SPECIAL DISCUSSIONS OF THE FARADAY SOCIETY
NO. 1   1970

Thin Liquid Films and Boundary Layers

Published for
THE FARADAY SOCIETY
by
ACADEMIC PRESS
LONDON AND NEW YORK

ACADEMIC PRESS INC. (LONDON) LTD
24–28 Oval Road,
London, N.W.1. 7DD.

U.S. Edition published by

ACADEMIC PRESS INC.
111 Fifth Avenue,
New York, New York 10003

Library of Congress Catalog Card Number: 70–141730
ISBN: 0-12-694550-0

Printed in Great Britain by
THE ABERDEEN UNIVERSITY PRESS
ABERDEEN, SCOTLAND

Thin Liquid Films and Boundary Layers

28th, 29th and 30th September, 1970

A SPECIAL DISCUSSION on Thin Films and Boundary Layers was held at the University of Cambridge on the 28th, 29th and 30th September, 1970. It represented the first of a new series of Discussion on physical chemistry topics of particular significance in industrial and technological research. Such Special Discussions are expected to be held on topics of scientific and industrial importance and timeliness at intervals and not necessarily on a yearly basis.

The President, Prof. Geoffrey Gee, C.B.E., F.R.S. opened the meeting and welcomed 220 members and others. Among the visitors from overseas were:

Mr. C. M. Allen, *U.S.A.*
Dr. R. De Backer, *Belgium*
Mr. Bongrand, *France*
Dr. E. Drauglis, *Germany*
Mr. F. Dumont, *Belgium*
Mr. J. G. J. Egberink, *Netherlands*
Mr. J. A. de Feijter, *Netherlands*
Dr. G. H. Findenegg, *Austria*
Dr. G. Frens, *Netherlands*
Mr. C. Guestaux, *France*
Dr. M. L. Hair, *U.S.A.*
Mr. H. Hasmonay, *France*
Dr. T. Hendrikx, *France*
Dr. E. P. Honig, *Netherlands*
Mr. J. Hougardy, *Belgium*
Dr. A. M. Joseph-Petit, *Belgium*
Prof. H. Lange, *Germany*
Mr. J. W. Lichtenbelt, *Netherlands*
Dr. A. Lucas, *Belgium*
Prof. J. Lyklema, *Netherlands*
Ir. A. M. Michels, *Netherlands*
Dr. K. J. Mysels, *U.S.A.*
Prof. Dr. H. van Olphen, *Netherlands*
Prof. J. Th. G. Overbeek, *Netherlands*
Dr. G. Peschel, *Germany*
Dr. A. Prins, *Netherlands*

Dr. H. E. Ries, Jr., *U.S.A.*
Mr. J. B. Rijnbout, *Netherlands*
Dr. K. Roberts, *Sweden*
Prof. J. R. Robinson, *New Zealand*
Prof. Dr. E. Ruckenstein, *U.S.A.*
Prof. Dr. G. Schay, *Hungary*
Prof. A. Scheludko, *Bulgaria*
Dr. G. Schreier, *Germany*
Dr. M. Schwuger, *Germany*
Dr. J. A. N. Scott, *Netherlands*
Prof. L. E. Scriven, *U.S.A.*
Dr. R. Senkus, *U.S.A.*
Dr. J. M. Serratosa, *Spain*
Dr. H. Sonntag, *Germany*
Dr. J. H. Stark, *Switzerland*
Dr. W. Stone, *Belgium*
Dr. B. Stuke, *Germany*
Dr. G. Szasz, *Switzerland*
Dr. M. van den Tempel, *Netherlands*
Dr. L. Ter-Minassian-Saraga, *France*
Mr. J. Thorin, *France*
Mr. P. Viaud, *France*
Prof. Miss M. Vignes, *France*
Prof. A. Vrij, *Netherlands*
Prof. A. Watillon, *Belgium*
Prof. K. G. Weil, *Germany*

CONTENTS

5

General Introduction

By B. A. Pethica

Unilever Research Laboratory, Port Sunlight, Cheshire, England

Received 14th October, 1970

These introductory remarks are intended to serve two purposes. In the first place, the reasons for holding this Special Discussion of the Faraday Society are set out. Secondly, a review of current problems in research into thin liquid films and boundary layers is given.

During the past year, the Council of the Faraday Society decided to form an Industrial Sub-Committee of the Standing Committee on Conferences. This decision was a recognition of the fact that research in industry is making a substantial contribution to the advance of those basic sciences which the Faraday Society is concerned to foster. It was also recognized that the motivation of industrial research towards the direct use of its results constitutes a powerful current stimulus to the advance of fundamental science in a variety of areas. The principal purpose of the Industrial Sub-Committee is to propose topics for Discussions or Symposia which are of scientific merit and timeliness and of industrial significance. These topics may be adopted as part of the Society's customary programme, or may be the basis for Special Discussions or Symposia, of which this meeting is the first example.

This decision by the Society marks no lessening of concern with its traditional objectives and standards; rather it marks the Society's reassertion of the professional objectives common to academic and industrial scientists alike and its responsiveness to developments on the borders of physical chemistry. One such border is that between physical chemistry and the engineering sciences, as demonstrated at this Discussion.

The physico-chemical problems of industry have always been of interest to the Society, as is evident from its contributions in catalysis. This new venture is in some ways a return to an early tradition, as an examination of the titles of the Society's Discussions back to the 1920's will show. In 1920, the Society discussed *Basic Slags*, and in 1921 the subject for debate was *The Failure of Metals*. There followed Discussions on *The Physical Chemistry of the Photographic Process* (1923) and *Textile Fibres* (1924). Coming to more recent times, one of the topics for 1954 was *The Physical Chemistry of Dyeing and Tanning*; and of direct interest to us at this meeting, the Society in 1948 discussed *The Interaction of Water and Porous Materials*. That Discussion contained some remarkable foretastes of our debates at this 1970 meeting, and I can hardly do better to illustrate the joint scientific and industrial value of the *Thin Liquid Film* topic than to quote from the 1948 Discussion a remark by the late Prof. D. H. Bangham, then Director of the British Coal Utilization Research Association, and formerly Professor of Physical Chemistry in Cairo. "In view of the large number of papers written on the subject of adsorbed water it is astonishing how little experimental work is directed towards ascertaining its *properties*. Most workers are content with being able to give a self-consistent account of a very small range of facts of which more than one explanation is possible. There are, however, awaiting solution, a number of technical problems which turn upon the behaviour of these films, and it is important that their nature should not remain merely a matter of conjecture ".[1]

Bangham went on to give a highly topical example of these " technical problems ", relating to the adhesiveness of dust particles. We can easily add examples to make a formidable list of industrial problems which depend for their solutions, in part at least, on our understanding the properties of thin liquid layers—examples drawn from such areas as lubrication, corrosion, flotation, foaming, emulsion formation, colloid stability, wetting etc. Bangham's remarkable pioneering work on poly-molecular liquid layers on solids is not as well known as it should be, and it is sad to note that after his death the subject has languished among physical chemists in Britain until recently. There were many earlier indications that liquid layers on solids can have unexpected properties—for example, in the work of Hardy [2] in 1913—but we may regard Bangham and Razouk's [3] analysis of the contact angle equilibrium in 1937 as a turning point, showing as it did that for non-zero contact angles, the polymolecular layer adsorbed on the solid beyond the boundary of a liquid drop at saturation vapour pressure does not simply have the properties of the bulk liquid. At almost the same time Frumkin [4] was drawing very similar conclu-sions in the U.S.S.R. The proposal that the formation of structured liquid layers near to solid surfaces is of importance in lubrication processes also has a long history. Griffiths [5] in 1920 suggested that liquid surface layers possess a special rigidity, Bastow and Bowden [6] in 1931, among others, taking the opposite viewpoint. Here again, workers in the Soviet Union, notably Fuks,[7] have interested themselves deeply in the mechanical properties of liquid boundary layers.

The current broad interest in the properties of liquid boundary layers and thin films owes much to the development of quantitative theories of colloid stability, particularly in the Soviet Union and Holland by Deryaguin and Landau,[8] and Verwey and Overbeek.[9] These theories have facilitated a vast volume of quantitative research into surface forces, and to some extent the present debate on the reality and significance of special structural properties in thin liquid films and boundary layers is the result of a growing realization that the accumulated data will not be covered by the interplay of two sets of forces alone (the electrical double-layer interactions and van der Waals' forces), and of a certain impatience among some colloid scientists with the opacity of the now complicated corrections to a formerly elegant theory, particularly so far as the electrical double-layer is concerned. On the other hand, despite the many quantitative studies directed to elucidating the special properties of liquids in boundary layers and thin films, particularly by Deryaguin and his co-workers, considerable doubt remains as to the generality and extent of the alteration of liquids in these films. It must also be admitted that *quantitative* predictions of the effect of these layers, in influencing colloid stability for example, are not yet available. Progress towards such quantitative prediction will require further phenomenological investiga-tions and many exact studies of molecular and thermodynamic behaviour in films and boundary layers.

To the extent that, in dilute electrolyte systems, the long-range nature of surface effects due to electrical double-layers and van der Waals' forces is commonly accepted, the issue is not simply one of the distance over which surface forces act in liquid layers, but rather as to whether or not the liquid itself can play a more direct role than that of providing a fluid dielectric with convenient dispersion characteristics. In principle, the situation can be covered by using more refined two-force models. Such models will take account of saturation effects, ion volumes, dipole terms, asymmetry of polarizability, etc., and hence they will include more directly the involvement of the solvent. The corresponding calculations of disjoining pressure and other measurable quantities will then include the energy and entropic contributions due to the solvent. Nevertheless, the role of solvent orientation is sufficiently distinct for Deryaguin to introduce the concept of " forces of the third kind ".[10] From the experimentalist's point of view, the importance of this concept is that it suggests new experiments designed to reveal the molecular situation at the liquid boundary, thereby guiding further theory with a wider range of measurements than hitherto have been available. It is remarkable, for example, that until recently there has been so little interest in temperature effects in lyophobic colloid systems. An early exception is again in the work of Bangham's group (Bond, Griffiths and Maggs).[11] Temperature variations play a minor role in the classical DLVO theory but a major role in any model explicitly involving entropic layers and long-range ordering in the liquid. Without doubt, some of these recent experiments on temperature effects have been inspired by the " ordered liquid " theories, and represent attempts to break away from the two-force theory. The time has come to give up our addiction to precise thermostatting at 25°C. If the phenomena are not very temperature sensitive, why bother? And if they are sensitive, it is the temperature coefficient that is the most interesting variable.

In this Discussion we have a variety of measurements of temperature effects of direct relevance to deciding the existence and role of solvent layers and liquid structure changes in boundary layers. In the paper by Prins and van den Tempel, the study of the effect of temperature changes on a free " equilibrium " soap film suggests that disjoining pressure equilibrium may not be the controlling factor in apparent film stability. The paper of Clunie and his co-workers on temperature and salt effects on a non-ionic stabilized foam system raises substantial doubts as to the meaning of the derived Hamaker constants, which appear to vary by a factor of two over a 10 K range of temperature. The effects of temperature on liquid layers on quartz, as described by Adlfinger and Peschel, suggest a strong correlation between the disjoining pressure and ordering in the liquids, as do the effects of temperature on viscosity in thin capillaries recorded by Churayev, Sobolev and Zorin. In the simple, well-defined graphon+liquid alkane system, described in the paper of Ash and Findenegg, the heats and excess volumes of wetting show temperature coefficients that indicate significant liquid structural changes at the boundary, and the data for the graphon/water interface suggest extensive liquid structuring. The paper by Vincent and Lyklema deserves the special attention of colloid chemists, since it gives evidence, through temperature effects, of some structuring of water at the surface of silver iodide sols—perhaps the most exhaustively studied colloid system. On the other hand, the n.m.r. experiments over a range of temperatures on the aqueous polymer latex system reported by Clifford, Oakes and Tiddy, and those on the water+vermiculite system reported by van Olphen and his co-workers strongly suggest tight binding of small amounts of water with little evidence of extended alteration in the Brownian motion deep into the liquid layer. Clifford's results show that the extended structural effects in water near to polyvinyl acetate latex particles, reported earlier from the same laboratory,[12] are most likely caused by the fibrillar and porous nature of the surfaces of those latices.

The polymer latex data show clearly the importance of the precise characterization of the solid surfaces contacting the liquid layers. Of the surface layers discussed in the various papers at this meeting, the mica surface studied by Bailey, Price and Kay, the layered vermiculites and clays discussed in several papers, the silver iodide sol surface and the graphon interface are perhaps the best characterized of the solid+liquid systems. The free-standing liquid films have been customarily regarded as a reliable model system, but Prins and van den Tempel may cause us to revise this opinion. Surfaces of solid quartz, steel and rubber are less well characterized. Quartz has been widely studied, but the surface characterization is ambiguous in most cases.

An interesting illustration of this point is that the heat of immersion into water of un-annealed Aerosil silicas with a range of surface hydrations shows a striking dependence on the temperature of immersion, suggesting strongly that un-annealed silica can induce significant structural effects in the local water. When the silica powders are annealed and rehydrated at the surface, the temperature variation of the heat of immersion is no longer present for silicas annealed in air (in the presence of a small vapour pressure of water) but partially remains for silicas annealed *in vacuo*.[13] Results of this kind show that short-range surface effects have a profound influence in " triggering-off " structural changes in the local liquid layers, and that we should be cautious of supposing that where these effects occur they result from long-range forces entirely. This same point comes out clearly from numerous results on the thermodynamic functions for the adsorption of vapours on solid surfaces. It is well known that the B.E.T. equation, as commonly applied to vapour adsorption, assumes that the second and successive layers of adsorbed vapour have bulk-liquid properties, and to the extent that the B.E.T. equation is successful, it gives support to the view that polymolecular liquid layers on many solids have no unusual structure. Even for systems in which the B.E.T. equation gives a good fit, however, the experimentally determined differential heats and entropies of adsorption do not usually become indistinguishable from the bulk-liquid values until several layers are adsorbed. It is almost certain that increased accuracy in the determination of these parameters would show that very many layers are necessary before true bulk-liquid properties are reached, and we should never forget that even if the deviations in molar functions are small, the liquid concentrations are high. The detailed preparation of the solid surface has a profound effect on the energetics of the physical adsorption of vapours, and necessarily therefore on the behaviour of polymolecular layers [see, e.g., Holmes [14]].

Another set of considerations are provoked by taking the view that the solid/liquid and fluid/liquid interfacial systems discussed at this meeting represent extreme examples of solutions in the liquid phase. The properties of the solvent take pride of place in this approach, and many suggestive correlations come to mind from considerations of solution and liquid-state thermodynamics. This approach is illustrated in part by Adlfinger and Peschel and by Ash and Findenegg. The copious data on the properties of solutions of ions, inert gases, polar and amphipathic molecules suggest that a variety of structure-making and -breaking effects will be manifest at surfaces. Extrapolating from thermodynamic and n.m.r. studies of surfactant micelles,[15] for example, one would not expect extensive long-range solvent orientation changes at the surface of oil in water emulsions stabilized by highly charged surfactants. The variations of water-structuring effects with surface charge reported by Vincent and Lyklema are strongly reminiscent of short-range solvent polarization effects in aqueous electrolyte solutions.

Perhaps most interesting of the relevant speculations that arise from considerations of bulk liquid thermodynamic properties is that the term " liquid " includes " liquid crystals " and a variety of systems involving the reversible aggregation of amphipathic molecules. The phase-rule relationships in these bulk nematic and smectic systems have been worked out for many examples, and the use of expressions such as " surface liquid phase " may in some instances be well justified. In this meeting Drauglis, Lucas and Allen develop the smectic phase analogy for lubricating films of fatty acids in hydrocarbon oils, and Clifford and his co-workers remind us of the relationships between lamellar soap phases and thin foam films. The paper from Haydon's laboratory is a valuable contribution on several scores—it provides data on thin oil layers stabilized by a chemically defined and thermodynamically characterized solute in the oil. A striking result is that this stabilizer, glyceryl mono-oleate (which is, incidentally, related to well-known emulsion stabilizers used in the food industry), shows reversible micellar aggregation in hydrocarbon solutions. Furthermore, these black oil films form typically at concentrations near to or above the micelle point of the stabilizer, which is a common situation also in the formation of black films from aqueous surfactant solutions. These considerations suggest that the sharp salt-induced transition (recorded by Clunie and co-workers) between the first and second black films stabilized by a non-ionic surfactant is, in fact, a phase change which should have a direct parallel in bulk solutions of the stabilizer.

These suggestions take us directly to the point that, even if the forces controlling the thinning of free-standing thin films are correctly represented by the interaction of electrical double layers and van der Waals' forces in planar (or laterally extended) geometric arrangements, we must still enquire as to the causes of the planar arrangements. Smectic and nematic phases provide one of the keys to answering this question. The thermodynamic stability of liquids which are either themselves molecularly asymmetric and flexible, or contain such asymmetric solutes, involves a balance of conformational entropy and energy terms deriving in part from geometrical considerations and in part from " chemical " factors (asymmetric polarizabilities of different groups, etc.). This balance is necessarily modified in the vicinity of the asymmetry we call a surface, and our Discussion will go far in showing whether and where these structural changes are significant in a wide range of experimental situations.

[1] Bangham, *Disc. Faraday Soc.*, 1948, **3**, 102.
[2] Hardy, *Proc. Roy. Soc. A*, 1913, **88**, 313.
[3] Bangham and Razouk, *Trans. Faraday Soc.*, 1937, **33**, 1459.
[4] Frumkin, *Zhur. Fiz. Khim.*, 1938, **12**, 33.
[5] Griffiths, *Phil. Trans. A*, 1920, **221**, 163.
[6] Bastow and Bowden, *Proc. Roy. Soc. A*, 1931, **134**, 404.
[7] Fuks, *Research in Surface Forces*, ed. Deryaguin, (Consultants Bureau, New York, 1964), vol. 1, p. 79.
[8] Deryaguin and Landau, *Acta physicochim.*, 1941, **14**, 633.
[9] Verwey and Overbeek, *Theory of the Stability of Lyophobic Colloids*, (Elsevier, Amsterdam, 1948).
[10] Deryaguin, *Research in Surface Forces*, ed. Deryaguin, (Consultants Bureau, New York, 1964), vol. 1, p. 3.
[11] Bond, Griffiths and Maggs, *Disc. Faraday Soc.*, 1948, **3**, 29.
[12] Johnson, Lecchini, Smith, Clifford and Pethica, *Disc. Faraday Soc.*, 1966, **42**, 120.
[13] Tyler, Taylor, Pethica and Hockey, *Trans. Faraday Soc.*, in press.
[14] Holmes, *The Solid-Gas Interface*, ed. Alison Flood, (Edward Arnold & Marcel Dekker, 1967), vol. 1, p. 127.
[15] Clifford and Pethica, *Trans. Faraday Soc.*, 1965, **61**, 182.

Bursting of Soap Films

Part 4.—The Behaviour of Ions on a Crowded Surface

By G. Frens,* Karol J. Mysels† and B. R. Vijayendran ‡

R. J. Reynolds Tobacco Co., Research Dept., Winston-Salem,
North Carolina 27102, U.S.A.

Received 13th *April*, 1970

The growing hole of a bursting soap film is preceded by an aureole of accelerating, contracting, and thickening film. Measurements of the cross-sectional profile of such aureoles are reported and provide the basis for an interpretation in terms of surface tension changes accompanying the contraction of the film which is complete in less than a millisecond. The surface tension of sodium dodecyl sulphate solution decreases to below 15 mN/m (dyn/cm) under these conditions. As desorption is negligible, the data can be interpreted in terms of an extension of surface pressure-area per molecule curves towards high pressures and low areas. This extension shows an unexpected decrease of slope at low areas and indicates a rapid increase in intermolecular attractions as the surfactant ions become crowded on the surface.

Once a thin liquid film, specifically a soap film, develops a tiny perforation, the latter will keep growing and the whole film disappears rapidly. In soap films the edges of a hole recede at speeds of the order of 10^3 cm s^{-1}. The driving force is the surface tension of the two faces of the film, which exerts an uncompensated force on the perimeter of the hole.

It was assumed that the growing hole is surrounded by undisturbed film except for a very narrow rim, which contains all the receding material. Recently, however, it has been found [1, 2] that the expanding hole is preceded by a wide zone (the aureole) of moving film material. The width of the aureole is comparable to the radius of the hole, and the expansion of the hole and the aureole are approximately proportional. The thickness δ of the film increases in the aureole region from the thickness δ_0 to approximately $2\delta_0$ at the rim of the hole. Flash photographs, such as fig. 1, which show some of the complicated features of aureoles have been published.[1, 3]

A theoretical analysis of the aureole phenomenon was given by Frankel and Mysels.[4] They showed that the existence of aureoles can be explained if there is a gradient in the surface tension of the film in the vicinity of the hole. Such a gradient is closely related to the thickness variations in the aureole. As a film element increases in thickness, it decreases in surface area. If the relaxation of the surface through the desorption of surfactant molecules is slow as compared with the rapid compression of the collapsing film structure, then the film surface resembles an insoluble monolayer. The surface tension of a film element becomes smaller as its surface area decreases, i.e., as its thickness δ increases. It is the purpose of the present paper to show that an interesting extension of the classical Π—A curves (surface pressure against area per molecule) becomes available when this theory is applied to experimental data obtained for the thickness profile of the aureoles.

* present address: Philips Research Laboratories, N.V. Philips' Gloeilampenfabrieken, Eindhoven, Netherlands.

† present address: Research Dept., Gulf General Atomic Inc., San Diego, California 92112, U.S.A.

‡ present address: Research Dept., Pitney-Bowes, Stamford, Conn., U.S.A.

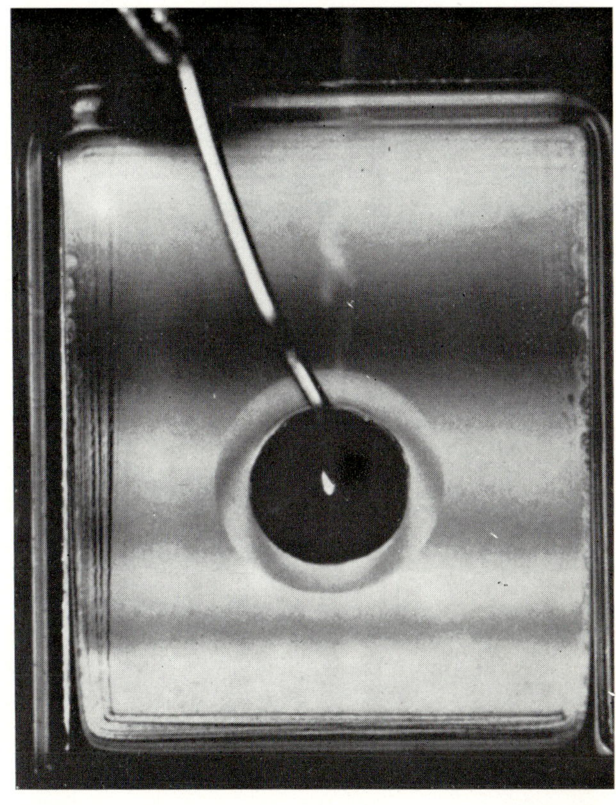

FIG. 1. Bursting soap film, photographed using a 0.5 μs flash.   The aureole surrounding the hole is relatively narrow and shows a large frontal shock.

[*To face page* 12.

THEORETICAL

In principle, it is possible to calculate values for the surface tension as a function of the increase in thickness and of the time during a burst for any arbitrary behaviour of an aureole, provided that the history of the evolution of the aureole is sufficiently well known.[4]   In practice, one would need too many and too accurate data to make such an analysis feasible.

The interpretation of data is much simpler if the physical situation in a bursting soap film corresponds to a pseudo-equilibrium in which there is no desorption from the surface and where the surface pressure depends only on the area per molecule. In such a system there is no intrinsic (relaxation) timescale and the bursting becomes self-similar, i.e., all features of the aureole and of the hole expand from the origin, each at its own constant velocity.

This assumption means that the surface tension σ depends only on the relative shrinkage α, or thickening β, of the film, defined by $\alpha = 1/\beta = \delta_0/\delta$, and not on time, so that a single (σ,α) curve, such as that of fig. 2, characterizes the system.

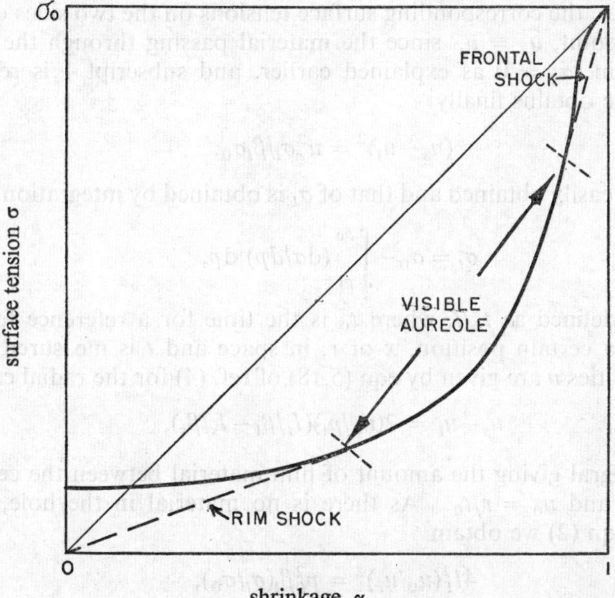

FIG. 2.—Schematic (σ,α) curve.   The dashed lines indicate portions corresponding to shock waves in unidimensional bursting.

If the burst originates from a straight line—is unidimensional—a complete analysis is then possible.[4]   If it originates from a point, which is the case in our experiments, only certain features of the unidimensional analysis can be extended to this radial case [4] and others have to be computed numerically.

A fundamental velocity in this analysis is Culick's velocity, $u_c = (2\sigma_0/\rho\delta_0)^{\frac{1}{2}}$, where σ_0 is the surface tension of the film at rest and ρ its density.   Culick [4] showed that this is the velocity of the rim of the hole, but in his model there was no aureole. In the presence of an aureole it can be shown that u_h is always smaller than u_c in the unidimensional case [4] and is smaller or slightly larger [6] in the radial case.   As indicated in fig. 2, only certain parts of the (σ,α) curve which are convex downwards are represented by the visible smooth slopes of the aureole; those that have the opposite curvature are hidden in shock waves, the most important of which is the rim itself.[4]   The rim may be considered as the shock wave due to the zero surface tension within the hole.   Generally, there is a shock wave at the outer edge of the aureole, but there may be others.[1]   In the unidimensional case, the shock waves correspond to chords exactly tangent to the (σ,α) curve.   In the radial case, chords are not exactly tangent.[6]   A special case is that of a (σ,α) curve lying entirely above the diagonal joining the initial state to the origin.   In that case, the entire aureole is compressed into the shock wave represented by that diagonal and the width of the aureole is reduced to zero.[4]

A useful extension of the theory is an expression for the limiting properties of the aureole at its thickest point where it joins the final shock wave, i.e., the rim.   We denote these by the subscript l.   Conservation of mass and momentum gives for a shock wave moving with a velocity u_s and separating film elements moving with u_1 and u_2, respectively,

$$(u_s - u_1)(u_s - u_1)\rho\delta_1 = 2(\sigma_1 - \sigma_2), \tag{1}$$

where σ_1 and σ_2 are the corresponding surface tensions on the two sides of the wave.[4] For the limiting point, $u_2 = u_s$, since the material passing through the shock wave remains in the rim, $\sigma_2 = 0$ as explained earlier, and subscript 1 is replaced by l. Introducing u_c one obtains finally

$$(u_s - u_l)^2 = u_c^2 \sigma_l / \beta_l \sigma_0. \tag{2}$$

The value of σ_0 is easily obtained and that of σ_l is obtained by integration :

$$\sigma_l = \sigma_0 - \int_{pl}^{po} (\mathrm{d}\sigma/\mathrm{d}p)\,\mathrm{d}p, \tag{3}$$

where p may be defined as t_0/t, where t_0 is the time for a reference feature of the aureole to reach a certain position, x or r, in space and t is measured at the same point.   The velocities u are given by eqn (5.18) of ref. (4) for the radial case and yield

$$u_s - u_l = 2(u_0/p_l)(I_l/\beta_l - I_s/\beta_s), \tag{4}$$

where I is an integral giving the amount of film material between the centre and the point considered and $u_0 = r/t_0$.   As there is no material in the hole, $I_s = 0$, and combining with eqn (2) we obtain

$$4I_l^2(u_0/u_c)^2 = p_l^2\beta_l(\sigma_l/\sigma_0), \tag{5}$$

which is satisfied only at the rim.   As this is derived on the assumptions of self-similarity, but involves only information concerning the profile of a single aureole, it can provide a criterion of self-similarity for each aureole.

For the unidimensional case, the similar expression

$$I_l^2(u_0/u_c)^2 = \beta_l(\sigma_l/\sigma_0) \tag{6}$$

is derived similarly, using eqn (5.6) of ref. (4).

EXPERIMENTAL

Experimentally, the velocity of the rim and that of any shock waves is probably best estimated from photographs taken at different time intervals after initiation of the burst by an electric spark.[1]   Details of the profile of an aureole are difficult to obtain precisely in this way and methods [7] measuring either the variation of thickness with time at a fixed point, as the aureole moves by, or the time required for a feature to proceed from one point to another, are preferred.   The results reported here were obtained with an instrument using a laser and measuring simultaneously the thickness of the film at two points along a horizontal radius, as indicated in fig. 3, which also shows a typical oscilloscope result and its interpretation.

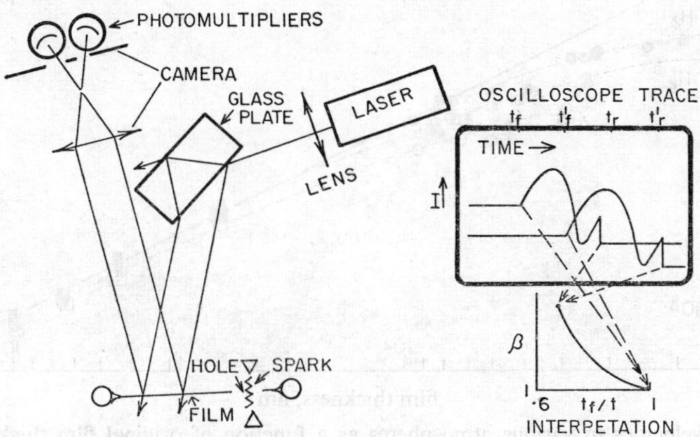

FIG. 3.—Schematic diagram of the apparatus used.   The laser beam is split and concentrated at two points of the film.   The reflected interference pattern, changing as the aureole passes these points, is recorded through photomultipliers and a two-beam oscilloscope and interpreted as shown on the right.

RESULTS

A major difficulty, discussed in detail earlier,[1] was that whereas the theory predicts that all velocities should vary with Culick's velocity, i.e., with $1/\sqrt{\delta}$, gross deviations were observed for thin mobile films as well as for rigid films of all thicknesses.   The explanation offered earlier,[1] that even the apparently undisturbed film is slightly compressed, was supported by direct evidence for rigid films, and may be significant when applied to these, but we have not been able to support it experimentally for mobile films.

On the other hand, we have found that frictional resistance of the atmosphere— the windage of the moving aureole and rim—greatly affects the velocity of the rapidly moving rim of thinner films and seems to account for the deviations observed with first black and thicker thin films.   Fig. 4 shows rim velocities observed in air, in helium and in hydrogen for films of various thicknesses of the same solution.   The viscosities of these gases are 185,198, and 89 μP and their densities 1.20, 0.165 and 0.083 g/l., respectively.   The deviations from the $1/\sqrt{\delta}$ behaviour indicated by the straight line become less as the inertia and viscosity of the atmosphere decreases.   For thin films, the deviations in hydrogen are still significant though much smaller than in air, but above some 100 nm they cease to be perceptible.   Hence, further discussion will be restricted to films having greater thicknesses and bursting in hydrogen.

That bursting is self-similar to a first approximation is shown by the constancy of rim velocities as the aureole grows, which has been reported earlier [1] for films in air.   This has been supported now by experiments in hydrogen.   The velocity of the frontal shock is also constant and obeys the $1/\sqrt{\delta}$ relation closely.   A more sensitive criterion is given by the shape of the aureoles, which should be unchanging after correction for growth as a function of time and the corresponding spreading in space.   A convenient form is to represent the thickness of the aureole as a function of t_f/t, where t_f is the time for the frontal shock to reach the point of observation and t corresponds to the relative thickening $\beta = \delta/\delta_0$, as indicated schematically in fig. 4. The solid line of fig. 5 and the corresponding large points show the results obtained

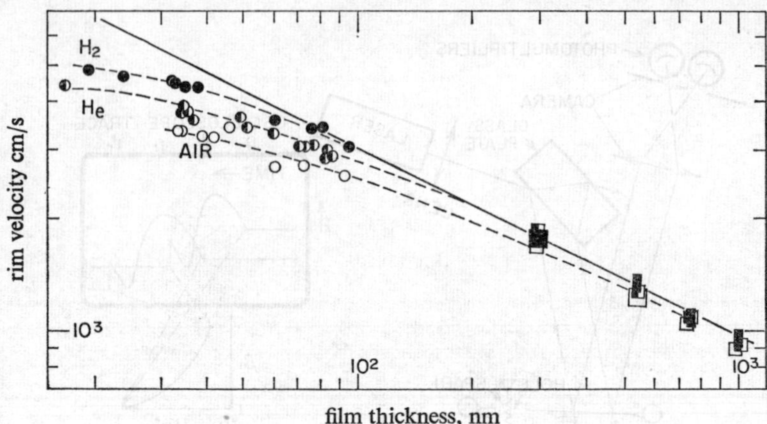

FIG. 4.—Rim velocities in various atmospheres as a function of original film thickness δ_0. The line shows the slope of $\delta_0^{-\frac{1}{2}}$.

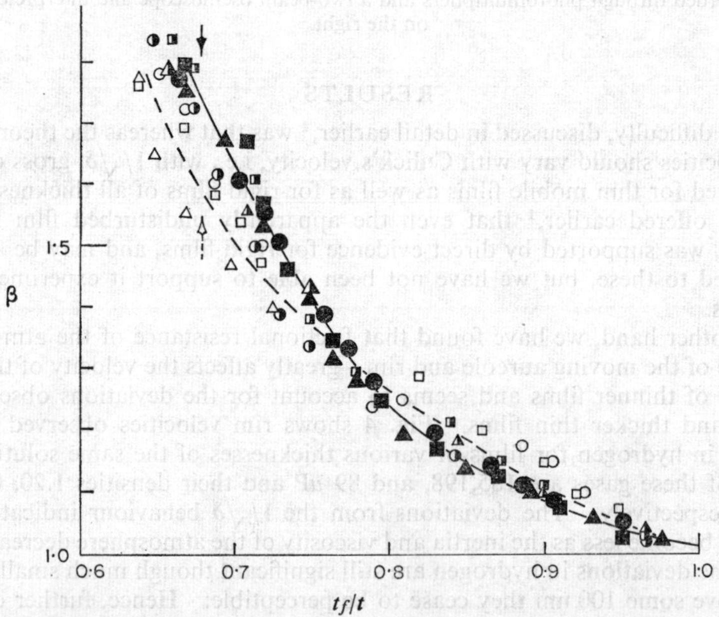

FIG. 5.—Reduced aureole profiles of 535 ± 15 nm thick films drawn from a 33.7 mM solutions of NaLS. (Results for 10.4 mM are indistinguishable). Large shaded points and solid line: $t_f = 770\pm10$ μs; open points and dashed line: $t_f = 370\pm10$ μs; half shaded points: $t_f = 540\pm10$ μs.

in three experiments for the later stages of an aureole. The precision of the measurements defines the line quite well. The position where the rim should lie under assumptions of self-similarity according to eqn (5), is indicated by an arrow and is close to that observed. For younger aureoles, the precision gradually decreases as shown by the smaller points and differences in profile appear as indicated by the dashed line. These differences, although systematic, are not sufficiently larger than all the uncertainties of the measurement to be accepted as definitely real. Furthermore, the criterion of eqn (5) indicates a lack of self-similarity for the dashed line.

Fig. 6 shows the (σ,α) curves corresponding to the profiles drawn in fig. 5. These were obtained by computer integration, using Simpson's approximation, of the

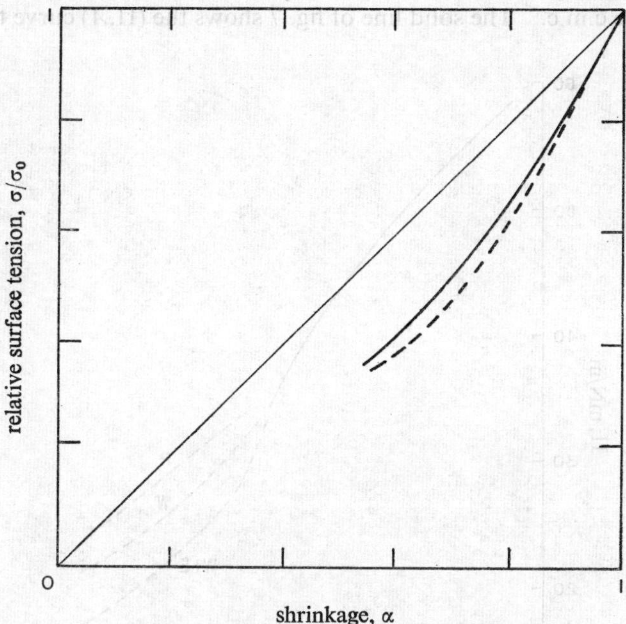

FIG. 6.—(σ,α) curves derived from the lines of fig. 5.

differential expression (5.22) of ref. (4). The difference between these curves is relatively minor so that the uncertainty in σ is much less than in the profile itself. This is generally true for (σ,α) curves for aureoles differing only in their thicker areas and results from the complex relation [4] between the two.

Deviations such as those of fig. 5 between young and old aureoles could stem from two sources : windage which presumably is greatly reduced by operating in hydrogen, or relaxation of the surface by desorption of surfactant molecules under surface pressures much above the equilibrium values, leading to higher σ values for a given α for older aureoles. The curves of fig. 6 do indicate such an effect upon σ and, if the differences are real, they could be used to study such desorption. On the other hand, extrapolation to zero age of the aureole could then also be used to correct for any relaxation and serve as a basis for interpretation in terms of the self-similar theory.

As shown in fig. 6, the original surface tension of the solution is reduced by about 63 % in the last stages of the aureole. In absolute terms, this corresponds to a reduction from 38 to only 14 mN/m, considerably below values normally encountered at the air-aqueous solution interface.

DISCUSSION

The (σ,α) curve is closely related to the surface pressure-area per molecule, (Π,A), curve, since $\Pi = \sigma_{H_2O} - \sigma$ and $\alpha = A/A_0$, where A_0 is the area per molecule in the original film provided that there is no desorption. Whereas σ_0 is readily measured, the direct determination of A_0 requires relatively delicate experiments with foaming or with tracers. Indirect determination of A_0 can be based on highly accurate surface tension and activity measurements through the Gibbs equation. Hence, in practice, literature data, which are often conflicting, have to be used. For sodium dodecyl sulphate (NaLS) in water, a value of 0.415 nm² (41.5A²), based on foaming experiments,[8, 9] seem most likely to be correct for solutions at, and slightly above, the c.m.c.   The solid line of fig. 7 shows the (Π,A) curve thus obtained

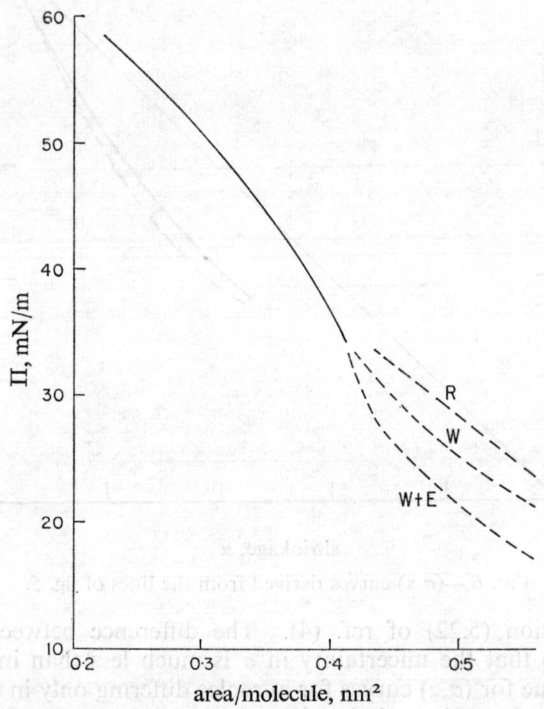

Fig. 7.—The solid line is the (Π,A) curve corresponding to the solid (σ,α) curve of fig. 6.   The dashed lines are some of the available literature data for equilibrium adsorbed monolayers of NaLS solutions : R, from equations of Rehfeld [10] ; W, data of Weil [9] ; W + E, areas of Weil [9], surface tensions of Elworthy and Mysels.[11]

from our data and based on this value.   It is essentially a mirror image of the (σ,α) curve of fig. 6 and provides an extension of (Π,A) curves for ionic monolayers into a new region.

If the interpretation of a (σ,α) curve in terms of (Π,A) is justified, it should give, to a first approximation, a smooth extension of the classical (Π,A) curve for higher areas. The interrupted lines of fig. 7 show some of the literature data [9-11] in this region. The values at higher areas are obtained on solutions of NaLS alone of varying concentrations and, therefore, varying ionic strengths. Our data pertain to a single solution and, therefore, presumably to a constant ionic strength of the subjacent liquid. This difference could lead to slightly different slopes. Reducing ionic strength at constant A should lead to an increase of interionic repulsions and, therefore, to an increased surface pressure. Hence, the interrupted lines should tend to be steeper than if they had been taken at the constant ionic strength corresponding to the c.m.c.   As it is, the data of Weil [9] seem to give the smoothest fit with ours. Literature data, particularly the surface tension values below the c.m.c., show considerable disagreement among themselves despite the precautions taken by each author to ensure purity and accuracy.[9-12]   Our data are obtained above the c.m.c. to reduce the effect of surface-active impurities by solubilizing them.   Hence, any conclusions from the intercomparison of these data must be highly tentative but is indicative of the possibilities of this approach.

The striking feature of fig. 7 is that the heretofore known curves for ionic surfactants, as exemplified by the interrupted lines, are convex to the abscissa, whereas most of our curve is concave.   If the change of curvature did occur at the point where the two sets of data meet, this would be a strange coincidence.   In fact, however practically all aureoles examined until now (the only exception [3] being a complex mixture of alkylbenzene sulphonates) show a frontal shock.   This shock corresponds to a part of the (σ,α) curve which is not directly accessible because it includes a curvature opposite to that of the accessible part (fig. 2).   Hence, the frontal shock wave indicates an extension of the curvature observed by conventional techniques into the region where $A < A_0$.   The surface compression and, therefore, the change in A during this frontal shock varies in different solutions and can reach 10 %.

The aureole which follows after the frontal shock represents still lower values of A.   Fig. 2 indicates that the reversal of the curvature in this part of the (Π,A) curve is intrinsic to the existence of the aureole phenomenon.[4]   Hence, the reality of the observation that (Π,A) plots have concave portions at $A < A_0$ requires only a qualitative validity of our experimental results and their interpretation.   It does not depend on the precision of the measurements, nor on the detailed outcome of the calculations.

The curvature deduced from bursting experiments would be reduced by any relaxation effects since the shrunken area, as measured by α, would be attributed to the smaller number of molecules remaining in the surface.   It does not seem possible, however, that any small relaxation effects occurring in our experiments could account for the observed result.

An equation of state for the surface would describe the (Π,A) curve for A below as well as above A_0.   Such a quantitative interpretation of the (Π,A) curve in terms of intermolecular forces should become more meaningful now that the aureoles in bursting soap films give access to the metastable states of a surface where the van der Waals attraction between the molecules becomes strong enough to reverse the curvature of the (Π,A) plot.   Progress along these lines awaits the improvement of the experimental data, including those concerning the surface tension and the area per molecule below the c.m.c.

[1] W. R. McEntee and K. J. Mysels, *J. Phys. Chem.*, 1969, **73**, 3018.
[2] I. Liebman, J. Corry and H. E. Perlee, *Science*, 1968, **161**, 373.
[3] K. J. Mysels and J. Stikeleather, *J. Colloid Interface Sci.*, submitted for publication.
[4] S. Frankel and K. J. Mysels, *J. Phys. Chem.*, 1969, **73**, 3028.
[5] F. E. C. Culick, *J. Appl. Phys.*, 1969, **31**, 1128.
[6] K. J. Mysels and B. R. Vijayendran, unpublished work.
[7] A. T. Florence, G. Frens, B. R. Vijayendran and K. J. Mysels, unpublished work.
[8] A. Wilson, M. B. Epstein and J. Ross, *J. Colloid Sci.*, 1957, **12**, 345.
[9] I. Weil, *J. Phys. Chem.*, 1966, **70**, 133.
[10] S. J. Rehfeld, *J. Phys. Chem.*, 1967, **71**, 738.
[11] P. H. Elworthy and K. J. Mysels, *J. Colloid Interface Sci.*, 1966, **21**, 331.
[12] (a) B. A. Pethica and A. V. Few, *Disc. Faraday Soc.*, 1954, **18**, 258.
   (b) A. P. Brady, *J. Phys. Chem.*, 1949, **53**, 56.
   (c) G. D. Miles and L. Shedlovsky, *J. Phys. Chem.*, 1944, **48**, 57.
   (d) E. J. Clayfield and J. B. Matthews, *Proc. 2nd Int. Congr. Surface Activity*, 1957, p. 172.

Response of an Equilibrium Film to External Disturbances

By A. Prins and M. van den Tempel *

Unilever Research Laboratories Vlaardingen/Duiven, The Netherlands

Received 2nd April, 1970

An equilibrium film situated in air saturated with water vapour is subjected to a disturbance consisting of a rapid change of the temperature of the surrounding atmosphere. The resulting large change in film thickness is found to be due to exchange of water between film and atmosphere, and not to expansion or contraction of the film. Pseudo-equilibrium films of widely varying thickness can be formed by means of this process. For a film with a sufficiently large area, the thickness of an element far from the border is determined by the water vapour pressure equilibrium rather than by the disjoining pressure equilibrium. The thickness profile of a large, vertical film in a state of apparent rest is explained on the basis of the different time scales associated with the various equilibrium processes.

It was noticed by Gibbs [1] that a thin liquid film in air need not necessarily be in complete thermodynamic equilibrium, even when it is in a state of apparent rest. In a small element of the film (i.e., having dimensions comparable to the film thickness), several of the equilibrium conditions will be rapidly satisfied. In particular, after a disturbance the uniformity of chemical potentials and of temperature will be re-established in a time which is very short compared to the usual observation time, and therefore these equilibrium conditions may be regarded as being permanently satisfied in a film element. The values of the chemical potentials in the film element may, however, differ from those in other parts of the system. Equilibration by transport along the film or through the adjoining bulk phases may then require a time comparable to the usual observations time, or even much longer.

The equilibrium thickness of the film element is determined by mechanical equilibrium of the forces acting in a direction perpendicular to the plane of the film.[2] The mere existence of thick films slowly draining to their equilibrium thickness shows that the time required for this equilibration process must be measured in minutes. Measurements of film thickness can be carried out before *this* mechanical equilibrium has been established, and such measurements can be used to obtain information about the rate and the nature of the other slow equilibration processes. For example, the experiment of Plateau, in which a soap bubble is thinned locally and reversibly by the warmth of a finger, indicates the effect of small temperature variations. Recent work [3, 4] has shown that films become thinner when exposed to an atmosphere with reduced water vapour pressure. Measurements of equilibrium film thickness [2, 5] require extreme caution to prevent evaporation, and even then they must be carried out in close proximity to the border, where complete equilibration with the contacting bulk solution may be expected to be fairly rapid.

The present investigation is concerned with a more detailed investigation of the effect of small changes in temperature on the thickness of a film which is in a state of apparent rest, in particular in regions well away from the film borders.

* authors' address : Olivier van Noortlaan 120, Vlaardingen, The Netherlands

EXPERIMENTAL

In this work, it was necessary to take elaborate precautions to prevent uncontrolled temperature fluctuations in the system. The films were made and investigated in a completely closed glass bottle, placed in a large box of double-walled glass. Water of constant temperature was circulated between the bottom and side-walls of the box. The box was covered with sheets of foamed plastic, and placed into a constant temperature room. All the necessary manipulations with the film took place from outside the box, using magnetic coupling across the wall of the glass bottle. Under these conditions, the temperature in the glass bottle fluctuated over 0.02°C in a period of 10 h.

After the glass bottle had been provided with about 400 cm³ of the solution, the system was left to equilibrate for at least 24 h. Then a film was produced by slowly raising a frame, originally submerged in the solution. The film of 4 × 4 cm, was suspended between nylon wires of about 10 microns thickness, weighted with a piece of glass tubing to ensure a perfectly flat film. The thinning process of the film was observed by following the downward motion of the interference fringes. Fast-draining films were used in most experiments. After a few minutes, a " black " region developed at the top of the film, and covered the whole area in about 30 min. Measurements of film thickness were started before the development of a black film, and continued for at least several days.

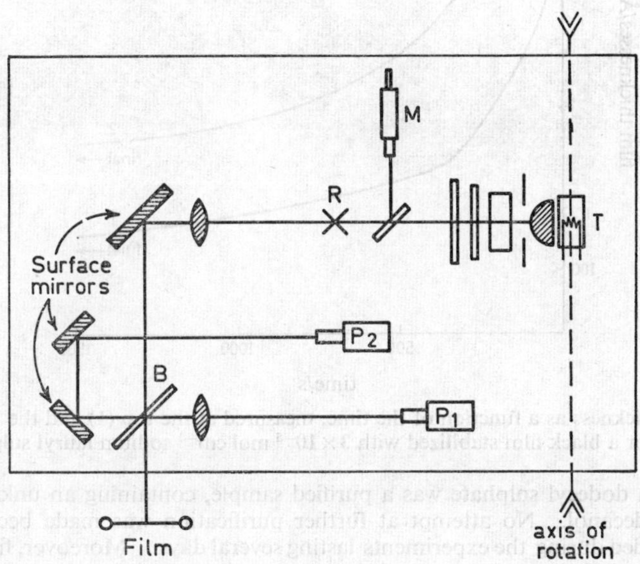

FIG. 1.—Schematic top view of the optical apparatus used for the scanning reflectometric film thickness measurements.

Film thickness was monitored by recording the intensity of reflected light. A light beam from a halogen lamp T (see fig. 1) passed through a layer of water, a heat filter, and an interference filter (5460 ± 80 Å). From the beam-splitter B, a part of the light moves on to the film and is reflected in the photomultiplier P_1. The remaining part of the original beam is reflected in the photomultiplier P_2, where it serves as a reference. The ratio of the output of the two photomultipliers was recorded. The maximum intensity of the light reflected by the film having the optical thickness of one-quarter wave-length was used for calibration. The equivalent water thickness of the film was calculated from the amount of reflected light by the usual procedure (e.g., ref. (2)). The optical apparatus was fitted on a rigid frame that was made to carry out small-amplitude oscillations around an axis perpendicular to the film. The distance between axis and film was about 1 m. In this way the area of the film seen by the photomultiplier P_1 (0.3×0.02 cm^2) was made to oscillate over a nearly vertical line of about 2 cm length, in 19 s.   Black surfaces were arranged in suitable locations to absorb stray light and to reduce background illumination.

Meaningful results can only be obtained if the film is perfectly flat and exactly perpendicular to the axis of rotation of the optical apparatus.   Alignment was carried out before each measurement by means of cross-wire R and microscope M, using the autocollimation principle.   It was verified that the presence of the light beam did not affect the film thickness. After several hours (see fig. 2), the film came to a state of apparent rest as indicated by a thickness profile remaining constant for at least many hours.   Then the temperature of the circulating water was increased by 1°C in about 2 min, and the response of the film was observed.

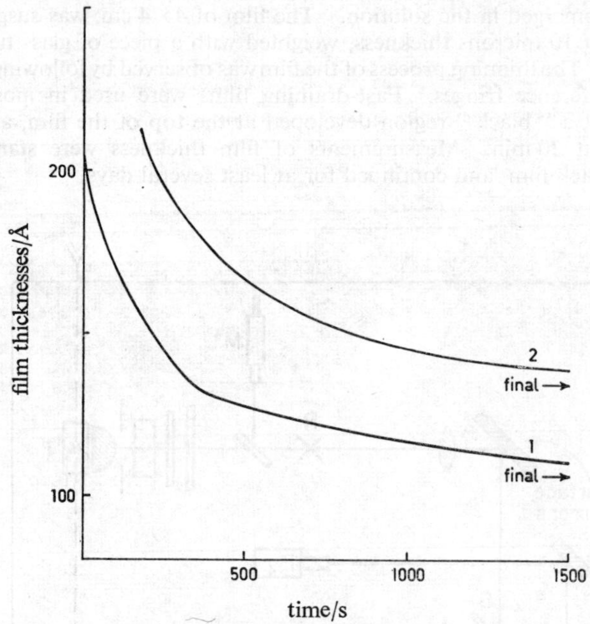

Fig. 2.—Film thickness as a function of the time, measured at the top (1) and the bottom (2) of the scanned area, for a black film stabilized with 3×10^{-5} mol cm^{-3} sodium lauryl sulphate at 25.0°C.

The sodium dodecyl sulphate was a purified sample, containing an unknown but small amount of dodecanol.   No attempt at further purification was made because hydrolysis cannot be avoided during the experiments lasting several days.   Moreover, films of sufficient stability could only be obtained in the presence of some dodecanol and/or inorganic electrolyte; even then the concentration of the main surfactant had to be in excess of the c.m.c. The cetyl trimethyl ammonium bromide was the same sample used in an earlier investigation.[6]   Water and other ingredients were of the highest purity available.

RESULTS AND DISCUSSION

Results of film thickness profile measurements are shown in fig. 3 and 4.   Even though elaborate precautions were taken to ensure that the system would be in complete thermodynamic equilibrium, the results show that this could not be achieved. For easy comparison, the " theoretical " profile has also been indicated in fig. 3 and 4. This profile was calculated on the assumption that the sum of the electrostatic repulsion between the charged film surfaces,

$$P_R = 64c\, RT\gamma^2 \exp\left(-\kappa\delta_w\right), \tag{1}$$

and the attraction due to Van der Waals' forces between the film molecules,

$$P_A = -A/6\pi\delta_A^3 \tag{2}$$

is just compensated by the hydrostatic head $P_H = -\rho g H$ of the film liquid. Here, the electrolyte concentration c in the film liquid is expressed in mol cm^{-3} and was assumed to be the same as in the bulk liquid. The aqueous core of thickness $\delta_w = (\delta_A - 20)$ Å is bounded on both sides by a layer of 10 Å thick, consisting mainly of the hydrocarbon parts of the surface-active material. Using this model, the equivalent water thickness $h = (\delta_A + 8)$ Å, which is the quantity plotted in the figures. The other quantities in the above equations have their usual meaning (see, e.g., ref. (2)), where $\gamma \approx \tanh 1$ and $A = 5 \times 10^{-13}$ erg.

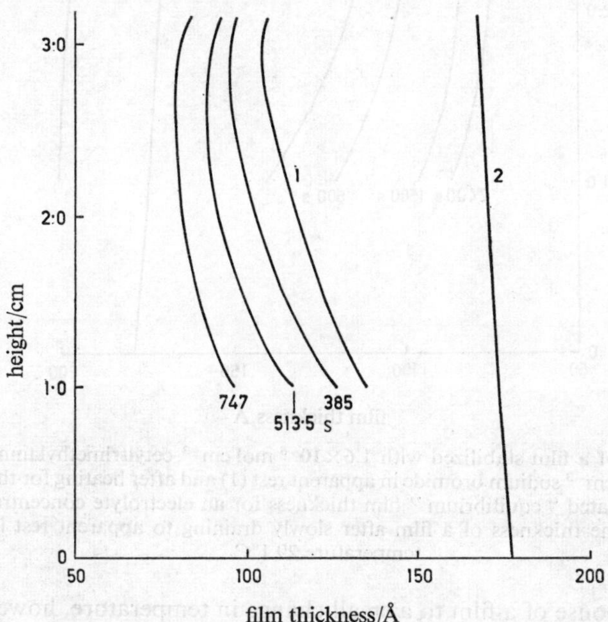

FIG. 3.—Thickness of a film stabilized with 3×10^{-5} mol cm^{-3} sodium lauryl sulphate in apparent rest (1) and after heating for the time indicated. Curve 2 is the calculated " equilibrium " film thickness for an electrolyte concentration of 3×10^{-5} mol cm^{-3}. Original temperature 25.0°C.

In all cases, the thickness profile of the " equilibrium " film differed appreciably from the profile expected on the basis of the usual criterion [2, 5] for determining equilibrium film thickness, i.e., the condition for mechanical equilibrium in a direction perpendicular to the plane of the film. This means that either this equilibrium condition is not satisfied, or the electrolyte content is much higher than in the bulk solution, in a film element at some distance from the lower border. Evidence has already been presented to show that electrolyte is accumulated in the film liquid,[7] and this is confirmed by the analysis of the present results.

If the equilibrium condition,

$$\Delta P = P_R + P_A + P_H = 0, \tag{3}$$

is satisfied in every element of the film at rest, it is possible to estimate the actual electrolyte concentration in the film liquid. This is because the thickness that satisfies eqn (3) is determined almost completely by the electrolyte concentration in the film liquid. Results of such calculations (fig. 5) show that a very steep concentration gradient in the film liquid is required to explain the thickness profile on this basis.

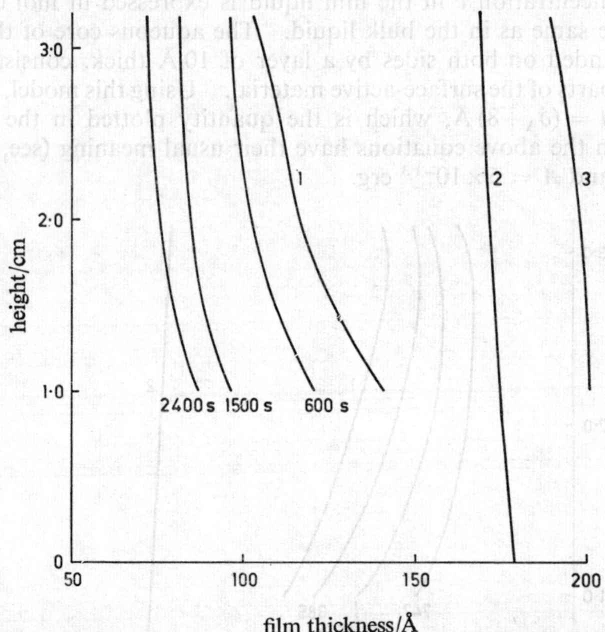

Fig. 4.—Thickness of a film stabilized with 1.6×10^{-6} mol cm^{-3} cetyltrimethylammonium bromide and 2.84×10^{-5} mol cm^{-3} sodium bromide in apparent rest (1) and after heating for the time indicated. Curve 2 is the calculated " equilibrium " film thickness for an electrolyte concentration of 3×10^{-5} mol cm^{-3}, curve 3 the thickness of a film after slowly draining to apparent rest in 7 h. Original temperature 29.1°C.

From the response of a film to a small change in temperature, however, it may be concluded that eqn (3) is *not* in general satisfied in every film element, and also that the actual concentration gradient in the film liquid is less steep in this case (see fig. 5). The important observation is here that small temperature differences between the film and the bulk liquid result in immediate and reversible variations of film thickness. The greater part of the measured rate of film thinning (fig. 6) must be due to evaporation of water from the film surfaces, which proceeds at a rate determined by the diffusion coefficient D in the gas phase and the distance x to the region of condensation. The resulting rate of film thinning is

$$-\frac{dh}{dt} = \frac{18D}{RTx} \Delta p, \tag{4}$$

where Δp is the excess water vapour pressure of the thin film. In a first approximation, the excess vapour pressure is determined by the temperature difference only :

$$\Delta p = (Qp_0/RT^2)\Delta T, \tag{5}$$

where Q is the heat of evaporation of water at room temperature and p_0 its equilibrium vapour pressure. The value of ΔT was estimated from temperature measurements by means of a set of thermistors, placed in the gas phase a few mm from the film. The actual temperature of the film element will be somewhat lower but even then application of eqn (4) and (5) shows that the observed rate of film thinning can be explained as a result of evaporation of water from the film, followed by diffusion of water vapour over a distance x of a few cm, and condensation on the surface of the bulk liquid.

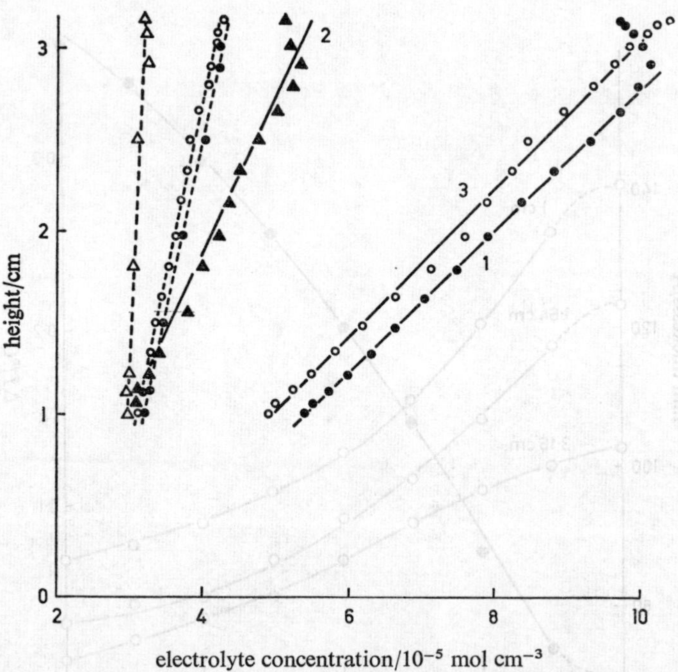

FIG. 5.—Electrolyte concentration as a function of the height for a film in apparent rest stabilized with 3×10^{-5} mol cm^{-3} sodium lauryl sulphate (curve 1), 1.5×10^{-5} mol cm^{-3} sodium lauryl sulphate and 1.5×10^{-5} mol cm^{-3} sodium bromide (curve 2), 1.6×10^{-6} mol cm^{-3} cetyltrimethylammonium bromide and 2.84×10^{-5} mol cm^{-3} sodium bromide (curve 3) for $\Delta P = 0$ (full lines) and for $\Delta P = -2g\Delta c$ (dashed lines).

A possible contribution from extension to the thinning of films on warming was studied by suspending a thick film from the black film, using a technique that has already been described in connection with film elasticity measurements.[6] The rate of descent of the border line between the black and the thick film was practically unaffected by the change in temperature. It is concluded therefrom that downward motion of the border line is only determined by the usual drainage processes in the film, and there is no appreciable contribution from an extension of the black film. Fig. 7 shows a typical result of such measurements.

Evaporation of water from a film element results in increased solute concentrations, and therefore in decreased surface tension and water vapour pressure. The thickness of the element, at any time during the establishment of the water vapour pressure equilibrium, is determined not only by its decreased volume but also by expansion or contraction resulting from other equilibration processes. In order to clarify the behaviour of the film element under these circumstances, it is necessary to realize that eqn (3) is not the only condition to be satisfied in complete thermodynamic equilibrium.   There are several other conditions, which may conveniently be distinguished according to the characteristic time associated with the corresponding equilibration process.   For any film element at any given time, some of these conditions may already be satisfied because the equilibration processes take less time than the age of the element.   Other processes may require much more time, in particular because an equilibration process may proceed at very different rates in the plane of the film and perpendicular to that plane.

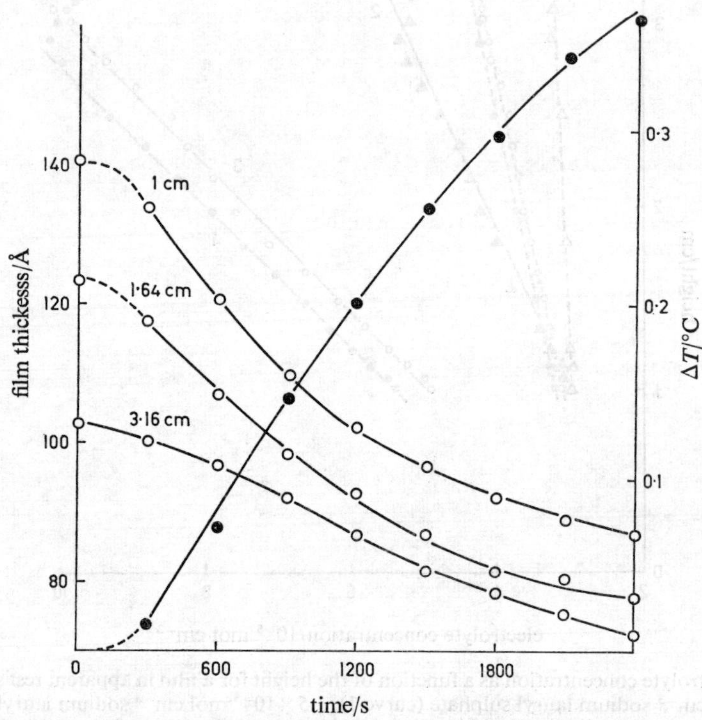

Fig. 6.—Increase in temperature (●) and the resulting decrease in film thickness (○) measured at three indicated heights in the film as a function of time.   The film is stabilized with 1.6×10^{-6} mol cm^{-3} cetyltrimethylammonium bromide and 2.84×10^{-5} mol cm^{-3} sodium bromide.   Original temperature 29.1°C.

For the present discussion, the most important equilibrium conditions and the corresponding equilibration processes are: (i) Uniformity of temperature and of chemical potentials of all components *in a film element*, including the contacting gas layer.   Equilibration occurs by diffusional transport between film liquid and surfaces [8]; the characteristic time is much less than a milli-second.

(ii) Uniformity of surface (or film) tension along the entire surface.   A simple experiment demonstrating the rapidity of this process was already described by Gibbs,[1] and later experience [6] has shown that the time required is of the order of milliseconds.   The mechanism by means of which a thin film satisfies these two equilibrium conditions simultaneously consists of adjusting the thickness of the film elements, by expansion or contraction.   Flow of film liquid with respect to its surface does not contribute significantly to changes in film thickness in so short a time.

(iii) Uniformity of temperature, and of chemical potentials of all volatile components, in *the whole system*.   This is achieved in a time of the order of seconds, by evaporation, diffusion in the gas phase and condensation. The process and the characteristic time for thermal equilibration are similar.

(iv) Mechanical equilibrium in the direction perpendicular to the plane of the film, i.e., between disjoining pressure and hydrostatic head. This is the condition expressed by eqn (3). The main process by means of which this equilibrium is established consists of flow of film liquid with respect to the surface,[5] and the characteristic time is of the order of many hours.

(v) Uniformity of chemical potentials of non-volatile components in the whole system. Diffusion of non-volatile components along the film proceeds about 10^5 times slower than diffusion in the gas phase, and therefore the chemical potentials of such components may not be uniform in a large film unless it has been left undisturbed for at least several days.

During equilibration, as required by conditions (iii) to (v), the equilibrium conditions (i) and (ii) remain *permanently* satisfied. Moreover, the thickness of an element of the film, of given composition, is already completely determined by conditions (i) and (ii), irrespective of whether any further equilibrium conditions are satisfied.

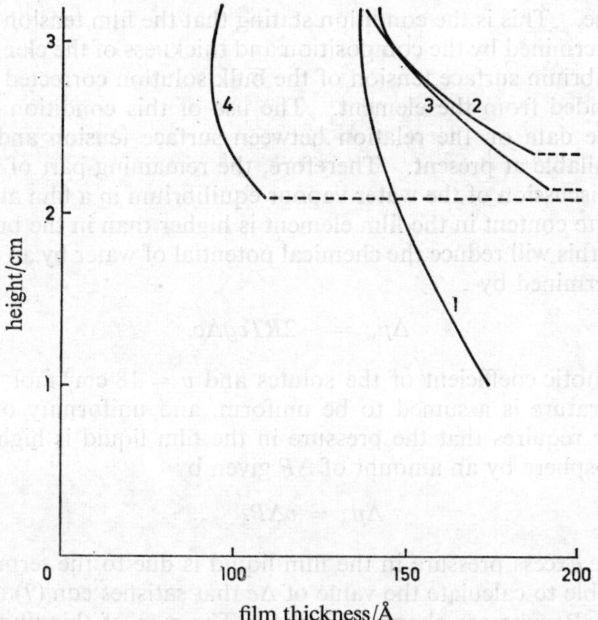

Fig. 7.—Thickness of a film stabilized with 1.5×10^{-5} mol cm^{-3} sodium lauryl sulphate and 1.5×10^{-5} mol cm^{-3} sodium bromide in apparent rest (1), 30 s after loading with a thick film (2) and after further draining for 760 s when the temperature is kept constant (3) or when the film is heated (4).

The interpretation of the film thickness measurements reported in fig. 3 and 4 must, therefore, be based on the supposition that condition (v) is, in general, *not* satisfied, even if the thickness profile remains unchanged during several hours. The liquid composition of the element may differ appreciably from that of the bulk liquid, and even from that of other film elements. The actual composition of the element at a given time is determined by the accumulated effects of processes (i)-(v) since the birth of the elements, and a full description of these effects is not possible at present.

That the composition of a film element depends upon the states through which it has passed during its production is shown by curve 3 in fig. 4. This curve relates to a slow-draining film, made from an aged solution, and which required 7 h to attain a state of apparent rest.

It is evident that the composition of a film element during the measurement reported in fig. 3 and 4 is determined by conditions (i)-(iii) being satisfied, and this composition will, in general, not be such that ΔP of eqn (3) will vanish. The excess pressure ΔP gives rise to flow of film liquid with respect to the surfaces.[5] The resulting rate of thinning in a part of the film with area a^2 is then

$$-dh/dt = 8\Delta P\delta_w^3/3\eta a^2. \tag{6}$$

An evaluation of the magnitude of this effect requires an estimate of ΔP for which we need the electrolyte content in the film liquid. The values shown in fig. 5 cannot be used for this purpose, because they are based on the supposition $\Delta P = 0$. If it is assumed that equilibrium condition (iv) need not necessarily be satisfied in a film element in a state of apparent rest, it is still possible to estimate the electrolyte concentration in the film element from condition (iii) expressing the uniformity of the chemical potential of water. In principle, it would also be possible to use condition (ii) for this purpose. This is the condition stating that the film tension in the element at height H, as determined by the composition and thickness of the element, should be equal to the equilibrium surface tension of the bulk solution corrected for the weight of the film suspended from the element. The use of this condition would require extremely accurate data on the relation between surface tension and composition, which are not available at present. Therefore, the remaining part of this paper will be confined to a discussion of the water vapour equilibrium in a film at apparent rest.

If the electrolyte content in the film element is higher than in the bulk solution by an amount of Δc, this will reduce the chemical potential of water by an amount that is approximtely determined by

$$\Delta\mu_w = -2RTvg\Delta c. \tag{7}$$

Here, g is the osmotic coefficient of the solutes and $v = 18 \text{ cm}^3 \text{ mol}^{-1}$. For a film at rest the temperature is assumed to be uniform, and uniformity of the chemical potential of water requires that the pressure in the film liquid is higher than in the surrounding atmosphere by an amount of ΔP given by

$$\Delta\mu_w = v\Delta P. \tag{8}$$

Assuming that the excess pressure in the film liquid is due to the terms appearing in eqn (3), it is possible to calculate the value of Δc that satisfies eqn (7) and (8) for any film thickness h. Results are shown in fig. 5. The rate of thinning due to liquid flow can now be found by inserting the resulting ΔP in eqn (6). For a reasonable value of the surface area (several cm²) this gives a rate of thinning of one Å in a day.

It is concluded that the thickness of a film element at some distance from the border is not determined by disjoining pressure equilibrium but by vapour pressure equilibrium of volatile components. It is hoped that further work along these lines will provide information about the meaning of " pressure " in a thin film.

Thanks are due to Mr. Th. C. Kouters for skilful execution of the experiments.

[1] J. W. Gibbs, *The Scientific Papers*, (Dover Publ. 1961), vol. 1, p. 305.
[2] J. Lyklema and K. J. Mysels, *J. Amer. Chem. Soc.*, 1965, **87**, 2539.
[3] J. S. Clunie, J. F. Goodman and P. C. Symons, *Nature*, 1967, **216**, 1203.
[4] M. N. Jones and D. A. Reed, *J. Colloid Interface Sci.*, 1969, **30**, 577.
[5] A. Scheludko, *Adv. Colloid Interface Sci.*, 1967, **1**, 39.
[6] A. Prins, C. Arcuri and M. van den Tempel, *J. Colloid Interface Sci.*, 1967, **24**, 84.
[7] J. S. Clunie, J. F. Goodman and J. R. Tate, *Trans. Faraday Soc.*, 1968, **64**, 1965.
[8] J. Lucassen and R. S. Hansen, *J. Colloid Interface Sci.*, 1967, **23**, 319.

Aqueous Foam Films Stabilized by a Non-Ionic Surface-Active Agent

By J. S. Clunie, J. M. Corkill, J. F. Goodman and B. T. Ingram

Procter & Gamble Limited, Newcastle Technical Centre, Basic Research Department, Newcastle-upon-Tyne, England

Received 10*th April*, 1970

The thicknesses and tensions of black films formed by aqueous solutions of a pure, non-ionic surface-active agent, n-decyl methyl sulphoxide (DMS), have been measured at 298 and 308 K as a function of sodium chloride concentration. In the $DMS+NaCl+H_2O$ system, second black films are formed at low ionic strengths whereas first black films are formed at higher ionic strengths (>100 mol m^{-3}). Estimated values for the composite Hamaker constant for first and second black films have been obtained and compared with theoretical values.

In most studies on aqueous foam films the surface-active agents used have been ionic in character.[1] By comparison, studies using non-ionic surface-active agents are fewer in number and mostly confined to films formed by aqueous solutions of commercial alkyl phenol polyoxyethylene ethers (OP-7 to OP-20).[2-4] Nevertheless, some observations have been made on films formed by aqueous solutions of pure n-dodecyl hexaoxyethylene glycol monoether,[5] and a few results have also been reported for films stabilized by another pure non-ionic surface-active agent (a partially-fluorinated n-alkyl dimethylamine oxide [6]).

In the present investigation, a study has been made on films formed by aqueous solutions of n-decyl methyl sulphoxide (n-C$_{10}$H$_{21}$.SO.CH$_3$.DMS). This non-ionic surface-active agent has a compact hydrophilic group and, unlike the n-alkyl dimethylamine oxides,[7] is protonated only under extremely acid conditions (pK$_a<0$).[8] We have determined the thicknesses and excess tensions of the foam films formed by solutions of this material in water and in the presence of added sodium chloride (to 5 kmol m^{-3}). From these measurements estimated values for the composite Hamaker constant for first and second black films have been obtained and compared with calculated values.

EXPERIMENTAL

MATERIALS

n-Decyl methyl sulphoxide (DMS) was prepared and purified as described previously.[9]

The water used for preparing solutions was twice distilled in a silica vessel after initial distillation from aqueous alkaline potassium permanganate. At 298 K this water had a specific conductivity of less than $10^{-4}\,\Omega^{-1}\,m^{-1}$ and a surface tension of 72.0 mN m^{-1}.

A.R. sodium chloride was roasted in a platinum crucible at $\sim$900 K. A 1 kmol m^{-3} solution raised the surface tension of water by 1.6 mN m^{-1} at 298 K,[10] indicating the absence of any surface-active impurities.

SURFACE TENSION MEASUREMENTS

Solutions were contained in a double-walled thermostatted cell with a thermistor for temperature monitoring.   Surface tensions were determined using a du Noüy tensiometer and the usual corrections were applied.[11]

FILM THICKNESS MEASUREMENTS

The apparatus consisted of a glass cell (3.5×10^{-4} m³ capacity) made vacuum tight by demountable seals, liquid seals and greaseless vacuum taps (fig. 1).   An auxiliary reservoir

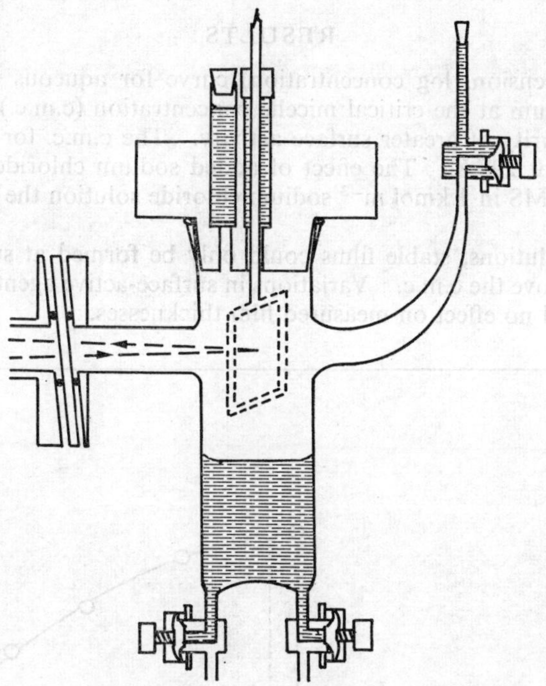

FIG. 1.—Cell used for optical reflectance measurements.

(2.5×10^{-4} m³ capacity) was connected to the cell by a siphon through which the surface-active solution was introduced into the evacuated cell.   Both cell and reservoir were totally immersed in a constant temperature water bath controlled to ± 0.01 K by a mercury-toluene regulator with two 100 W heaters and a cooling coil.   The water bath was housed in a light-proof enclosure but was fitted with a window for viewing the film when necessary.

Films were formed by totally withdrawing a rectangular glass frame ($50 \times 20 \times 5.0$ mm) from the surface-active solution.   The frame was supported vertically from a Perspex vessel top through a liquid seal filled with the solution under investigation.   Film thickness was determined by an optical reflection method similar to that described previously,[12] with the addition of a beam splitter to allow continuous monitoring of the intensity of the incident monochromatic light beam.   Throughout the course of any film's lifetime, the intensity of the incident beam was constant to within ± 2 %.   The measured background light intensity was ~20 % of the reflectance from the thinnest films studied.   The uncertainties in reflection coefficient measurements lead, on the basis of a single homogeneous layer model for the film,[12] to a corresponding imprecision in film thickness of less than 2 %.

Film thicknesses were calculated from the reflectance measurements using the symmetrical three-layer model previously adopted.[6]   The parameters of this model are the refractive indices of the surface monolayers (1.41) and the aqueous core (1.34), and also the thickness of the surface monolayers (1.4 nm). The uncertainty in the calculated thickness for possible variations in these parameters is unlikely to exceed 0.4 nm.

FILM TENSION MEASUREMENTS

The apparatus and technique were essentially those described previously [13] but to increase the sensitivity in the present series of measurements an 80 mm wide glass frame made of 1 mm thick glass was used to support the film. The electronic microbalance was calibrated with standard weights at 298 and 308 K. The estimated error in the measurement of excess tension was $\pm 2.5 \ \mu N \ m^{-1}$.

RESULTS

The (surface tension, log concentration) curve for aqueous solutions of DMS showed no minimum at the critical micelle concentration (c.m.c.) indicating the absence of any impurity of greater surface activity. The c.m.c. for aqueous solutions was 2.0 mol m^{-3} at 298 K. The effect of added sodium chloride was to lower the c.m.c. (e.g., for DMS in 1 kmol m^{-3} sodium chloride solution the c.m.c. was 1.0 mol m^{-3} at 298 K).

With DMS solutions, stable films could only be formed at surface-active agent concentrations above the c.m.c. Variations in surface-active agent concentrations up to $1.5 \times$ c.m.c. had no effect on measured film thicknesses.

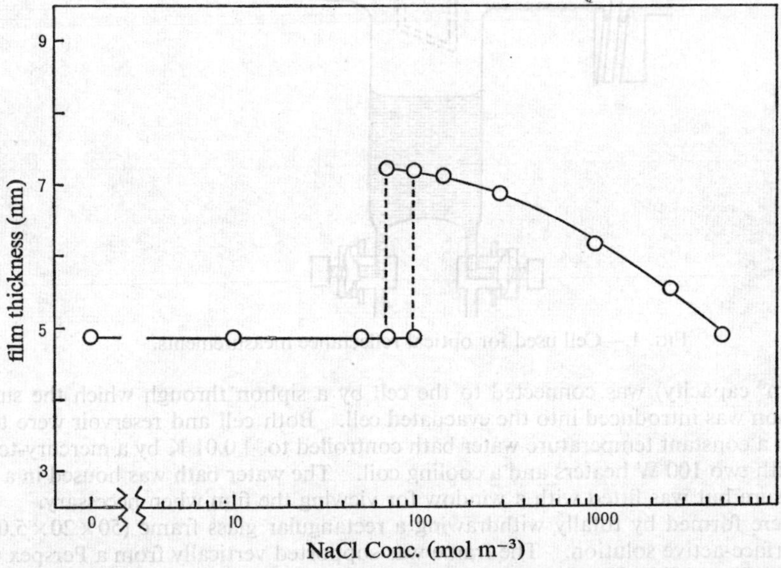

FIG. 2.—Equilibrium film thickness in the DMS+NaCl+H$_2$O system as a function of NaCl concentration at 298 K. Results at 308 K are identical.

FILM THICKNESS

Measurements were made at 298 K and 308 K. All films showed mobile drainage behaviour and rapidly reached constant thicknesses which were maintained indefinitely and which were reproducible to ± 0.2 nm. Fig. 2 shows the equilibrium thicknesses of films formed from 2.2 mol m^{-3} solutions of DMS containing varying amounts of sodium chloride. For any given film the measured reflection coefficients were the same at both temperatures. Since any variation in the parameters of the optical model in this small temperature interval will be very small, the calculated film thicknesses are independent of temperature.

Equilibrium film thicknesses were low (4.9 nm) and independent of electrolyte concentration up to a sodium chloride concentration of 70 mol m^{-3} where a sharp increase in film thickness to 7.3 nm was observed. With further increases in sodium chloride concentration, the equilibrium film thickness decreased continuously, returning to a value of 4.9 nm at the highest concentration studied (5 kmol m^{-3}). The abrupt transition in film thickness from 4.9 to 7.3 nm occurred over a fairly narrow range of sodium chloride concentrations (70-100 mol m^{-3}). In this concentration range two co-existing black films could be formed in the vertical film holder, viz., an upper film of thickness 4.9 nm in equilibrium with a lower film of thickness 7.3 nm.

FILM TENSION

Measurements were made at 5 K intervals in the temperature range 298-308 K. Fig. 3 shows the excess film tension, $\Delta\sigma$ ($\Delta\sigma = \sigma^f - 2\gamma$, where σ^f is the total film tension

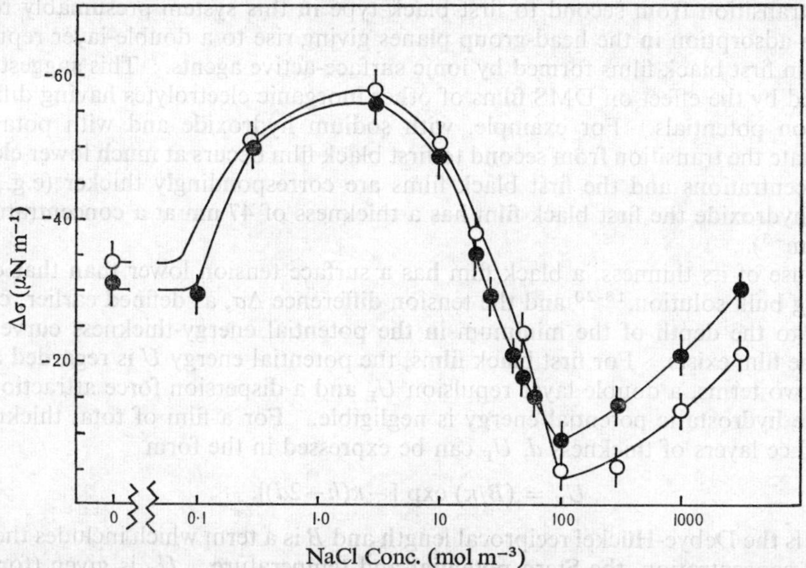

Fig. 3.—Excess film tension $\Delta\sigma$ in the DMS+NaCl+H$_2$O system as a function of NaCl concentration at 298 K ($\bullet$) and 308 K ($\bigcirc$).

and γ is the surface tension of the bulk solution), as a function of sodium chloride concentration at 298 and 308 K. The sharp minima in the curves at 100 mol m^{-3} corresponded to the abrupt transition in film thickness. $\Delta\sigma$ appeared to change linearly with temperature, the extreme values of the temperature coefficient being -0.4 and $+1.1$ μN m^{-1} K^{-1} at sodium chloride concentrations of 50 mol m^{-3} and 1 kmol m^{-3} respectively.

DISCUSSION

Mysels et al.[14, 16] have defined " first " and " second " black films in terms of the behaviour of the equilibrium thickness with respect to the ionic strength of the bulk solution. First black films show a monotonic decrease of thickness with increasing

SP1—B

ionic strength, whereas with second black films the thickness is independent of the ionic strength. It is generally considered [15] that the equilibrium thickness of the first black film is governed by the balance between the van der Waals attractive forces and the repulsion due to the overlap of the electrical double layers associated with a surface charge in the head-group planes. For the second black film, the repulsion force is less well defined and is short range in nature.

For films stabilized with non-ionic surface-active agents one might expect that since the head-groups are uncharged, only second black films would be observed. In the DMS system, second black films are indeed found at sodium chloride concentrations less than 70 mol m^{-3} but, after a restricted concentration region in which co-existing first and second black films are observed, only first black films are formed. Due to salting out of the DMS, the highest electrolyte concentration at which films could be formed was 5 kmol m^{-3}, and at this composition the film thickness had returned to a value close to the electrolyte-free second black film thickness of 4.9 nm. In films formed from ionic surface-active agents,[16, 17] increasing electrolyte results in a transition from the first to the second black type; in the present system the converse is the case.

The transition from second to first black type in this system presumably results from ion adsorption in the head-group planes giving rise to a double-layer repulsion force as in first black films formed by ionic surface-active agents. This suggestion is supported by the effect on DMS films of other inorganic electrolytes having different adsorption potentials. For example, with sodium hydroxide and with potassium thiocyanate the transition from second to first black film occurs at much lower electrolyte concentrations and the first black films are correspondingly thicker (e.g., with sodium hydroxide the first black film has a thickness of 47 nm at a concentration of 0.1 mol m^{-3}).

Because of its thinness, a black film has a surface tension lower than that of the adjoining bulk solution,[18-20] and the tension difference $\Delta\sigma$, as defined earlier, can be equated to the depth of the minimum in the potential energy-thickness curve [21] in which the film exists. For first black films, the potential energy U is regarded as the sum of two terms, a double layer repulsion U_E and a dispersion force attraction U_A, when the hydrostatic potential energy is negligible. For a film of total thickness h and surface layers of thickness d, U_E can be expressed in the form

$$U_E = (B/\kappa) \exp[-\kappa(h-2d)], \tag{1}$$

where κ is the Debye-Hückel reciprocal length and B is a term which includes the bulk solution concentration, the Stern potential and temperature. U_A is given (for non-retarded forces) by

$$U_A = -A^*/12\pi h^2, \tag{2}$$

where A^* is the composite Hamaker constant for the film.

Since at equilibrium the derivative of the potential energy with respect to thickness is zero, it follows from (1) and (2) that, if A^* is independent of h,

$$\kappa U_E + (2U_A/h) = 0 \tag{3}$$

The potential energy $(U_E + U_A)$ can consequently be expressed in the form,

$$U = U_A(1 - 2/\kappa h), \tag{4}$$

or
$$\Delta\sigma = (-A^*/12\pi h^2)(1 - 2/\kappa h). \tag{5}$$

Since both h and $\Delta\sigma$ can be directly measured eqn (5) affords a method for determining A^*.

In fig. 4, the values of A^* obtained from $\Delta\sigma$ and h for the first black films studied are shown as a function of h at 298 and 308 K.   The large variation of A^* with temperature

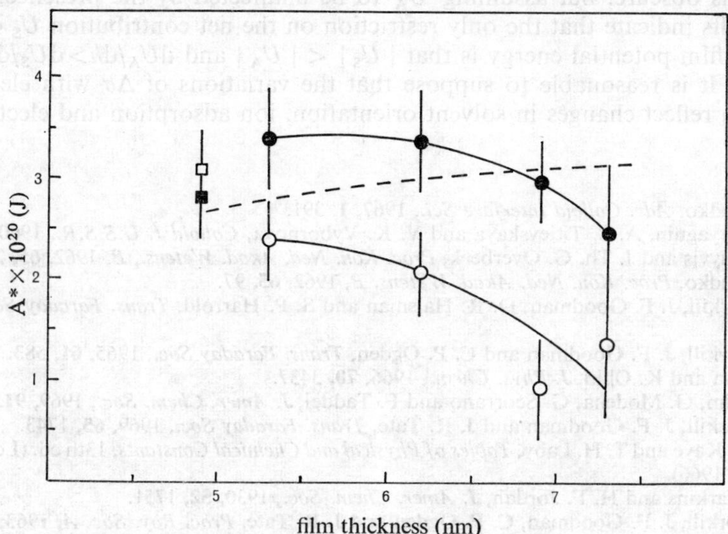

Fig. 4.—Hamaker constant A^* for first ($\bigcirc$) and salt-free second ($\square$) black films as a function of film thickness at 298 K (closed symbols) and 308 K (open symbols).   Dashed line represents the calculated Hamaker constant for the non-retarded van der Waals forces.

is most unexpected since the densities and polarizabilities of the film components should show only a small change for a 10 K temperature difference.   Taking the value of $A^* = 2.5 \times 10^{-20}$ J into eqn (3) leads to a value for U_E and hence the Stern potential can be calculated.   At the transition concentration, a potential of 20 mV is obtained which is similar to that obtained [3] for first black films stabilized by the non-ionic surface-active agent OP-20.

Approximate values of A^* may be calculated by the method of Duyvis [22] from the Hamaker constants and thicknesses of the film surface layers and aqueous core.   The Hamaker constants A for water and decane can be calculated from a form of the London equation,[23]

$$A = \frac{27}{64} h v_L \left(\frac{n^2 - 1}{n^2 + 2} \right)^2 \tag{6}$$

where h is Planck's constant, n the refractive index and v_L the characteristic frequency. Using the values of v_L given by Gregory,[24] we obtain A values for water and decane of 3.9×10^{-20} and 5.8×10^{-20} J respectively.   Inserting these into the Duyvis equation leads to A^* values that are in moderate agreement with those calculated from the experimental data using eqn (5).

An alternative method of estimating A^* is to assume that for the second black films, in the *absence* of electrolyte, the short-range repulsion is represented by a cut-off potential and hence that $U = U_A$ at the equilibrium thickness.   The value of A^* for the electrolyte free films is comparable to that calculated from eqn (5) for the first black films (fig. 4).   However, for these films, although h is independent of electrolyte concentration, $\Delta\sigma$ shows large changes with electrolyte concentration passing through a maximum at ~ 1.0 mol m^{-3}.   It seems improbable that compositional changes could lead to variations in A^* large enough to account for these effects, and hence some other forces besides the van der Waals attraction and a steric (cut-off) repulsion must operate in these films. The nature of these additional short-range forces remains obscure, but assuming U_A to be unaffected by the presence of ions then the results indicate that the only restriction on the net contribution U_S of these forces to the film potential energy is that $|U_S| < |U_A|$ and $dU_A/dh > dU_S/dh$ at all thicknesses. It is reasonable to suppose that the variations of $\Delta\sigma$ with electrolyte concentration reflect changes in solvent orientation, ion adsorption and electrostatic screening.

[1] A. Scheludko, *Adv. Colloid Interface Sci.*, 1967, **1**, 391.

[2] B. V. Deryaguin, A. S. Titievskaya and V. K. Vybornova, *Colloid J. U.S.S.R.*, 1960, **22**, 407.

[3] E. M. Duyvis and J. Th. G. Overbeek, *Proc. Kon. Ned. Akad. Wetens.*, B, 1962, **65**, 26.

[4] A. Scheludko, *Proc. Kon. Ned. Akad. Wetens. B*, 1962, **65**, 97.

[5] J. M. Corkill, J. F. Goodman, D. R. Haisman and S. P. Harrold, *Trans. Faraday Soc.*, 1961, **57**, 821.

[6] J. M. Corkill, J. F. Goodman and C. P. Ogden, *Trans. Faraday Soc.*, 1965, **61**, 583.

[7] F. Tokiwa and K. Ohki, *J. Phys. Chem.*, 1966, **70**, 3437.

[8] D. Landini, G. Modena, G. Scorrano and F. Taddei, *J. Amer. Chem. Soc.*, 1969, **91**, 6703.

[9] J. M. Corkill, J. F. Goodman and J. R. Tate, *Trans. Faraday Soc.*, 1969, **65**, 1743.

[10] G. W. C. Kaye and T. H. Laby, *Tables of Physical and Chemical Constants*, 13th ed. (Longmans, London, 1966).

[11] W. D. Harkins and H. F. Jordan, *J. Amer. Chem. Soc.*, 1930, **52**, 1751.

[12] J. M. Corkill, J. F. Goodman, C. P. Ogden and J. R. Tate, *Proc. Roy. Soc. A*, 1963, **273**, 84.

[13] J. H. Clint, J. S. Clunie, J. F. Goodman and J. R. Tate, *Nature*, 1969, **223**, 291.

[14] K. J. Mysels, K. Shinoda and S. Frankel, *Soap Films, Studies of their Thinning and a Bibliography* (Pergamon Press, London, 1959).

[15] J. Lyklema and K. J. Mysels, *J. Amer. Chem. Soc.*, 1965, **87**, 2539.

[16] M. N. Jones, K. J. Mysels and P. C. Scholten, *Trans. Faraday Soc.*, 1966, **62**, 1336.

[17] G. Ibbotson and M. N. Jones, *Trans. Faraday Soc.*, 1969, **65**, 1146.

[18] B. V. Deryaguin, G. A. Martinov and Yu. V. Gutop, *Colloid J. U.S.S.R.*, 1965, **27**, 298.

[19] K. J. Mysels, H. F. Huisman and R. Razouk, *J. Phys. Chem.*, 1966, **70**, 1339.

[20] A. Scheludko, B. Radoev and T. Kolarov, *Trans. Faraday Soc.*, 1968, **64**, 2213.

[21] F. Huisman and K. J. Mysels, *J. Phys. Chem.*, 1969, **73**, 489.

[22] E. M. Duyvis, *The Equilibrium Thickness of Free Liquid Films* (Thesis, Utrecht, 1962).

[23] D. Tabor and R. H. S. Winterton, *Proc. Roy. Soc. A*, 1969, **312**, 435.

[24] J. Gregory, *Adv. Colloid Interface Sci.*, 1970, **2**, 396.

GENERAL DISCUSSION

Dr. B. A. Pethica (*Unilever Res., Port Sunlight*) said: Frens, Mysels and Vijayendran suggest that their derived (surface-pressure, apparent-molecular-area) curves for a collapsing film stabilized by sodium dodecyl sulphate indicate a rapid increase in intermolecular attractions as the surface becomes very crowded. The form of the curves is, however, typical of those for collapsing insoluble monolayers, and it is probable that a combination of solution and stacking effects in the collapsing film will explain the results.

Dr. K. J. Mysels (*R. J. Reynolds, Winston Salem*) said: Pethica has raised the question whether the unexpected reversal of curvature in the (Π,A) curves at low areas per molecule might not be due to a partial collapse of the surface monolayer. There is no way to answer that question with complete certainty on the basis of our observations. However, the surfactant layer is highly ionized and not compact in our experiments (*ca.* 30 Å^2/ion). This, and the fact that variations in the rate of concentration—due to differences in thickness of the bursting film—by a factor of 10, do not produce significant deviations from the measured (σ,α) curve, are arguments against monolayer collapse as an explanation for the inversion of curvature. The reversal of curvature is also not to be explained as a time effect due to desorption. If the reverse curvature were not intrinsic for the film surface there could not be a beginning to the propagation of the aureole, and therefore no such phenomenon to produce relaxation effects. In that case only a single shock-wave at the rim of the hole would propagate into an undisturbed film. In our opinion the reversal of curvature might indicate something like the incipient formation of a closely-packed, vertically-oriented monolayer of long-chain ions. In such a layer, important cohesive forces between chains would come into effect and act in the direction of inversion of the curvature.

Dr. G. Frens (*Philips Res. Lab., Eindhoven*) said: One would like to relate the surface tensions of super-saturated surfaces as obtained from the interpretation of the aureoles in bursting soap films to data for surface tension as a function of surface coverage below the c.m.c. We have pointed out that a problem here is the lack of agreement between the data of the different authors who have determined (Π,A) curves for NaLS below the c.m.c.

Mrs. Lucassen-Reynders of Unilever Research Laboratories has tackled this problem, and some of her results were made available to me. She had shown earlier that two equations,

$$-\frac{\sigma_0-\sigma}{RT\Gamma^\infty} = \ln\left(1-\frac{\Gamma}{\Gamma^\infty}\right) + \frac{H^s}{RT}\left(\frac{\Gamma}{\Gamma^\infty}\right)^2, \tag{1}$$

and

$$\frac{c}{a} = \frac{\Gamma/\Gamma^\infty}{1-\Gamma/\Gamma^\infty} \exp\left(-\frac{2H^s}{RT}\frac{\Gamma}{\Gamma^\infty}\right), \tag{2}$$

relate surface pressure $(\sigma_0-\sigma)$ to adsorption (Γ) and Γ to the surfactant concentration c for ionized surfactant.[1] These equations have three constants (a, Γ^∞ and H^s)

[1] E. H. Lucassen-Reynders, *J. Phys. Chem.*, 1966, **70**, 1777.

38   GENERAL DISCUSSION

whereas $(\sigma_0 - \sigma)$ and c are the experimentally measured variables.   Mrs. Lucassen
computed which values of the constants a, Γ^∞ and H^s gave the best fit between theory
and experimental data.   The best set of constants is that for which the parameter

$$D \equiv \sum_{\text{all exp}} (\sigma_{\text{expt.}} - \sigma_{\text{calc.}})^2 \tag{3}$$

has its minimum.   She analyzed some twenty sets of experimental data from the
literature.   Of these the surface tensions of NaLS as measured by Elworthy and
Mysels [1] gave satisfactory results.   The other data led to unreasonable values for
a, Γ^∞ and H^s, probably because there was too large a scatter in the experimental
points.

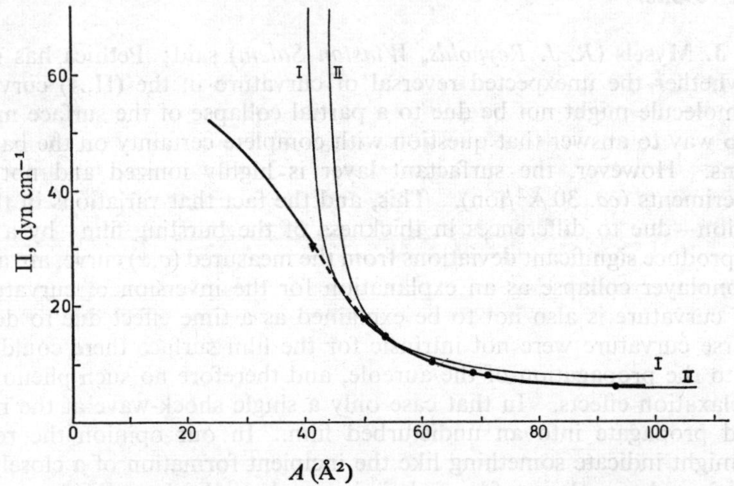

FIG. 1.—(Π, A) curve for NaLS ; I and II calculated by Lucassen-Reynders ; ■, experimental points
(ref. (2)); ▼, " anchoring " point at c.m.c. (ref. (2)).

Mrs. Lucassen's procedure will not only sort out the best set against experimental
data, it also produces a (Π, A) curve (or $\sigma_0 - \sigma$ against Γ) through eqn (1).   In fig. 1
we have combined this (Π, A) curve below the c.m.c. with the curve at high surface
coverages, which we obtained from our bursting experiments.   The values $\Pi = 30.3$
mN m^{-1} and $A = 0.415$ nm^2 at the c.m.c. (where the soap film data are anchored
in the (Π, A) plane) were taken from the literature.[1, 2]   The value of Π agrees with
our own measurements.

One would expect eqn (1) to break down near the c.m.c. as it predicts that the
surface pressure would become infinite on saturation $(\Gamma = \Gamma^\infty)$.   It is seen, however,
that only a short interpolation is necessary in order to connect smoothly the two
parts of the curve in fig. 1 into one (Π, A) plot, which might then be meaningfully
interpreted in a more complete equation of state than eqn (1).

Dr. A. T. Florence and **Dr. K. J. Mysels** (*Strathclyde University* and *R. J. Reynolds,
Winston Salem*) said : In reply to Goodman, we have been concerned with the question
whether the existence of the aureole may not be due to the presence of impurities or
hydrolysis in the sodium-dodecylsulphate used.   As mentioned in our paper we used

[1] P. H. Elworthy, K. J. Mysels, *J. Colloid Interface Sci.*, 1966, **21**, 331.
[2] I. Weil, *J. Phys. Chem.*, 1966, **71**, 738.

solutions well above the c.m.c. so as to solubilize any surfactant impurities and thus reduce any such effects. These effects, particularly those of dodecyl alcohol, cannot be very significant. This is shown by some yet unpublished experiments of Dr. A. T. Florence at the R. J. Reynolds Lab. He combined the purification by foaming of a solution below the c.m.c.[1] with photographic studies of bursting. Enough dodecyl alcohol was initially added to the solution to give rigid films. These show a peculiar bursting behaviour.[2] Foaming rapidly reduced the alcohol concentration to the point where the films become mobile and bursting produces the usual aureole. This aureole persisted in the experiments without qualitative change through several days of further purification by foaming. Razouk and Mysels[3] used a similar purification method. They obtained indistinguishable results in solutions of sodium dodecyl sulphate and sulphonate. This indicates that hydroylsis of the sulphate is not a limiting factor under these conditions.

Prof. A. Vrij (*Utrecht, Netherlands*) said: I would ask Prins the following questions: (i) in fig. 3 of his paper he has plotted a calculated equilibrium thickness (graph 2) for an electrolyte concentration of 3×10^{-5} mol cm^{-3}, which is the total concentration of sodium lauryl sulphate. In my opinion, however, he should have used the concentration of free soap ions (i.e., the c.m.c.) which is about a factor of 4 smaller. (ii) He proposes that the pressure in the film could be higher than in the surrounding atmosphere. Does this mean that in that case the van der Waals' attraction forces exceed the double-layer repulsion forces? (iii) The plots of calculated electrolyte concentrations as a function of height in fig. 5 are linear. Does this have any physical significance?

Dr. A. Prins (*Unilever Res. Lab., Netherlands*) said: In reply to Vrij, the activity of surfactant molecules is indeed affected by micelle formation. However, since it is not known how the counter ion activity changes with the concentration in the presence of micelles, we have assumed that for the calculation of the film thickness, the nominal concentration of the counter ions has to be used. This is confirmed by the agreement between the measured and the calculated film thickness at the bottom plateau border. In addition, as appears from fig. 4 of our paper, exchange of most of the surfactant by salt results in exactly the same profile of the film. Moreover, a calculation based on a salt concentration equal to the c.m.c. would result in an even thicker film, making the discrepancy between theory and experiment even bigger.

The suggested explanation of the water vapour pressure equilibrium is indeed that the pressure in the film exceeds atmospheric pressure in order to compensate for the decrease in water vapour pressure caused by the higher salt concentration. The linear relation between the salt concentration and the height in the film suggests that it is perhaps possible to give a simple explanation for the observed phenomena. The physical significance, however, is not known at present.

Dr. M. N. Jones (*University of Manchester*) said: With regard to the paper by Prins and van den Tempel, Reed and I have attempted to measure the thickness profile of films up to 10 cm in height drawn from sodium n-dodecylsulphate solutions at low concentrations (~ 0.009 M) and found that while a given film had a thickness which remained constant within ~ 0.5 nm over a period of several hours there was considerable variation (± 10 nm) in thickness between different films drawn from

[1] K. J. Mysels and A. T. Florence, in *Clean Surfaces*, G. Goldfinger ed., (New York, 1970).
[2] W. R. McEntee and K. J. Mysels, *J. Phys. Chem.*, 1969, **73**, 3018.
[3] Razouk and K. J. Mysels, *J. Amer. Oil Chem. Soc.*, 1968, **45**, 381.

the same solution.   At such low ionic strength the film thicknesses were very dependent on the method used to humidify the atmosphere in the film chamber and it was possible to observe, even with the naked eye, that different elements of the film had different thicknesses.   The addition of electrolyte, however, had a marked effect on the thickness profiles and at higher ionic strengths (~ 0.15 M 1 : 1 electrolyte) we found that for a given film although the thickness decreased with time, this change was only small ~ 0.5 nm and when the system was left to thermostat for a long period (~ 70 h) the decrease was even smaller, which implied that it was due to evaporation. What is perhaps more significant is that under these conditions the initial black film thickness of freshly drawn films was always constant within ± 0.05 nm and we concluded that these are true equilibrium thicknesses.   At higher salt concentrations (~ 0.2 M 1 : 1 electrolyte) we always obtained uniform films of constant thicknesses. It would seem from these observations that the diffusion of non-volatile components along the film is dependent on ionic strength and I would like to know whether they have made any observations on thinner films.   Eqn (6) of this paper predicts that the rate of thinning should be much *slower* for thin films and this is contrary to our observations.

Dr. A. Prins (*Unilever Res. Lab., Netherlands*) said: In reply to Jones, a measurable thickness profile as described in our paper appears only when films are used which contain not too much salt.   By increasing the salt concentration the film profile becomes more and more parallel and a Perrin film has a uniform thickness which, however, is determined by other forces than are discussed in our paper.   Under the usual conditions of observation the rate of thinning of large films is always determined by evaporation and not by the process described by eqn (6).

In reply to Lyklema, Bruil carried out his film thickness measurements at 0.5 cm above the bottom plateau border.   It might well be that his results were affected by the phenomena discussed in our paper, which should result in slightly thinner films than predicted by the DVLO theory at that level.

Dr. H. Sonntag (*Deutsche Akademie der Wissenschaften, Berlin*) said: Clunie *et al.* expect that films stabilized with non-ionic surface-active agents only might form second black films.   I cannot agree with this opinion, because, (i) we only found first black films in polar systems with various non-polar surface-active agents and never obtained an alteration of film tension; (ii) aqueous films between oil droplets stabilized with nonylphenol ethylene oxide (20) formed second black films after adding K_2SO_4 or $MgSO_4$ as electrolyte.   With $Ca(NO_3)_2$ or $BaCl_2$ e.g., we obtained only first black films at all concentrations.   I think the experimental results are still insufficient to give a theoretical explanation of the formation of second black films.   Moreover I would ask whether they are quite sure that there is no building-up of adsorbed multilayers of surface-active agents above the critical micelle concentration?

Dr. J. M. Corkill (*Procter & Gamble Ltd., Newcastle-upon-Tyne*) said: In reply to Sonntag, it is generally accepted that the stability of first black films is determined by a balance between van der Waals' attraction and electrostatic double-layer repulsion.   For aqueous foam films stabilized by a non-ionizable surface-active agent, the existence of any diffuse double layers can only result from ion adsorption at the film-core/surfactant-monolayer interface.   In the absence of ion adsorption one would therefore not expect the formation of first black films.   Any stable films that are formed in such systems are likely to be of the second black type.   Recent radio-tracer experiments [1] using ^{35}S-labelled decyl methyl sulphoxide have shown the equilibrium black film radioactivity to be virtually independent of electrolyte concentration in bulk solution (up to 4 kmol m^{-3}).   The measured activity corresponds to an area of 0.34 ± 0.02 nm^2 per surfactant molecule in the film surface.   This value is identical to the area per surfactant molecule at the bulk-solution/air interface obtained from surface tension measurements on the same system.

Dr. S. Levine (*Manchester University*) and **Prof. G. M. Bell** (*Chelsea College*) said: In the paper by Clunie, Corkill, Goodman and Ingram, eqn (1) for the double-layer repulsion, which is based on the DLVO theory, is applied to electrolyte concentrations as high as 5 mol/l.   However, this is derived from the Poisson-Boltzmann (Gouy-Chapman) equation, which in fact should not be used for concentrations of 1 : 1 electrolytes greater than say 0.3 mol/l. at the rather low value of the Stern potential (20 mV) quoted by the authors.   The corrections to the Poisson-Boltzmann equation have been considered by many authors.[2]

Dr. J. M. Corkill (*Procter & Gamble Ltd., Newcastle-upon-Tyne*) said: In reply to Levine, we agree with his comment concerning the limitation of the Gouy-Chapman theory to electrolyte concentrations below ~ 300 mol m^{-3}.   However, we have found good agreement between experimental and theoretical A^* values both below and above the strict limits of applicability of double-layer theory.   In these film calculations, the contribution to the potential energy from electrical double-layer repulsion is much smaller than that due to van der Waals' attractive forces (maximum of 40 % for 7.2 nm films).   The influence on A^* of errors in calculating the double-layer repulsion term for high ionic strength films will probably be small.   This may explain the consistency between calculated A^* values for films at high and low ionic strengths and their agreement with the A^* value calculated for the electrolyte-free second black film.

Prof. A. Scheludko (*Sofia*) (*communicated*): Exerowa and Kolarov have measured the thickness of microscopic horizontal films of $C_{10}H_{21}SOCH_3$ aqueous solution (the compound was supplied by Dr. Goodman), the concentration of the surfactant being the same as in the paper of Clunie *et al.* (2.2×10^{-3} M) and electrolyte NaCl 10^{-4} M.   White films have been obtained with thickness of 460-520 Å instead of 48 Å measured by Clunie *et al.* at low electrolyte concentration.   The outer capillary pressure was 290 dyn cm^{-2} which yields for the potential of the diffuse electric layer $\phi_0 = 17$ mV as calculated according to the DLVO theory and with a van der Waals-Hamaker constant $K = A/4\pi = 4 \times 10^{-14}$.[2]   The calculation has been carried out with the complete formulae and has account of all the necessary conditions as described in ref. (2).   The certainty of the value obtained for ϕ_0 is diminished because of the considerable effect of the van der Waals component of the disjoining pressure with such a low value of the potential and thicknesses; the van der Waals constant is known only with limited accuracy.[1]   It is certain, however, that the potential is considerably lower than that of water (30 mV), which is evident from the fact that the equilibrium thickness is smaller with the same electrolyte concentration (700 Å without the surfactant [2] against ~ 500 Å here).   The significant difference in thickness compared to 48 Å of Clunie *et al.* is completely explained by the higher outer pressure of the cell in our work.

[1] J. S. Clunie, J. M. Corkill and B. T. Ingram, unpublished results.
[2] see e.g., S. Levine and G. M. Bell, *Disc. Faraday Soc.*, 1966, **42**, 69.

In fig. 1 the curve disjoining pressure Π against film thickness h at 10^{-4} M NaCl is presented, Π being calculated from the approximate expression

$$\Pi = 64cRT\gamma^2 \exp(-xh) - K/h^3,$$

with $\gamma = 0.164$, corresponding to $\phi_0 = 17$ mV.

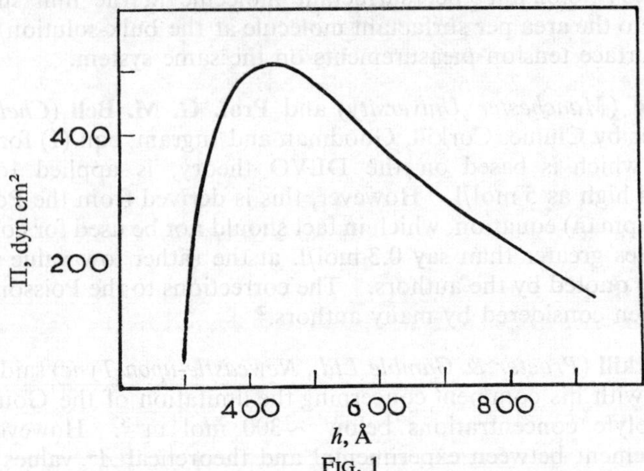

FIG. 1

As is seen, the force barrier corresponds to some 500 dyn cm^{-2} so that with a higher outer pressure this barrier would be overcome and the film would either collapse or form a more stable film (Perrin's film), the latter not being described by the DLVO theory.   The outer pressure in the vertical frame of Clunie *et al.* also includes the hydrostatic pressure which at the upper end of the frame is ~ 5000 dyn cm^{-2}

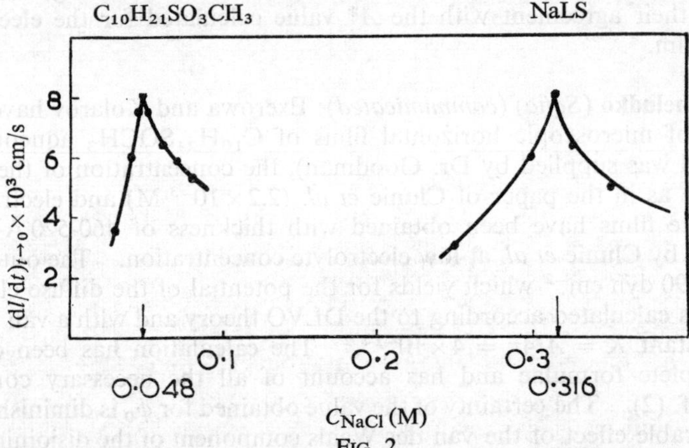

FIG. 2

so that the force barrier is definitely exceeded and a thermodynamically more stable Perrin film is formed which then covers the whole frame.   Up to this point, therefore, the films described behave normally.

Exerowa and Kolarov have also studied the transition first/second (Perrin's) black film with the same surfactant by means of the new method of foam destruction by a stream of α-particles.   As shown in ref. (3) the rate of foam column reduction (dl/dt) subjected to α-particle bombardment possesses a sharply outlined maximum near the electrolyte concentration corresponding to the first/second black film transition with microscopic films.

In fig. 2 the results obtained with $C_{10}H_{21}SOCH_3$ (2.2×10^{-3} M) and for comparison with NaLS (5×10^{-4} M) are shown. As an α-ray source Pu^{239}, with intensity 558 α-particles/s was used. In the second case the maximum is at a NaCl concentration of 0.316 M, very close to the value obtained using the contact angle method by us [4] and by Mysels.[5] For the surfactant used by Clunie *et al.* a maximum at 5×10^{-2} M NaCl has been observed. Therefore the first/second black film transition does not show at first sight any peculiarity compared to the above results.

The striking and very interesting moment in the paper of Clunie *et al.* is the thickening of the film from 40 to 70 Å at a NaCl concentration of 7×10^{-2} M. They interpret this effect as a result of the increase in ϕ_0-potential. This explanation is based mainly, we think, on the fact that the ϕ_0-potential with this surfactant has, as we have shown, an especially low value and therefore can easily increase on changing the conditions, e.g., the electrolyte concentration. Quantitative interpretation on the basis of the DLVO theory, however, seems to be groundless in this case. As is well known, the DLVO theory gives a maximum and a minimum for the energy and disjoining pressure isotherms when the film thickness decreases. The maximum explains the first black films but is not able to explain Perrin films. In order to describe the latter, another minimum with a consequent rise of the disjoining pressure above the outer pressure should exist at very small thicknesses. Therefore at thicknesses close to those of the Perrin films, significant deviations from the DLVO theory are expected in the direction of positive disjoining pressure. Such a large deviation has been observed in the measurements of the contact angle black film/bulk liquid in the region of first/second black film transition.[4] Therefore the application of the DLVO theory when calculating the constants at thicknesses exceeding Perrin's film thickness less than 30 Å is quite uncertain.

It is appropriate to recall the measurements on films of concentrated aqueous solutions of fat acids.[6] With 2 M butyric acid solution, very thick equilibrium films of some 1000 Å have been obtained. The thickening at such high electrolyte concentration is not due to the diffuse electric layers. The high positive values of the disjoining pressure in aqueous solution of butyric acid have been confirmed by Voropaeva using the method of crossed platinum fibres,[7] i.e., with different phase surfaces. This suggests a volume origin for the effect. Maybe in the work of Clunie *et al.* a similar situation appears. Other possible reasons for the deviations from the DLVO theory are discussed in ref. (8). As far as there is no theory to explain Perrin films, we suggested [9] that all these deviations from the DLVO theory should be combined in a third component of the disjoining pressure Π_{bl}. This component must rise sharply at small thicknesses in order to explain the Perrin films. The object should be a study of this Π_{bl}, and, in this sense we find the results in the paper of Clunie, Corkill, Goodman and Ingram of considerable interest.

Dr. J. M. Corkill (*Procter & Gamble Ltd, Newcastle-upon-Tyne*) said: In reply to Scheludko, we agree with the explanation advanced by Scheludko for the occurrence of thick (~ 50 nm) *microscopic* films below the transition concentration found by us

[1] A. Scheludko and D. Exerowa, *Kolloid Z.*, 1960, **168**, 24.

[2] D. Exerowa, *Kolloid-Z.*, 1969, **232**, 703 ; *Proc. 5th Int. Congr. Surface Activity*, 1968, **2**, 153.

[3] D. Exerowa and D. Ivanov, *Compt. Rend. Acad. Bulg. Sci.*, 1970, **23**, 547.

[4] T. Kolarov, A. Scheludko and D. Exerowa, *Trans. Faraday Soc.*, 1968, **64**, 2864.

[5] F. Huisman and K. Mysels, *J. Phys. Chem.*, 1969, **73**, 489.

[6] A. Scheludko and D. Exerowa, *Ann. Univ. Sofia, Fac. Chim.*, 1959/60, **54**, 205.

[7] T. Voropaeva, B. Deryaguin and B. Kabanov, *Kolloid-Z.*, 1962, **24**, 398.

[8] A. Scheludko, *Adv. Colloid Interface Sci.*, 1967, **1**, 391.

[9] A. Scheludko, *Ann. Univ. Sofia Fac. Chim.*, 1967/68, **62**, 47.

for large vertical films. For second black (Perrin) films, i.e., films with thicknesses independent of electrolyte concentration, we agree that the dominant repulsive force Π_{bl} does not behave simply as a cut-off potential. However, we find it difficult to invoke a Π_{bl}, which is not a double-layer repulsion, to explain the occurrence in the decyl methyl sulphoxide system of first black films, i.e., films where the thickness decreases with increasing electrolyte concentration. This difficulty becomes greater when we consider the lower concentrations and higher thicknesses involved in the transition from second to first black films with different added electrolytes. For example, with NaOH a first black film of thickness 47 nm is formed at a concentration of 0.1 mol m^{-3}. The thickness of this first black film then decreases progressively with increasing NaOH concentration.

Dr. B. A. Pethica (*Unilever Res., Port Sunlight*) said: Clunie *et al.* suggest that the transition from second to first black type in their films stabilized by a non-ionic surfactant is due to ion adsorption in the head-group plane. It is highly improbable that an adsorption of ions governed by a Stern adsorption potential will give such a sharp transition as is shown on their fig. 2. The adsorption of ions can be tested by electrophoretic methods using, e.g., oil drops stabilized by the same non-ionic, or by surface potential measurements. The film transitions are probably associated with salt effects on smectic phase transitions in solution. Bearing in mind that the film ionic composition is quite possibly different from that of the bulk associated solution, it would be useful to have related data on the solution properties of the stabilizer. The determination of the apparent Hamaker constants in their paper is open to objection. The set of equations (1)-(5) rest on the improbable assumption that the term B (which contains a Stern potential) does not vary with film thickness. Assuming that the electrical repulsion is a simple exponential function of thickness, the electrical term disappears from the argument, leaving $\Delta\sigma$ a direct function of the Hamaker constant A, thickness and the Debye-Hückel reciprocal length. The resulting values of A vary by a factor of two over a 10 K temperature interval, which leaves grave doubts as to the physical significance of the calculations.

Dr. J. M. Corkill (*Procter & Gamble Ltd., Newcastle-upon-Tyne*) said: In answer to Pethica's comment concerning the abruptness of the film transition, we refer to our reply to Lyklema; it is not necessary to postulate a corresponding lamellar mesomorphic phase transition in the *bulk solution*. The assumption that the Stern potential is independent of the separation between the charged surfaces is commonly employed in double-layer repulsion calculations, although some authors have suggested that the surface charge is more likely to remain constant. For weakly overlapping double layers, the difference between the two treatments leads to a very small difference in repulsion energy.[1,2] The variations in A^* with temperature that we have observed are hence unlikely to originate in a variation of the B term. We return to the point made in our paper that the temperature dependence of $\Delta\sigma$ for the first black films cannot be explained in simple terms, such as a variation in Stern potential.

Prof. J. Lyklema (*Wageningen*) said: In fig. 2 of their paper, Clunie *et al.* observe a steep rise in film thickness in the salt concentration range 0.07-0.1 M. An explanation is offered invoking ion absorption in the head-group plane of the surfactant.

[1] E. J. W. Verwey and J. Th. G. Overbeek, *Theory of the Stability of Lyophobic Colloids*, (Elsevier, Amsterdam, 1948).
[2] J. E. Jones and S. Levine, J. *Colloid Interface Sci.*, 1969, **30**, 241.

If this were true one would also expect an analogous irregularity in the micelle formation behaviour of the used surfactant as a function of the salt concentration around the 0.07-0.1 M region.   Was anything like that observed?

Dr. J. M. Corkill (*Procter & Gamble Ltd., Newcastle-upon-Tyne*) said: In reply to Lyklema, the critical micelle concentration (surface tension data) of decyl methyl sulphoxide decreases with addition of salt but no abrupt discontinuities are observed. The dependence of critical micelle concentration on ionic strength is similar to that reported for other non-ionic surface-active agents.[1]   To account for the transition in film thickness it is not necessary to postulate an abrupt rise in ion adsorption. It is generally accepted that a charged film can exhibit two minima in its potential energy-thickness relationship.   The secondary minimum, at larger thicknesses, results from superposing a double-layer repulsion term on the gravitational and van der Waals' energy contributions.   If the hydrostatic pressure is sufficiently large, a secondary minimum does not occur and either a primary minimum (second black) film is formed or rupture takes place.[2]   In our system, the hydrostatic pressure remains constant at the point of film thickness measurement (-300 N m^{-2}) as the solution composition is changed.   The transition occurs when the surface charge of the film has risen sufficiently to create a secondary minimum, rather than an inflection in the energy-thickness diagram.   The observation that further increasing the ionic strength leads to a deepening of the secondary minimum in addition to moving it to lower thickness is consistent with the predictions of the DLVO theory.

Prof. R. J. Good (*Bristol University*) said: With regard to the paper by Clunie *et al.*, it may be misleading to refer to film tension σ^f, for a second black film, without distinguishing it physically from the film tension of a thick film and from surface tension.   The state of stress of a second black film is a Hookean state, characterized by an elastic modulus, i.e., the tension in the film, for small elongations, is proportionate to the elongation.   In this, it is unlike the surface of a bulk liquid or of a thick film ; for these cases, the static stress is independent of elongation.   The dynamic stress in the latter two cases may be dependent on elongation, through the Gibbs elasticity ; however, this is a strongly time-dependent condition of stress.   This difference between the second black film and the other surfaces exists because of the lack of a reservoir of fluid in the interior of the black film.   So the black film acts, mechanically, simply as a linkage for transmitting the pull of the bulk-liquid surface to the Plateau border at the top of the frame, and thence to the top of the frame itself. There is a contact angle formed by the surfaces of the plateau border and the film, just as there is a contact angle at the bottom.

I would ask whether Clunie *et al.* measured the contact angles directly, and in particular, the angle at the top of the frame?   This set of measurements should enable them to tell whether the plateau border has a higher surface tension than the bulk liquid, indicating appreciable depletion of surfactant.

Dr. J. M. Corkill (*Procter & Gamble Ltd., Newcastle-upon-Tyne*) said: In reply to Good, we have directly measured the contact angle between the bottom of the film and the bulk solution at 298 K using the refraction method.[3]   We found no significant difference between this measured, equilibrium contact angle and that calculated from the film tension measurements.

[1] P. Mukerjee, *J. Phys. Chem.*, 1965, **69**, 4038.
[2] J. Th. G. Overbeek, *J. Phys. Chem.*, 1960, **64**, 1178.
[3] S. Frankel and H. M. Princen, *J. Phys. Chem.*, 1970, **74**, 2580.

Composition and Energy Relationships for Some Thin Lipid Films, and the Chain Conformation in Monolayers at Liquid-Liquid Interfaces

By D. M. Andrews,* E. D. Manev † and D. A. Haydon‡

Laboratory of Biophysical Chemistry and Colloid Science,
University of Cambridge, Free School Lane, Cambridge CB2 3RT, England

Received 2nd April, 1970

Optically black films have been formed in aqueous media from solutions of glyceryl mono-oleate in aliphatic hydrocarbons. The thicknesses of the hydrocarbon cores of the films were estimated from electrical capacitance measurements and the compositions from interfacial tension data. The thicknesses and the compositions were found to be interrelated in a simple way and were markedly dependent on the chain length of the hydrocarbon solvent.

An electrical potential applied across a liquid film subjects it to a large compressive force under which most types of film became significantly thinner. From thickness measurements in applied fields the strengths of the steric interactions which stabilize the films were calculated. From a knowledge of the steric interaction, together with an estimate of the London-van der Waals forces from contact angle measurements, the curve of potential energy against film thickness has been calculated for one system. The magnitude of the steric interaction at a given film thickness varies considerably for films of different solvent content. As a consequence, a general picture of the time-average conformations of the hydrocarbon chains of glyceryl mono-oleate in the black films and at different hydrocarbon/water interfaces may be deduced.

Detailed investigations of optically black lipid films in aqueous media have so far been inhibited by the difficulty of making simple stable films of well-characterized substances. However, solutions of glyceryl mono-oleate in aliphatic hydrocarbons and other nonpolar solvents form relatively stable films in aqueous solutions.[1] The thickness of these films may be measured by optical [2] and electrical [3, 4] methods and the adsorption of the oleate in the film may be estimated from interfacial tension measurements. From these data the composition of the films may be calculated. The free energy of formation of the films may be found from the contact angles between the thin film and bulk interface and may be interpreted to give the magnitude of the London-van der Waals forces.[5]

A uniform electric field normal to the surfaces of the black film exerts a compressional force, which may be considerably larger than the London-van der Waals forces, and which may produce appreciable thinning of the film.[6] This thinning is opposed by the steric interaction of the chains of the adsorbed oleate and, from the dependence of film thickness on applied field strength, it is possible to deduce the magnitude of the steric interactions and to infer qualitatively the conformation of the oleate chains both in the black film and in the adjacent oil-water interfaces.

The free energy against thickness relationship has been deduced for films of n-decane in saturated sodium chloride, stabilized by glyceryl mono-oleate. Films

---

* present address : Research Dept., Unilever Limited, Port Sunlight, Cheshire.
† present address : Institute of Physical Chemistry, University of Sofia, Sofia, Bulgaria.
‡ present address: Physiological Laboratory, Univ. of Cambridge, Downing Street, Cambridge.

of glyceryl mono-oleate in other non-polar solvents have also been examined and, although these studies have been less detailed, they have revealed some important respects in which the solvent may influence the oleate chain conformation and the composition of the films.

EXPERIMENTAL

METHODS

The films were formed in a 1 mm hole in a Fluon vessel, as described previously.[3]   For experiments with volatile solvents the top of the cell was sealed by means of a glass lid. Capacitances were measured as described previously.[3]   When a d.c. bias was applied across the film (from a potentiometer) a 1 μF capacitor was used to isolate the capacitance bridge.

Interfacial tensions were determined by the drop-volume technique.[7]   The activity coefficients of the glyceryl mono-oleate in the more volatile solvents were obtained by vapour pressure osmometry, using a Hewlett Packard type 302 Vapour Pressure Osmometer.   The sensitivity of this instrument was increased by a factor of *ca.* 3 beyond the makers' specification.   In order to obtain good reproducibility for both tension and osmometer results it was necessary to equilibrate the apolar solutions with the appropriate aqueous phase for at least 24 h prior to the experiment.

MATERIALS

The glyceryl mono-oleate was obtained from Sigma and was found by thin layer chromatography (kindly carried out by Dr. H van Zutphen) to be >99 % pure 1-isomer.   No significant ageing of the interfacial tensions was observed except at low concentrations.

The solvents were all of *puriss* grade and were $\geqslant$99 % by g.l.c.   Before use they were passed through an alumina column to remove trace surface active impurities.   A.R. NaCl was roasted at 700°C to remove organic impurities.   The water was twice distilled, first from a commercial still and then from a Pyrex still fitted with a quartz column, condenser and receiver.   All experiments were carried out at 20°C.

RESULTS

The two effects with which this paper is primarily concerned are illustrated in fig. 1 and 2.   First, the specific capacitance of the film increases considerably when

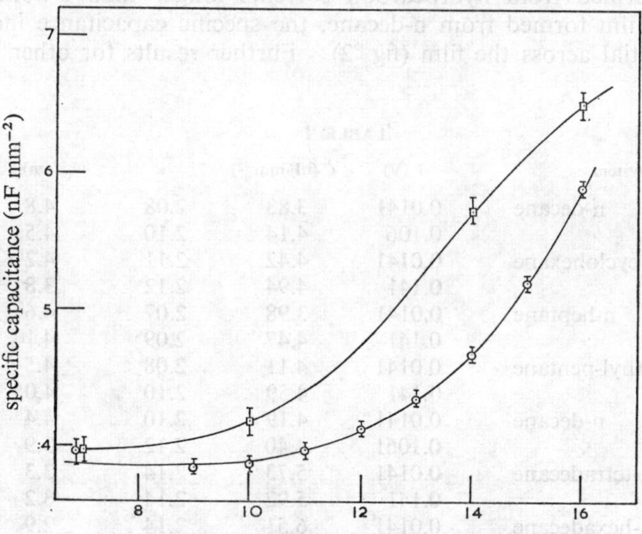

FIG. 1.—The specific capacitance of black films formed from solutions of glyceryl mono-oleate (*ca.* 12 mM) in normal alkanes. $\odot$, aqueous phase 0.1 M NaCl; $\square$, aqueous phase saturated NaCl.

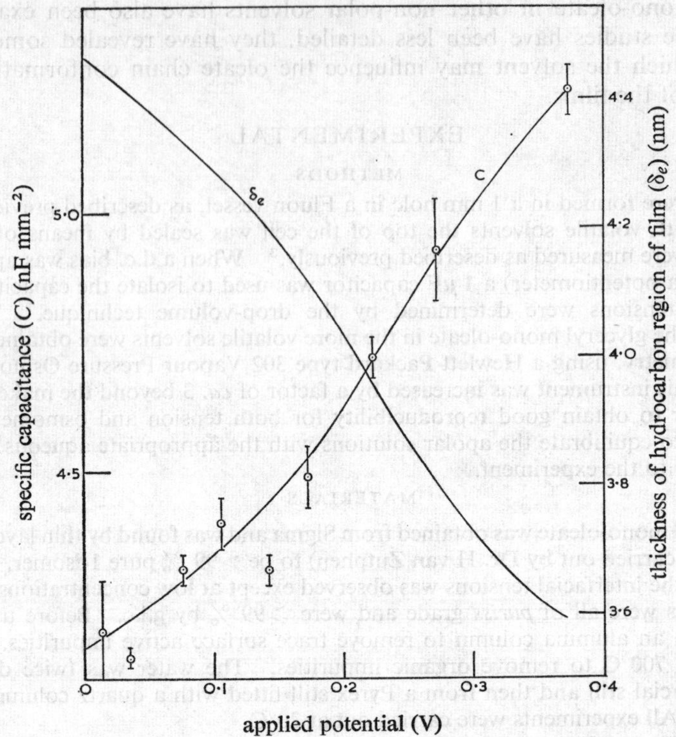

FIG. 2.—The specific capacitance (C, ⊙) and thickness of the hydrocarbon region (δ_e) of black films under applied potentials. The films were formed from glyceryl mono-oleate in n-decane, and the aqueous phase was saturated NaCl.

the films are formed from hydrocarbon solvents longer than n-nonane (fig. 1). Secondly, for a film formed from n-decane, the specific capacitance increases with increasing potential across the film (fig. 2). Further results for other solvents are given in table 1.

TABLE 1

system		V (V)	C (nF mm⁻²)	ε	δ_e (nm)	$\overline{\varphi}$
0.1 M NaCl	n-decane	0.0141	3.83	2.08	4.8	0.53
		0.106	4.14	2.10	4.5	
satd. NaCl	cyclohexane	0.0141	4.42	2.11	4.2	0.72
		0.141	4.94	2.12	3.8	
	n-heptane	0.0141	3.98	2.07	4.6	0.73
		0.141	4.47	2.09	4.1	
	2,2,4-trimethyl-pentane	0.0141	4.11	2.08	4.5	0.70
		0.141	4.59	2.10	4.05	
	n-decane	0.0141	4.19	2.10	4.4	0.73
		0.1061	4.40	2.12	3.9	
	n-tetradecane	0.0141	5.73	2.14	3.3	1.03
		0.141	5.92	2.14	3.2	
	n-hexadecane	0.0141	6.51	2.14	2.9	1.18
		0.141	6.38	2.14	2.95	
	CCl₄	0.0141	5.63	2.15	3.4	0.85
		0.141	5.63	2.15	3.4	

All the systems had capacitances and conductances which were frequency dependent. The dispersions were simple in form and corresponded accurately to those which would be expected for two parallel-sided isotropic layers in series. The two layers concerned are the hydrocarbon core of the film and the aqueous phase, respectively.[3, 8, 9] Above and below the dispersion region the capacitance does not vary with frequency and the data here reported were obtained in the low frequency, constant-capacitance region (at *ca.* 1 kHz). It was confirmed that a given r.m.s. a.c. potential had a similar influence on the capacitance to the equivalent d.c. potential.

When the electric field across a film was changed, the re-establishment of equilibrium often took of the order of minutes or hours. This was especially so for films formed from solvents of higher molecular weight, and arose evidently from the difficulty of escape of such solvents from the films. Thus, when the potential was suddenly increased, bright spots of surplus liquid appeared in the film after a period of seconds or minutes, presumably from a squeezing-out or disproportionation process. The specific capacitance of the film was constant once the spots were visible, but the diffusion of the spots to the edge of the film, where they coalesced with the meniscus, took much longer.[10]

In many of the systems the films became markedly less stable under potential differences of more than *ca.* 150 mV (r.m.s.), and precise capacitance measurements could not be made. The use of saturated NaCl as the aqueous phase greatly helped to minimize inaccuracies in the capacitances for films formed from low-molecular-weight solvents. Unless the apolar and aqueous phases were in perfect equilibrium, the capacitances became anomalously high. In saturated NaCl this effect was considerably reduced, owing presumably to the lower solubility of the solvents, and a wider range of films could be examined.

A film thickness was obtained from the specific capacitance C by means of the equation

$$C = \varepsilon_0 \varepsilon / \delta_e \tag{1}$$

where ε_0 is the permittivity of free space, ε is the dielectric constant and δ_e is the thickness of the hydrocarbon core of the film. The validity of this procedure, and in particular the justification for ignoring contributions to the capacitance from the polar group region and the electrical double layers in the aqueous phases, has been examined in previous papers.[3, 8, 9, 11] The hydrocarbon core is regarded as bulk liquid hydrocarbon composed of a mixture of oleyl chains and the appropriate solvent.[12] The proportions of the two components have been estimated from the adsorption data as described below. The dielectric constants of the two components have been assumed additive on a volume fraction basis. Anisotropy in the hydrocarbon has been ignored. Even in hydrocarbon crystals this effect is small and the film structure more closely resembles a liquid than a crystal. The resulting dielectric constants are all close to 2.1 and the calculated thicknesses are shown in fig. 2 and table 1. From measurements of molecular models the thickness of the glyceride polar groups layer was estimated to be 0.45 nm. The total film thickness δ_L was therefore taken to be $(\delta_e + 0.9)$nm. Such optical measurements as have been reported (e.g., films of glyceryl distearate and n-hexane[13]) are consistent with the present estimated thickness.

In order to estimate the composition of the black films, the colligative properties of glyceryl mono-oleate and its adsorption at the various oil-water interfaces were studied. Glyceryl mono-oleate aggregates strongly in apolar media at concentrations above *ca.* 10^{-3} mol l^{-1}. When the apolar medium is equilibrated with 0.1 M

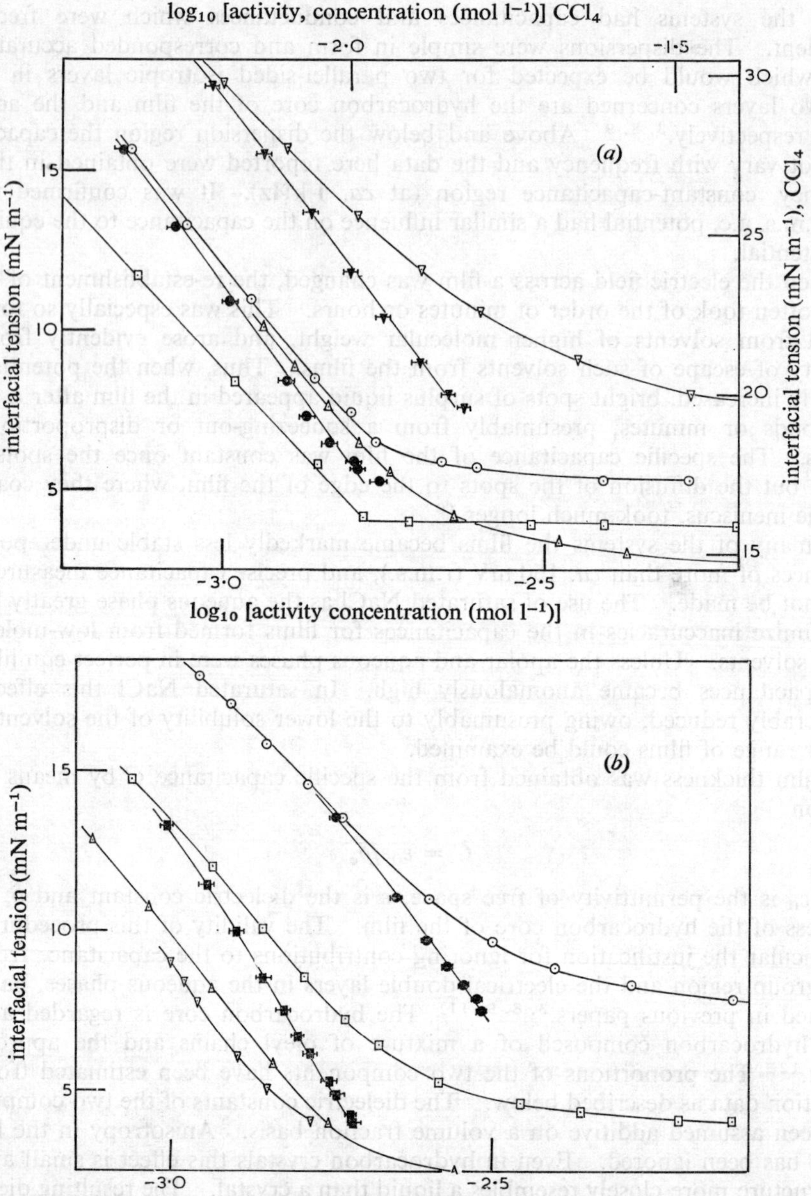

Fig. 3.—Interfacial tensions of glyceryl mono-oleate in various solvents as a function of $\log_{10}$ (activity), (Closed points) and, or, $\log_{10}$ (concentration), (open points). (a) $\square$, n-decane+0.1 M NaCl; $\triangle$, n-decane+saturated NaCl; $\odot$, $\bullet$, 2,2,4-trimethylpentane+saturated NaCl; $\triangledown$, $\blacktriangledown$, carbon tetrachloride+saturated NaCl (top and right hand axes). (b) All aqueous phases saturated NaCl. $\triangledown$, n-hexadecane; $\triangle$, n-tetradecane; $\square$, $\blacksquare$, n-heptane; $\odot$, $\bullet$, cyclohexane.

NaCl the onset of the aggregation is sufficiently sharp to be described as a critical micelle concentration. When equilibrated with saturated NaCl, however, the aggregation is less sharp and there is no clear c.m.c. This point is illustrated by the interfacial tension curves for decane in fig. 3(a). Black films tended to be stable only at concentrations above the c.m.c. or aggregation region, although films were occasionally sufficiently stable just below this concentration range for some types of measurement to be made. All the films of table 1 were formed above the aggregation region.

The adsorption of the glyceryl mono-oleate at the oil/water interface was found by means of the Gibbs equation which, as the two solvents were effectively insoluble in each other, and the oleate was very strongly adsorbed, reduces to [14]

$$-\mathrm{d}\gamma = \Gamma_2 RT \mathrm{d} \ln a_2,\qquad(2)$$

where γ is the interfacial tension, Γ_2 is the surface concentration of the glyceryl mono-oleate and a_2 is the activity of the oleate in the oil phase. Curves of γ against $\log_{10}$ (concentration) are shown in fig. 3, (a) and (b), for each of the systems of table 1. For only four systems, however, was it possible to determine the activity co-effiicients f_2. The latter were obtained from vapour pressure osmometer measurements, via the osmotic cocfficients g, by means of the equation

$$\int \mathrm{d} \ln f_2 = \int \mathrm{d}g + \int (g-1)\mathrm{d} \ln x_2.\qquad(3)$$

The results may be seen in fig. 3 (a) and (b). For the remaining systems the vapour pressure osmometer was insufficiently sensitive and there seemed no obvious alternative method of obtaining the activity coefficients. The adsorption in these instances was therefore estimated by assuming that the activity correction was identical to that for the heptane system. Both determined and estimated surface concentrations are shown in table 2.

TABLE 2

system		$-\left(\dfrac{\mathrm{d}\gamma}{\mathrm{d}\log_{10}a}\right)_{\max}$	$\Gamma_2 \times 10^{-18}$ (molecules m^{-2})
0.1 M NaCl	n-decane	21.8*	2.5†
satd. NaCl	cyclohexane	28.1	3.0
	n-heptane	31.0	3.3
	2,2,4-trimethyl-pentane	27.3*	2.9
	n-decane	22.9*	3.2†
	n-tetradecane	24.8*	3.4†
	n-hexadecane	24.5*	3.4†
	CCl$_4$	26.6	2.9

$*-(\mathrm{d}\gamma/\mathrm{d}\log_{10}c)_{\max}$; † based on activity correction for n-heptane system

The volume fraction $\bar{\varphi}$ of the oleate chains in the film was calculated as follows. The thickness δ_e of a film was estimated from the specific capacitance using an arbitrarily chosen dielectric constant (say, 2.1). Assuming that the partial molar volume of the oleate chain in the film was equal to the molar volume of 1-heptadecene in bulk, the first approximation to $\bar{\varphi}$ was deduced from Γ_2 the surface concentration of oleate (table 2). From this value of $\bar{\varphi}$, and on the assumption that the dielectric constants of the hydrocarbons in the film were equal to their bulk values, a new thickness was calculated and the procedure repeated. The second approximation to $\bar{\varphi}$ differed only slightly from the first. The assumption of bulk density for the hydrocarbon in the film was made by analogy with the interior of surfactant micelles in aqueous solution,[15-17] and is consistent with the electrical conductances of the films.[12] The assumption that the adsorption in the film was, for present purposes, identical to that at the bulk interface with which the film was in equilibrium has been justified by theoretical considerations,[14] and has also been substantiated experimentally for one system by contact-angle measurements. Thus, from the contact angles for systems below the c.m.c., the difference between the film tension and twice the bulk tension may be found, and it may be shown that the variation of this difference with glyceryl mono-oleate activity is extremely small compared to the variation of either of the individual tensions. It then follows [14] that the adsorption at the film and bulk interfaces differs only very slightly (<1 %).

DISCUSSION

FREE ENERGY AS A FUNCTION OF THICKNESS

In the absence of an electrical potential difference across a film, the forces acting normally to the film surface are assumed to be the London-van der Waals compression, and the steric repulsion which originates from the interaction of the oleate chains. All other forces, such as those from electrical double-layer overlap and from dipole-dipole interactions are assumed to be negligible. Experimental evidence so far available is entirely consistent with this assumption.[1, 18]

For unit area of film, the London-van der Waals forces F_L are given by

$$F_L = -A/6\pi\delta_L^3,\tag{4}$$

where A is the Hamaker constant. In the present systems the retardation correction should be less than *ca.* 15 % and will be disregarded. At equilibrium therefore

$$F_L+F_S = 0,\tag{5}$$

where F_S is the steric repulsion per unit area. When a potential V exists across the film there is an additional force F_e of compression given by

$$F_e = -CV^2/2\delta_e.\tag{6}$$

At equilibrium,

$$F_L+F_S+F_e = 0,\tag{7}$$

and therefore

$$F_S = (A/6\pi\delta_L^3)+(CV^2/2\delta_e).\tag{8}$$

The Hamaker constant for the present systems may be found from contact-angle measurements. At potentials of more than about 50 mV across the film, the electrical forces exceed the London-van der Waals forces, and for 370 mV applied potential the former are some 50 times the latter. From eqn (8) and a knowledge of the capacitance over a range of applied potential, the variation of the steric repulsion force with film thickness may be found. This has been done for all the systems, but only for the n-decane+saturated NaCl system have the measurements been taken to relatively high potentials. A plot of F_S against δ_L for this system is shown in fig. 4 (inset).

The change ΔA_S in free energy of the film due to the steric interaction may be calculated by means of the relationship,

$$\Delta A_S = -\int_\infty^{\delta_L} F_S d\delta_L.\tag{9}$$

The total free energy change ΔA of the system as the film thins may therefore be written

$$\Delta A = -\frac{A}{12\pi\delta_L^2}-\int_\infty^{\delta_L} F_S d\delta_L.\tag{10}$$

ΔA and its components are shown in fig. 4. The onset of the steric repulsion is so sharp as not to affect appreciably the depth of the minimum. This, incidentally, justifies the assumption which was made in the calculation of the Hamaker constant.[5] For the systems listed in table 1 the steric repulsion against film thickness is shown in fig. 5; the rise of the steric repulsion is as steep or more so than in the n-decane + saturated NaCl system.

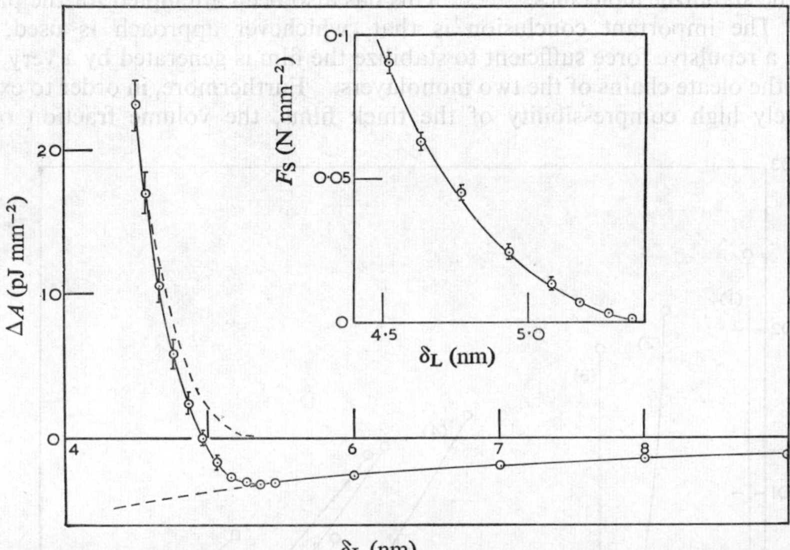

FIG. 4.—Free energy changes as a function of film thickness for glyceryl mono-oleate + n-decane films in saturated NaCl. The dashed curves represent the separate London-van der Waals and steric interaction contributions. The inset shows the steric repulsion force F_S, also as a function of film thickness. Hamaker constant $= 3.48 \times 10^{-21}$ J.

FILM COMPRESSIBILITY, COMPOSITION AND CHAIN CONFORMATION

It can be seen from table 1 that the thicker films contain a greater volume fraction of solvent, $(1 - \bar{\varphi})$, than do the thinner films. In fact, as the adsorption of the glyceryl mono-oleate is almost the same in each system (table 2), the film thickness is directly proportional to the amount of solvent in the film. (The finding that $\bar{\varphi} > 1$ for the n-tetradecane and n-hexadecane systems is attributed to the inaccuracy in the estimation of the activity correction.)

The slopes of the curves in fig. 5 are inversely proportional to the compressibility of the films. The thicker films, where $\bar{\varphi}$ is low, are thus much more compressible than the thinner films where $\bar{\varphi} \simeq 1$. There is therefore a direct relationship between oleate chain density and repulsive force.

The nature of this force, or of the related interaction free energy, has been discussed by a number of authors.[20] The Helmholtz free energy change of the system as the film thins is, for unit area of film,

$$\Delta A = A - A_\infty = (\sigma - 2\gamma) + \Sigma n_i(\mu_i - \mu_{i\infty}), \tag{11}$$

where σ is the film tension, γ is the interfacial tension of the interfaces between the equilibrium bulk phases, n_i is the number of moles of i in the system, and $\mu_{i\infty}$ and μ_i are the chemical potentials of i in the system before and after the film has thinned. In most instances, and certainly for the present systems, the second term on the right-hand side is negligibly small.[21] For a system in which adsorption equilibrium with the bulk phases is maintained, the first term is calculable in principle from the adsorption isotherms for the single interfaces and the thin films respectively.[21]   A crude attempt to do this for a thin lipid film was made previously.[9]   An alternative approach which is theoretically less satisfactory but which more readily yields an answer, is to assume the adsorption to be independent of film thickness and to estimate the osmotic pressure changes in the film produced by the overlap of the chains of the stabilizer molecules.[22, 23]   This has also been attempted for the present systems.[1]   The important conclusion is that, whichever approach is used, it is found that a repulsive force sufficient to stabilize the film is generated by a very small overlap of the oleate chains of the two monolayers.   Furthermore, in order to explain the relatively high compressibility of the thick films, the volume fraction of the

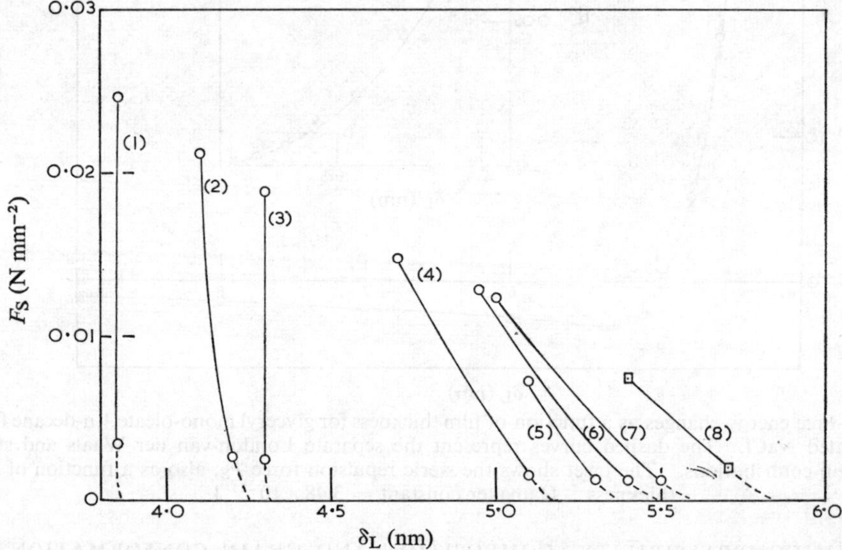

FIG. 5.—The steric repulsion force F_S as a function of film thickness for the systems of table 1. Saturated NaCl (Hamaker constant assumed to be 3.48×10^{-21} J [18]); (1), n-hexadecane; (2), n-tetradecane; (3), carbon tetrachloride; (4), cyclohexane; (5), n-decane); (6), 2,2,4-trimethylpentane; (7), n-heptane.   0.1 M NaCl (Hamaker constant assumed to be 6.76×10^{-21} J [5]); (8), n-decane.

oleate chain segments in the overlapping parts of the monolayers must be very small, compared to the average volume fraction $\bar{\varphi}$ in the film.   As, in these particular systems, the thickness of the hydrocarbon core of the film is closely similar to twice the extended chain length of the oleate, it is inferred that at any given time only a small fraction of the oleate chains are fully extended.

In the thinnest films $(\delta_e < 3.2 \text{ nm}) \bar{\varphi} \approx 1$, and the oleate chains of each monolayer are packed into a thickness of 1.5-1.6 nm.   Films thicker than this must be stabilized by the greater tendency of the oleate chains to extend themselves beyond *ca.* 1.5 nm from the interfaces although, as noted above, only a small proportion of them are required to do so.   The tendency to extend more fully is apparently related to the nature of the solvent.

From the foregoing arguments it is possible to construct a qualitative picture of the time-average volume fraction φ of oleate chain segments as a function of distance from the film interfaces (fig. 6).   The curve for the thin (e.g., hexadecane) films cannot be appreciably in error as the amount of solvent present is undetectable by the present methods.   For the thicker films the form of the curve is less certain, although as any segments located outside the minimum volume for the chains (i.e., the region 1.45 nm thick between $z = \pm 2.3$ nm and $z = \pm 0.85$ nm) necessarily leave a similar sized hole within this region, the shaded areas must be equal.

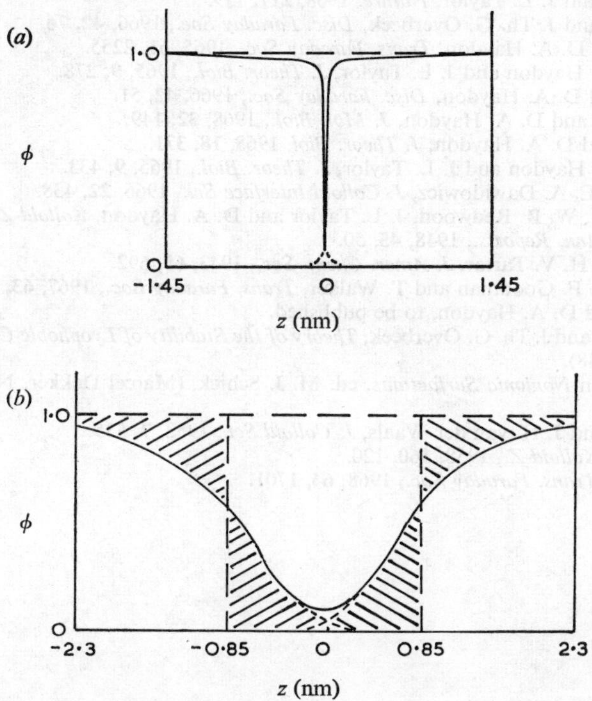

FIG. 6.—A schematic representation of the time-average volume fraction φ of oleate chain segments as a function of distance across the hydrocarbon region of a film (*a*) when n-hexadecane and (*b*) when n-heptane is used as the solvent.

While the above remarks apply to the thin film it is probable that if the curves of fig. 6 are correct for the film, they are also good indications of the situation in the soliated monolayers. Thus, for the thick films, where the individual monolayers are not compressed to significantly less than their maximum thickness, this must be so. For the thin films it is difficult to give a satisfactory argument in the absence of quantitative relationships. However, only a very small overlap of the monolayers at very small segment concentration is necessary to stabilize the film. If, therefore, the oleate chains in the monolayers were fully extended prior to film formation, it is almost certain that more work would have been required to compress them into the close-packed state than to stabilize the thick film, and a thick film would have resulted. As a thick film was not formed, it is concluded that the chains in the separate monolayers must have been already contracted into the nearly close-packed state.

D. M. Andrews was in receipt of a maintenance grant from Unilever Limited, Port Sunlight, during the course of this work. E. D. Manev thanks the British Council for the award of a Scholarship. The authors thank Mr. A. R. Taylor for considerable experimental assistance in the determination of the activity coefficients, and Dr. J. L. Taylor for the determination of the interfacial tension against concentration curve for the system hexadecane + saturated NaCl.

[1] D. M. Andrews, *Ph.D. Diss.* (Cambridge, 1970).
[2] R. J. Cherry and D. Chapman, *J. Mol. Biol.*, 1969, **40**, 19.
[3] T. Hanai, D. A. Haydon and J. L. Taylor, *Proc. Roy. Soc. A*, 1964, **281**, 377.
[4] H. Sonntag and H. Klare, *Kolloid-Z.*, 1964, **195**, 35.
[5] D. A. Haydon and J. L. Taylor, *Nature*, 1968, **217**, 739.
[6] D. A. Haydon and J. Th. G. Overbeek, *Disc. Faraday Soc.*, 1966, **42**, 76.
[7] R. Aveyard and D. A. Haydon, *Trans. Faraday Soc.*, 1965, **61**, 2255.
[8] T. Hanai, D. A. Haydon and J. L. Taylor, *J. Theor. Biol.*, 1965, **9**, 278.
[9] J. L. Taylor and D. A. Haydon, *Disc. Faraday Soc.*, 1966, **42**, 51.
[10] D. M. Andrews and D. A. Haydon, *J. Mol. Biol.*, 1968, **32**, 149.
[11] C. T. Everitt and D. A. Haydon, *J. Theor. Biol.* 1968, **18**, 371.
[12] T. Hanai, D. A. Haydon and J. L. Taylor, *J. Theor. Biol.*, 1965, **9**, 433.
[13] H. T. Tien and E. A. Dawidowicz, *J. Colloid Interface Sci.*, 1966, **22**, 438.
[14] G. M. W. Cook, W. R. Redwood, J. L. Taylor and D. A. Haydon, *Kolloid-Z.*, 1968, **227**, 28.
[15] G. S. Hartley, *Ann. Reports.*, 1948, **45**, 50.
[16] A. B. Scott and H. V. Tartar, *J. Amer. Chem. Soc.*, 1943, **65**, 692.
[17] J. M. Corkill, J. F. Goodman and T. Walker, *Trans. Faraday Soc.*, 1967, **63**, 768.
[18] D. F. Billett and D. A. Haydon, to be published.
[19] E. J. W. Verwey and J. Th. G. Overbeek, *Theory of the Stability of Lyophobic Colloids*, (Elsevier Amsterdam, 1948).
[20] R. H. Ottewill in *Nonionic Surfactants*, ed. M. J. Schick, (Marcel Dekker, New York, 1967), vol. 1, p. 649.
[21] E. L. Mackor and J. H. van der Waals, *J. Colloid Sci.*, 1952, **7**, 535.
[22] E. W. Fischer, *Kolloid-Z.*, 1958, **160**, 120.
[23] D. H. Napper, *Trans. Faraday Soc.*, 1968, **64**, 1701.

Interaction Energy of Emulsion Droplets and the Influence of Adsorbed Layers on it

By H. Sonntag, J. Netzel and B. Unterberger

Deutsche Akademie der Wissenschaften zu Berlin, Zentralinstitut für Physikalische Chemie, Berlin-Adlershof, Deutsche

Received 9th March, 1970

The equilibrium distance, contact angle, and formation velocity of black films between model droplets in aqueous solution of surface-active agent were measured. These variables produced information about the forces of interaction between the droplets. The concentration of the surface-active agent was increased to determine the effects of surface concentration upon the above forces, for small and wide particle spacings with reference to the thickness of the adsorbed layers. All experiments proved that by stepping up the concentration of the surface active agent the attraction between the particles was increased and, consequently the stability reduced.

The stability of colloids and the formation of coagulation networks in concentrated dispersions may be modified within wide limits by adsorbed layers of surface-active materials. They are capable of changing both the electrostatic repulsive forces and the dispersion forces. Other factors that have to be taken into account under conditions of small particle spacings are the dipole forces, the hydrogen bonds of water molecules adsorbed according to a given orientation, and the steric hindrance of the adsorbed layers, likely to occur in their mutual penetration or compression. While for practical purposes there is a common use of such modification of interaction forces by adsorbed layers, little quantitative work has so far been performed on this aspect. This paper will be limited to results obtained from investigations of non-ionic surface active materials which are less complex, since adsorption-dependent changes of the charge may be neglected.

The influence of non-ionic surface active agents on the stability of sols has already been described by some authors.[1-6] However, the influence of surface concentrations of surface-active agents upon interaction has not yet been studied in connection with model tests of foams and emulsions. An attempt is made to close this gap in this paper, which reports the effects undergone by the interaction forces between oil droplets which are separated by an aqueous film of surface-active agents.

EXPERIMENTAL

MATERIALS

Spectroscopically-pure n-octane and double-distilled water were used. The electrolytes were molten prior to use. The non-ionic surface-active agent used was nonylphenol with 20 moles of ethylene oxide (NP 20), a preparation made in the laboratory.

MEASUREMENT OF THE EQUILIBRIUM DISTANCE

The interaction forces between emulsion droplets may be characterized by the equilibrium spacing d_{eqn} resulting from an equilibrium of repulsive with attractive forces which is reached

"

with a certain electrolyte concentration. The distance between microscopic emulsion droplets were measured by interferometry. The radius of the plane-parallel contact zone was 0.01 cm. The apparatus used is described in detail in ref. (7). For emulsions, the calculation of droplet spacings from the intensity of light reflection proved to be easier than for foam laminae, since it could be based on the single-layer model. The alkyl groups of the surface-active material were found in the oil phase. with their refractive index to first approximation being identical with that of octane. The greatest change of the refractive index was found to take place at the boundary between the oily phase and the hydrated polar groups, so that the percentage of the adsorbed layers formed by the polar groups will be included simultaneously by the optical measurement of oil droplet spacing.

MEASUREMENT OF THE CONTACT ANGLE

A porous annular glass cell g as shown in fig. 1 was used to measure the contact angle between the black film and the volume phase. The surface of the glass frit was melt-sealed,

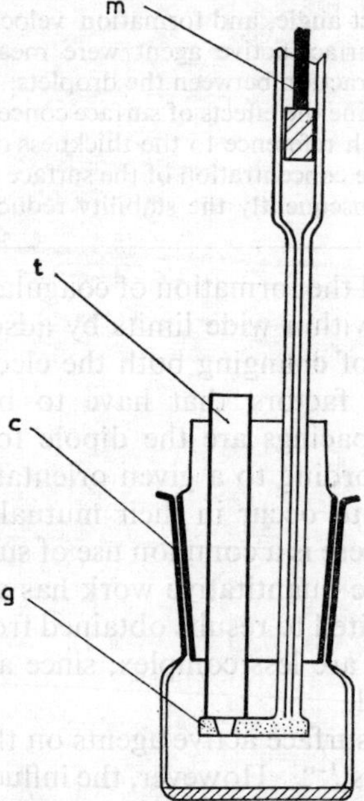

FIG. 1.—Apparatus for contact angle determination.

so that the drainage of the fluid took place only inside the conical bore, 0.08 cm in diameter. The drainage proper was effected by means of the micrometer screw m. The glass frit was inserted in a cuvette c with a plane-parallel bottom, which was filled with the oil phase. The glass tube t, shown to be fused on to the glass frit above bore level, was not needed for emulsion testing, but would be required to test the heterosystem air, water, oil not considered in this context.

The detector was mounted on an incident-light microscope, and the droplets were photographed prior to and after black film formation. The contact angle was calculated from the spacing of Newton's rings or from the growth of the contact area following black film formation, according to the method of Sheludko and co-workers.[8]

VELOCITY OF BLACK FILM FORMATION

The velocity of black, film formation was characterized by measuring the time that elapsed from the appearance of the first black spots to the completion of a homogeneous black film.[9]

RESULTS

The equilibrium spacings of octane droplets in NP-20 solutions of different concentrations were measured in electrolyte concentrations of 0.001-0.01 m KCl. The results are given in fig. 2.   It may be seem from the graph that up to the critical

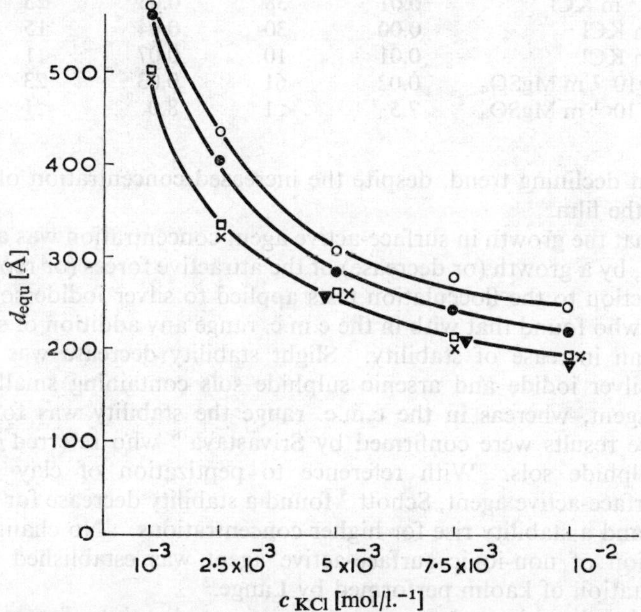

FIG. 2.—Equilibrium distance d_{equ}(Å) as a function of the KCl concentration, c (mol/l), for different surface-active agent concentrations: $\bigcirc$, 7.5×10^{-6} mol/l; $\bullet$, 7.5×10^{-5} mol/l; $\square$, 2.5×10^{-4} mol/l; $\times$, 7.5×10^{-4} mol/l; $\blacktriangledown$, 1×10^{-3} mol/l.

concentration of micelle formation (c.m.c.), which was reached at 1.4×10^{-4} mol/l, the equilibrium films decreased in thickness, with the electrolyte concentration remaining constant.   The influence of the surface-active agent grows as the droplet distances are reduced.   While above the c.m.c. level no further change of the equilibrium spacings takes place, the influence of the surface-active agent can be observed already at lower electrolyte concentrations.   These experimental findings might be explained by both a decrease of the repulsive forces or an increase of the attractive forces along with growing adsorption of surface active agent.   Equilibrium spacings were measured only under conditions were they were much greater than the layer thickness of the adsorbed polar groups.

Contact angle measurements were performed to find out whether, with reference to the sign, a further change in the influence of the surface-active agents on the interaction forces was possible even with small spacings.   The droplet distances chosen for that purpose were so small that the intensity of light reflection coincided with that of the minimum range.   They were smaller than 40 Å.

The experimental results are collected together in table 1.   While the influence of the surface active agent concentration upon the contact angle was small, a growth of the attractive forces rather than if the repulsive forces was evident, and it was even more obvious by comparing the time intervals t_A that elapsed from the appearance of the first black spots to the formation of homogeneous black films.   These intervals

TABLE 1.—CONTACT ANGLE θ AND TIME t OF FORMING BLACK FILMS IN EQUILIBRIUM, WITH SOLUTIONS CONTAINING 10^{-5} m AND 7.5×10^{-3} m NP 20 AS A FUNCTION OF ELECTROLYTE CONCENTRATION

electrolyte	10^{-5} m NP 20		7.5×10^{-3} m NP 20	
	$\theta°$	t (s)	$\theta°$	t (s)
10^{-1} m KCl	0.01	38	0.03	25
1 m KCl	0.00	30	0.04	15
2 m KCl	0.01	10	0.07	<1
5×10^{-2} m MgSO$_4$	0.02	61	0.03	23
5×10^{-1} m MgSO$_4$	7.5	<1	8.0	<1

show a clear-cut declining trend, despite the increased concentration of the surface-active agent in the film.

The result that the growth in surface-active agent concentration was accompanied, for all distances, by a growth (or decrease) of the attractive forces (or repulsive forces) is in a contradiction to the flocculation tests applied to silver iodide sols and latices by Ottewill [1, 2] who found that with in the c.m.c. range any addition of surface-active agent entailed an increase of stability. Slight stability decrease was observed by Glazman [4] in silver iodide and arsenic sulphide sols containing small amounts of surface-active agent, whereas in the c.m.c. range the stability was found to grow strongly, These results were confirmed by Srivastava [6] who referred to antimonic and arsenic sulphide sols. With reference to peptization of clay with smaller quantities of surface-active agent, Schott [5] found a stability decrease for these smaller concentrations and a stability rise for higher concentrations. No change in stability after the addition of non-ionic surface-active agent was established from experimental sedimentation of kaolin performed by Lange. [3]

Unfortunately, there have been no systematic studies into foam films as yet. However, it is believed that the authors' results provide an explanation for the different layer thicknesses found by van der Waarde [10] and Sheludko [11] who tested macroscopic and microscopic foam films with one and the same surface-active material. The macroscopic films were stabilized at surface-active agent concentrations above the c.m.c. and gave equilibrium thicknesses much smaller than those of microscopic films in which the surface-active agent concentrations were smaller by two orders of magnitude.

DISCUSSION

The only factors which have to be considered in calculating the interaction forces per unit area are the electrostatic repulsive forces (Π_{el}), the dispersion forces (Π_D), and the capillary pressure (Π_σ), since in all measurements the equilibrium spacings were bigger than the thicknesses of the adsorbed layers. The following equation has to be satisfied for the equilibrium spacings:

$$\Pi_{el} = \Pi_D + \Pi_\sigma. \tag{1}$$

The two forces may be analysed separately, if the influence of the surface-active agent is attributable mainly to the change of Π_{el} or to that of Π_D.

While Π_σ is also dependent on the given surface-active agent concentration, it can be determined separately by measurement of the interfacial tension. The interfacial tension and, consequently, the capillary pressure will decrease approximately by a factor of two, until the c.m.c. is reached. This will cause a reduction of the attractive forces and, therefore, cannot explain the reduction of the equilibrium spacings. Since the magnitude of the capillary pressure is identical with that of the dispersion forces for larger distances a change of the former could even lead to a compensation of the measured effect.

The dispersion forces are calculated according to the model proposed by Vold.[12] Two particles with the Hamaker constant A_1 interact in a medium with the Hamaker constant A_0. The adsorbed layer is subdivided into two parts,[13] the non-polar part of thickness δ_2 and the Hamaker constant A_2, on the one hand, and the polar groups of thickness δ_3 and the Hamaker constant A_3, on the other. In earlier work, evidence was produced to the effect that surface-active agents with equal alkane chain length but different polar group would affect the dispersion interaction due to the structure of this group and that the hydrocarbon chains (solute in oil) did not contribute to the energy of dispersion.[14] This would support the conclusion that, to a first approximation, A_2 is equal to A_1. The dispersion forces may then be expressed by the following equation:

$$\Pi_{\mathrm{D}} = \frac{1}{6\pi} \frac{(A_0^{\frac{1}{2}} - A_3^{\frac{1}{2}})^2}{(d - 2\delta_3)^3} + \frac{(A_3^{\frac{1}{2}} - A_1^{\frac{1}{2}})^2}{d^3} + \frac{2(A_0^{\frac{1}{2}} - A_3^{\frac{1}{2}})(A_3^{\frac{1}{2}} - A_1^{\frac{1}{2}})}{(d - \delta_3)^3}, \tag{2}$$

where d is the distance determined by interferometry.

If A_3 differs from A_0, smaller distances will always result in increased attraction, no matter whether A_3 is smaller or bigger than A_0 and A_1, since the first and third terms of eqn (2) will be very large. A decrease of dispersion energy may result from larger spacings,[13] provided that A_3 is smaller than both A_1 and A_0. The results obtainable from the above equation cannot be evaluated unless at least A_0 and A_1 are known from other measurements.

Let us substitute for A_0 the Hamaker constant of 3.5×10^{-13} erg found for microscopic foam films with low surface-active agent concentrations by Shelduko and co-workers[15]; $A_1 = 8.5 \times 10^{-13}$ erg is substituted for the oil phase, on the basis of measurements on latices[16] and emulsions with low surface-active agent concentrations.[17]

The adsorbed layers of surface-active agents will be capable of affecting the electrostatic forces in two ways, either by changing the structure of the double layer or by changing the potential of the diffuse double layer (ψ_δ). Since structural changes would be detectable mainly along with a close approach of the particles, this factor should be neglected for the purpose of equilibrium distance measurement ($d > 2\delta_3$).

The potential at the oil/water interphase may be generated by adsorption of OH' from the water or by adsorption of the surface active-agents proper. Under the first assumption, it would be possible that by increased adsorption of surface-active materials the potential-determining ions would be displaced, which would result in a reduction of the potential as such and, consequently, of the electrostatic interaction. This was tested by measuring equilibrium distances at constant surface-active agent concentration and variable pH values. The tests have shown that the equilibrium thicknesses remain constant in the measured pH interval of 3-7. Measurements of the interfacial tension, performed concurrently, have shown that within the above pH range the adsorption of surface-active agents remained unaltered.

The evaluation of the experimental results was to produce information as to the contribution made by each of the various factors (A_3, δ_3, and ψ_δ) to the influence of surface-active agents. The following equation is obtained for an 1 : 1-electrolyte [c(mol/l)] and 25°C by introducing, in addition, the various distance functions into eqn (1):

$$1.59 \times 10^9 c \left[\frac{\exp(19.83\,\psi_\delta) - 1}{\exp(19.83\,\psi_\delta) + 1} \right]^2 \exp(-0.329 \times 10^8 d_{eqn}\sqrt{c} =$$

$$\frac{1}{6\pi} \frac{(A_0^{\frac{1}{2}} - A_3^{\frac{1}{2}})^2}{(d_{eqn} - 2\delta_3)^3} + \frac{(A_3^{\frac{1}{2}} - A_1^{\frac{1}{2}})^2}{d_{eqn}^3} + \frac{2(A_0^{\frac{1}{2}} - A_3^{\frac{1}{2}})(A_3^{\frac{1}{2}} - A_1^{\frac{1}{2}})}{(d_{eqn} - \delta)^3} + \Pi_\sigma \qquad (3)$$

ψ_δ, δ_3 and A_3 may be calculated for different surface-active agent concentrations from the (d_{equ}, c) functions determined experimentally. The results are given in table 2.

TABLE 2.—INTERFACIAL TENSION σ, POTENTIAL OF THE DIFFUSE DOUBLE LAYER ψ_δ, HAMAKER CONSTANT A_3 AND THICKNESS δ_3 OF THE POLAR GROUPS AS A FUNCTION OF CONCENTRATION OF NP 20

surfactant concentration [mol/l.]	σ [dyn/cm]	ψ_δ [mV]	A_3 10^{13} [erg]	d_3 [Å]
7.5×10^{-6}	19.1	32	0.7	22
7.5×10^{-5}	10.5	19	0.7	22
2.5×10^{-4}	10.1	12	1.6	38
$7 \cdot 5 \times 10^{-4}$	9.0	12	1.6	38

The interfacial tension and consequently the capillary pressure, decreases with increasing surfactant concentration. This will cause a reduction of the attractive forces and therefore cannot explain the decline of equilibrium spacings. The double layer potential decreases up to the saturation concentration of the adsorbed surfactant molecules. On the other hand, the Hamaker constant and the thickness of the adsorbed layer increases up to the c.m.c. These results show, that the influence of non-ionics on the interaction energy is complex and that each of the various factors tend to decrease equilibrium spacings.

An interesting result would be obtained by comparison of the thickness of the polar groups with the structure of polyoxyethylene. Staudinger postulated two structures for polyoxyethylene, the zigzag and the meander configuration. Up to a degree of polymerization of about 9, the chain exhibits in a zigzag structure, whereas at higher degree of polymerization a meander structure is observed. For the oxyethylene unit in the meander-type chains a value of 2 Å is obtained and therefore the length of the oxyethylene chain in NP 20 should have a value of 40 Å. This value agrees well with the measured thickness of the polar groups of 38 Å in the saturated adsorbed layer.

More intricate problems were faced with regard to contact angle measurement, since with the small particle spacings some additional forces of interaction, with their orders of magnitude and distance functions still being unknown, had to be taken into account. Therefore, an analysis of the contact angle measurements and formation velocity is impossible. The only possible statement is that these measurements, too, showed an increased attraction of the particles with increasing surface concentrations. The hydrate shells, often postulated as the cause of stability, did not play any role, at least in our example. In a theoretical study by Havemann [17a] evidence was produced to the effect that water molecules adsorbed by a specified orientation (the dipoles being orientated parallel and arranged vertically relative to the interphase) cause the appearance of repulsive forces almost equal to that of the dispersion forces only if the degree of orientation is above 65 %. Such a high degree of orientation was not reached in our experiments.

The fact that, a finite particle distance is obtained may be attributed to the steric hindrance of the adsorbed layers which cannot be displaced from the interphase due to the insolubility of the given surface-active agent in the oil phase. Compression or mutual penetration is encountered by the adsorbed layers producing a force which can be compensated by the dispersion forces. This force was measured directly in non-polar media by compression of monomolecular and multimolecular adsorbed layers on mercury droplets.[18] The reversible change of distance between two flocculated mercury droplets was determined under conditions under which the droplets were pressed against each other by different pressures. The elasticity (vertical relative to the interphase!) derived therefrom for adsorbed layers of oleic acid, 54 Å in thickness, was 10^6 dyn cm^{-2}. Such a value was sufficient for compensation of the dispersion forces. The " mechanical " strength of the adsorbed layers under load, relative to the interphase in a vertical manner, is believed by the authors to be the cause of the coalescence stability. The latter has often been attributed to the strength in the adsorbed layer (mechanical properties of the layer, measured parallel with the interphase). Measurements of both the shear viscosity and elasticity, by the method of wave damping, have revealed what has become established knowledge, viz., that in monomolecular adsorbed layers of surface-active agent agreement between the coalescence stability, on the one hand, and the mechanical properties, measured parallel with the interface on the other, would occur only in few selected instances. This seems to support the conclusion that a breaking-up of the adsorbed layer, i.e., the tearing open of a hole, is not the decisive step towards coalescence.

The question why the stability may be both increased or reduced by non-ionic surface-active agents, depending on the nature of the disperse systems studied, still remains unanswered. The studies conducted by Cockbain [19] and Lemberger [20] may be considered as confirmations of our own results. In emulsions, 1-2 μ in particle size, increased adsorption of ionic surface-active agents will lead to stronger aggregation, as supported by the creaming volume quoted in ref. (19) and direct coulter counter determination reported in ref. (20). The same result was achieved by using macrodisperse solid dispersions.[3, 21] Yet it seems not plausible that particle size alone should be responsible.

[1] K. G. Mathai and R. H. Ottewill, *Trans. Faraday Soc.*, 1966, **62**, 759.
[2] R. H. Ottewill and T. Walker, *Kolloid-Z. Z. Polymere*, 1968, **227**, 108.
[3] H. Lange, *Kolloid-Z. Z. Polymere*, 1966, **211**, 106.
[4] M. Glazman, Jr., *Disc. Faraday Soc.*, 1966, **42**, 255.
[5] H. Schott, *J. Colloid Interface Sci.*, 1968, **26**, 133.
[6] K. L. Daluja and S. N. Srivastava, *Indian J. Chem.*, 1969, **7**, 790.
[7] H. Sonntag, J. Netzel and H. Klare, *Kolloid-Z. Z. Polymere*, 1966, **211**, 121.
[8] A. D. Scheludko, B. Radoev and T. Kolarov, *Trans. Faraday Soc.*, 1968, **64**, 2213.
[9] H. Sonntag and J. Netzel, *Tenside*, 1966, **3**, 296.
[10] Van der Waarde, private communication.
[11] D. Exerowa and A. D. Scheludko, *Proc. IV. Int. Congr. Surface Active Substances*, (Brüssell, 1964), vol. 2, p. 1097.
[12] M. J. Vold, *J. Colloid Sci.*, 1961, **16**, 1.
[13] H. Sonntag and K. Strenge, *Koagulation und Stabilität disperser Systeme*, VEB Deutscher Verlag der Wissenschaften, Berlin 1970, p. 43.
[14] H. Sonntag, *Tenside*, 1968, **5**, 188.
[15] A. D. Scheludko and D. Exerowa, *Kolloid-Z.*, 1960, **168**, 24.
[16] A. Watillon and A. M. Joseph-Petit, *Disc. Faraday Soc.*, 1966, **42**, 143.
[17] J. Netzel, *Diss.*, (Humboldt-Universität Berlin, 1968).
[17a] U. Havemann, unpublished, see ref. (13).
[18] K. Strenge and H. Sonntag, *Tenside*, 1969, **6**, 61.
[19] Cockbain, E. G., *Trans. Faraday Soc.*, 1952, **48**, 185.
[20] A. P. Lemberger and N. Mourad, *J. Pharm. Sci.*, 1965, **54**, 229.
[21] Th. Steudel, *Forschung Fortschritt*, 1964, **38**, 201.

Cohesive Properties of Thin Films of Liquids Adhering to a
Solid Surface

By J. F. Padday

Research Laboratories, Kodak Limited, Wealdstone, Harrow, Middlesex, England

Received 6th April, 1970

The process of forming and rupturing a thin liquid film at a solid surface is described thermo-dynamically for both high and low energy solid surfaces. In part 1 the build-up of thin films on high-energy surfaces from the first monolayer is considered and reviewed. Components of the surface free energy of formation of the thin film (disjoining pressure) are defined. For curved surfaces the disjoining forces should be combined with the Laplace capillary pressure to give a correct form of the Kelvin equation. It is suggested from the early work of Bangham and Deryaguin that thin liquid layers have anomalous physical properties. These studies are discussed in relation to the thickness of the liquid films.

In part 2, new experimental evidence of the critical rupture thickness of thin liquid films on low energy surface is presented. A number of pure liquids rupture spontaneously on low energy surfaces such as wax or polytetrafluoroethylene at very great thicknesses (0.01 cm). The effects of aqueous salt and surfactant solutions suggest these long-range forces are electrical in origin.

We first consider the process of thinning of a thick uniform layer of liquid in contact with a pure smooth horizontal solid surface as shown in fig. 1. It is assumed

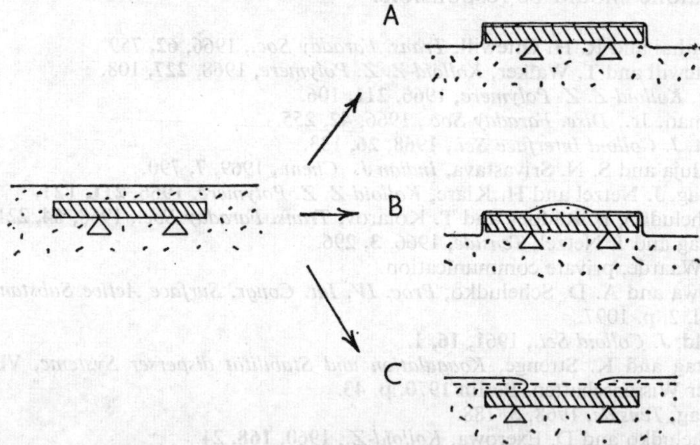

Fig. 1.—The processes of disjoining and rupture. A, formation of a thin stable layer; B, formation of a metastable layer with sessile drop or lens on a wetted surface; C, unstable thin layer with the formation of a dry patch.

that only vapour is present above the liquid surface and that the solid is completely wet when immersed below the surface of the liquid. If now the liquid thickness is successively reduced one of three phenomena occurs: either (A) the solid surface is

spread with and holds a uniform thin layer of liquid which is stable at all thicknesses ;
(B) or a thin layer is formed which when reduced to some critical thickness dis-
proportionates to form a very thin layer in contact with a thicker lens or sessile
drop ; (C) or at some relatively great thickness the liquid layer in contact with the
solid spontaneously breaks to leave a dry patch with a wetting meniscus in contact
with it as shown in fig. 1.

The phenomena of fig. 1A and B are invariably associated with high-energy surfaces
and are considered in part I of this paper.   The phenomenon of fig. 1C is associated
with low energy surfaces and is considered separately in part 2.

PART 1.—THIN FILMS ON HIGH ENERGY SURFACES

Hardy [1] appears to have been one of the earliest investigators of properties of
thin films of liquids in contact with solids.   He was concerned primarily with observations with lubrication but discovered the dual thickness films (fig. 1B) and likened
their behaviour to the formations of lenses of non-spreading oils on water.   It
became clear from this early work of Hardy and later Bangham and Fakhoury [2]
that to form stable or metastable thin films it was necessary to spread liquids on
high-energy surfaces such as mica, clean glass or silica, and charcoal.   Bangham [2-4]
and his coworkers investigated the build-up of successive layers of condensed vapour
to give a film of type A, fig. 1.   The partial vapour pressures recorded by Bangham
and Mosallam [4] did not exceed 0.7 of saturation, and the film thicknesses they obtained
were less than 20 Å.   They thus never made their films thick enough to investigate
the region of condensation on these surfaces, where the condensed vapour was of the
same thickness as a receding liquid film.

Bangham also attempted to estimate the surface tension of much thicker films
formed by a liquid receding on a solid surface.   He claimed that the thinner of the
two films in situation B of fig. 1 had a lower surface tension as measured from the
angle of contact between the thick and thin film.[5, 6]   However, these wetting
studies tended to be of a phenomenological nature and some of his experiments on
the break up of liquid films on freshly cleaved mica could not be confirmed by the
author.

Over the same period Deryaguin and his coworkers investigated the properties
of thin films of liquid on high-energy solids.[7-10]   Deryaguin introduced the term
" *wedging-apart pressure* " later known as " *disjoining pressure* " to describe the
equilibrium force required to remove a small increment of thickness of the thin liquid
layer.

DISJOINING PRESSURE OF THIN FILMS OF LIQUIDS ON SOLIDS

The process of joining a thin liquid film to a solid surface is shown in fig. 2.
Rupture in effect is the " reverse " of this process.   The disjoining process differs
from rupture in that only a small increment of liquid thickness is removed and not
the whole layer.   The steps of the process are taken as first the formation of the free
liquid film, (A–B) followed by the bringing into contact of the solid surface with
the thin film (B–C).   Thereafter specific interactions or solution may take place
between solid and thin film (C–D) and finally adsorption forces may lead to the
build-up of a electric potential at each interface of the thin layer (D–E).

Deryaguin and Shcherbakov,[11] Duyvis,[12] Kitchener [13] and Sheludko [14] have all
given thermodynamic accounts of the disjoining process.   They define the disjoining pressure as the change of free energy with thickness and break it down into

components.   Deryaguin and Obuchov [8] define this disjoining pressure Π by the expression,

$$\gamma = \gamma_0 + \int_h^\alpha \Pi \, dh, \tag{1}$$

where γ is the specific surface free energy of the thin liquid film, γ_0 is the specific surface free energy of an infinitely thick film and h is the thickness of the layer.

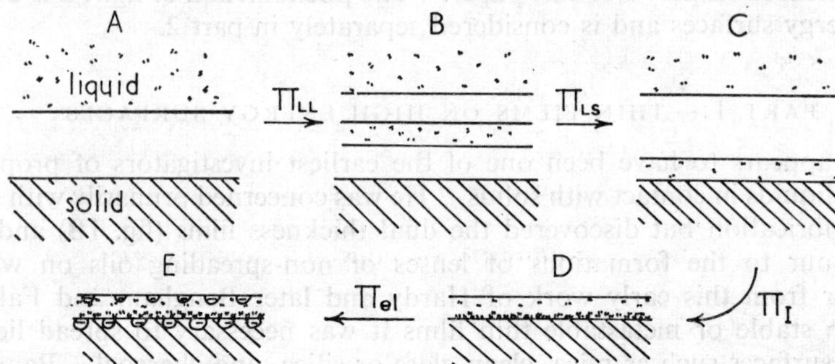

FIG. 2.—Steps in the formation of a thin film.   A–B, formation of thin liquid film; B–C, adhesion of thin films to solid surface; C–D reorientation of thin liquid layer by specific interactions; D–E adsorption of electric charge.

Provided that the structure of the bulk liquid is retained in the thin layer it is possible to inter-relate disjoining pressure with changes in surface tension, but Deryaguin and his coworkers have consistently maintained that thin liquid films adhering to solid surface are either totally oriented in the thin layer (Deryaguin's α-phase), or are partially oriented over a fraction of the total thickness of the film (Deryaguin's β-phase), the remainder of the layer containing unoriented or bulk liquid.   Hence, on Deryaguin's evidence, it is incorrect to consider disjoining pressure as involving a change in surface tension alone.   A more general definition of disjoining pressure is derived by Deryaguin and Shcherbakov [11] in terms of the chemical potential of the molecules forming the thin film and the total free energy w of the thin film.   Their equation was

$$RT \ln (P/P_s) = v_m(\delta w/\delta h), \tag{2}$$

where v_m is the molar volume of the substance forming the thin layer, P is the vapour pressure in equilibrium with the thin film, and P_s the saturated vapour pressure at the temperature T.   The term $-\delta w/\delta h$ is the disjoining pressure and is equivalent to Π of eqn (1).

Eqn (2) applies equally to the adsorption of the first monolayer of vapour on the solid surface and to successive build-up of condensed vapour into a true liquid layer. It thus allows the vapour pressure to be used to derive the disjoining pressure when direct measurement is not possible.

COMPONENTS OF THE DISJOINING PRESSURE

Sheludko [14] and Kitchener [13] split the disjoining pressure Π into two components; a term Π_v for the van der Waals interactions and a term Π_{el} for the component due to electrical double layer repulsion.   It is more convenient to split the van der Waals term into two components: one, Π_{LL}, for the disjoining pressure of the liquid film in the absence of any solid (equivalent to process A to B of fig. 2); and the other,

Π_{SL}, (equivalent to process B to C of fig. 2) for the effect of the solid on this film. In the absence of electrical charges and of specific interactions π_{LL} must always be negative and π_{SL} always positive.

To take account of the change in bulk properties in the film that is consistently claimed by many workers, a component of the disjoining pressure, Π_I is introduced. This component represents the change in free energy of the thin layer due to orientation, solvation or other specific interactions between the solid and the thin liquid film. Although measurement of this component may well prove difficult, its value must be positive for high energy surfaces.

The contribution of the electrostatic pressure Π_{el} has been derived by Langmuir,[15] Frumkin,[16] Verwey and Overbeek,[17] Deryaguin and Landau [18] and Sheludko.[14] Generally, Π_{el} was derived for two interacting electrical double layers of the same charge sign and density such as those that stabilize thin soap films. Read and Kitchener [19] calculated Π_{el} of a thin water film at a silica surface, taking into account the difference in potential between the solid-liquid and liquid-air interfaces. They pointed out that the usual assumption of constant potentials leads to physically improbable consequences because the negative charge at one surface was much greater than at the other. The counterions of the surface of high charge would render the surface of low charge positive and thereby induce attraction. However, it is certain that Π_{el} must be positive when the two interacting surfaces of the thin film are of like charge. Similarly, Π_{el} must be negative when the two surfaces are of opposite charge.

A curved liquid-air surface of radius r will possess a capillary pressure given by $2\gamma/r$ and a vapour pressure P different from the saturated value P_s. The capillary pressure is associated with the liquid-air-interface, and may be used as a direct measure of the disjoining pressure [10, 19] when the curved surface is free from thin film perturbations. However, when a curved liquid-air surface approaches a flat surface film, as with a Wilhelmy plate, then there will be a transition region, however small, where the capillary pressure diminishes and disjoining pressure increases. In this region we suppose in the first instant that the disjoining and capillary pressures are additive so that

$$\gamma\left(\frac{1}{r}+\frac{1}{r^1}\right) - \pi = \frac{\delta w}{\delta h} = \frac{RT}{v_m}\ln\frac{P}{P_s} \tag{3}$$

where r and r^1 are the principal radii of curvature, which possess a negative value when the liquid-air interface is concave and a positive value when convex. For thick films Π is zero and eqn (3) reduces to the Kelvin equation.

The contributions Π_{LL}, Π_{SL}, Π_I and Π_{el} are all complex functions of h and each will vary according to whether the thin film is flat or curved. Although Π_I and Π_{el} are probably interrelated, we assume additivity such that

$$\pi = \pi_{LL} + \pi_{SL} + \pi_I + \pi_{el}. \tag{4}$$

When Π is negative the thin film is unstable. For stability not only must Π be positive but $\delta^2 w/\delta h^2$ must be negative.

REGIONS OF STABLE THIN FILMS

The thin liquid film adhering to a long perfectly-wetted Wilhelmy plate will be considered following the model of Read and Kitchener.[19] Such a plate is shown in fig. 3 and here it will be used to distinguish certain zones of the thin films. If it is assumed that only pure vapour is in contact with the liquid and that the vapour is an ideal gas, the vapour pressure P at any height l above the free surface bulk liquid is given by

$$(RT/v_m) \ln (P/P_s) = -l\,g\rho. \tag{5}$$

The hydrostatic height may thus be used as a measure of the disjoining pressure. Combining eqn (3), (4) and (5),

$$\frac{RT}{v_m} \ln \frac{P}{P_s} = \gamma\left(\frac{1}{r}+\frac{1}{r^1}\right) - \pi_{LL} - \pi_{SL} - \pi_I - \pi_{el} = -l\rho g. \tag{6}$$

At flat films the first term is zero and disjoining pressure is the main component, whereas for thick films the first term predominates and all other terms are zero.

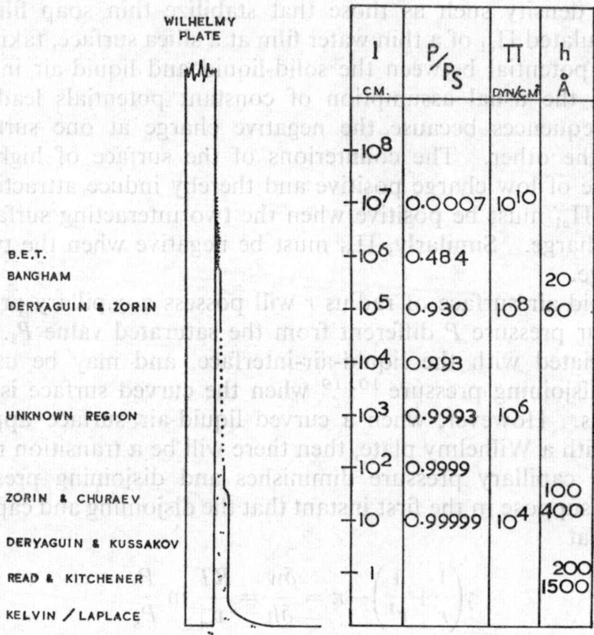

FIG. 3.—Regions of stable thin films on a Wilhelmy plate.

Where the Wilhelmy plate meets the free surface of liquid a cylindrical meniscus will be formed. At some height above the free surface equal to the capillary constant, a, the liquid meniscus will meet the flat thin film. The shape of the meniscus in this region is governed by the Laplace equation, because all Π terms of eqn (6) are too small to matter. This region is the meniscus zone of the " thin films " and may be located anywhere up the side of the Wilhelmy plate, merely by bringing a second Wilhelmy plate into close proximity so that capillary rise is obtained. At some height, say 10^8 cm, above this zone, the partial vapour pressure is so small that the solid Wilhelmy surface may be regarded as practically " clean ". Between these two zones lies a series of regions of different disjoining pressure in which all experimental data may be placed. The first region below the " clean " zone may be regarded as that region in which the first monolayer of adsorbed vapour condenses on the solid surface. Descending further, one enters the B.E.T. region where partial vapour pressure increases to a value of about 0.7 and multilayer adsorption sets in.

Bangham's [20] system of adsorbed water vapour on glass and mica covered a range of partial vapour pressures not exceeding 0.7 and therefore falls in to this B.E.T. region. He ascertained the thickness of his adsorbed layers as being no greater than 20 Å; hence the anomalous properties of condensed or adsorbed water claimed by Bangham refer specifically to the first few layers and not to the much thicker layers of Deryaguin and Zorin [21] and those by which anomalous water was formed as reported by Fedyakin [22] and by Deryaguin.[23] However, Bangham claimed anomalous properties for the much thicker layers he obtained with receding wetting films. The results of Deryaguin and Zorin [21] on thin films of water and other liquids on glass fall into the next region of partial vapour pressure 0.7 to 0.99. This is the region where capillary condensation usually takes place on porous solid surfaces.

In the B.E.T. region and the region of capillary condensation, the main contribution to the total disjoining pressure was believed by both Bangham and by Deryaguin to be Π_I, i.e., a specific but long-range interaction which is much larger than the sum of the van der Waals attractions, $\Pi_{SL} + \Pi_{LL}$.

The results of Read and Kitchener,[19] Deryaguin and Kussakov [10] and Zorin and Churayev [24] fall in a region of much lower disjoining pressures and refer to much thicker layers of 200 to 5000 Å. Read and Kitchener attribute their low disjoining pressure at these thicknesses to electrostatic repulsion Π_{el}, because the repulsion fell as ionic strength of added salt increased. The contributions of the other terms Π_{SL}, Π_{LL} and Π_I in Kitchener's experiments were regarded as zero at these much greater thicknesses.

Zorin and Churayev [24] measured the thickness of thin water films on quartz at low disjoining pressures. Their stable film was 100 Å thick and possessed a peripheral step 400 Å thick. This step could well correspond to the step found when a large clean sheet of glass is withdrawn from a basin of water and allowed to equilibrate.[25,26] The step on the plane glass sheet appears at about 10 cm above the free surface and corresponds to a disjoining pressure of about 10^4 dyn but is attributed by Satterly and Turnbull [25] to contamination. This step might mark a change from electrostatic to interactive disjoining forces.

Each of the terms of eqn (4) are a different function of the thickness h so that Π, the total disjoining pressure, is unlikely to fall away smoothly. Deryaguin and Zorin [21] showed that it did so for the non-polar liquids CCl_4 and C_6H_6 where the disjoining pressure decreased steadily to zero in a manner qualitatively predictable from London's theory of dispersion forces. For these liquids one may assume that

$$\pi = \pi_{SL} + \pi_{LL} = K(A_{SL} - A_{LL})/h^n, \tag{7}$$

where A_{SL} and A_{LL} are the Hamaker constants for intermolecular interactions between solid-liquid and liquid-liquid molecules respectively. K is a constant which depends on the value of n, which possesses a value of 3 when the dispersion forces are unretarded and a value of 4 when retarded. Approximate expressions for Π_{SL} and Π_{LL} are given by Kitchener [13] and Sheludko,[14] and these lead to negligible forces when h is greater than 500 Å.

Thin layers of polar liquid do not show the same trend. Water and n-alcohols are known to interact with polar surfaces, such as glass or mica, to produce positive value of Π_I (Deryaguins' α-phase or Bangham's aggregated phase). The stepwise increase or decrease of the stable film has been explained by Zorin [27] as due to the difference of Π_I for the two phases postulated by Deryaguin. In fig. 4A, the lower isotherm refers to the disjoining isotherm of Deryaguin's α-phase and the upper film on mercury. Scheludko [14] suggests that the same data may be explained as shown in fig. 4B without recourse to the two phases supposed by Deryaguin.

Between the thick or stepped region of low disjoining pressure and the capillary condensation region of very high disjoining pressure lies a large unknown region of disjoining pressures that has never, it seems, been investigated. In the region of capillary condensation, where $0.7 < P/P_s < 0.99$, the Kelvin equation may be tested experimentally. The Kelvin equation without disjoining effects leads to radii of curvature between 20 and 50 Å. Deryaguin and Zorin [21] show that disjoining effects extend to 100 Å in this region. This is also the region of partial vapour pressure appropriate to the formation of Deryaguin's anomalous water. Thus, any test of the Kelvin equation must take into account disjoining effects as given in eqn (3), and any application of disjoining pressure to curved surfaces must include the capillary pressure.

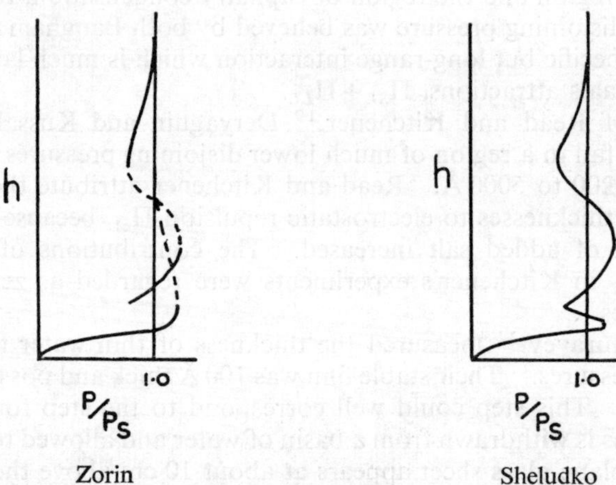

FIG. 4.—Disjoining isotherm of benzene on mercury. A, Zorin; B, Sheludko (taken from Sheludko [14]).

PART 2.—RUPTURE OF THIN FILMS AT LOW ENERGY SURFACES

A third type of liquid layer in contact with a solid surface, that of fig. 1C, is that which spontaneously ruptures when the thickness is reduced below some critical value. This type of thin liquid film is formed only at low-energy surfaces and it appears to be associated with large critical rupture thickness, i.e., 0.01-0.05 cm. Although negative disjoining pressures have been measured by Sheludko and Platikanov [28] the rupture process of fig. 1C, does not seem to be widely studied. The experiments by which the critical thickness of these films was observed are given here together with the principal results.

EXPERIMENTAL

The apparatus for measuring the critical rupture thickness, shown in fig. 5, is basically that described previously [29] for measuring spreading coefficients by the drop-height method. Instead of the solid surface E resting on the sample platform, it was supported inside a glass container G on three glass chips broken from a microscope slide. The surface of the plane solid which was either a waxed glass microscope slide or a flat piece of polytetrafluoroethylene (Teflon) was mounted level. A 10 ml glass syringe was used to raise or lower the level of liquid in the glass container G. All apparatus was carefully cleaned and washed.

Freshly distilled water, checked for purity by surface-tension measurement and by shaking, was poured into the vessel so as to cover the solid surface. The syringe was then used to lower the level of the liquid while at the same time its height was monitored with the pointer C. At some critical thickness the meniscus suddenly broke through to the Teflon. The critical rupture thickness was then measured as the distance between the pointer and the solid surface, with an accuracy of $\pm 1.0\,\mu$. The thickness at which breakdown took place was reproducible to within $\pm 3\,\%$ and was usually between 0.01 and 0.05 cm. Thicknesses below 0.003 cm were regarded as being too small to measure to this method. The surface tension of water after disjoining was measured and found to be unaltered (within ± 0.05 dyn/cm) thus dispelling the possibility of artifacts caused by impurities.

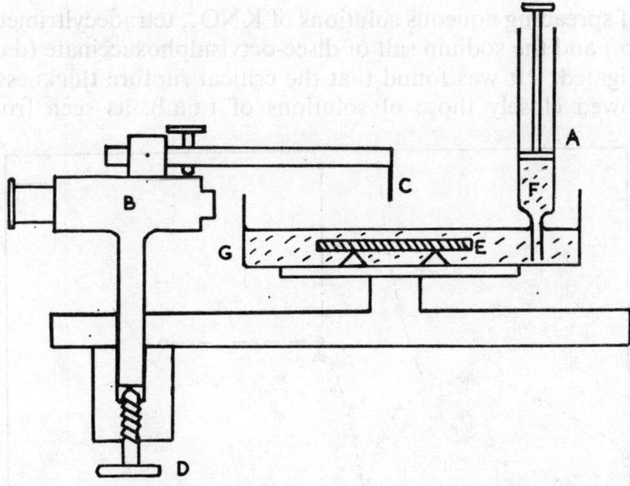

Fig. 5.—Apparatus for measuring the thickness of thin films on a low energy surface. A, syringe, B, microscope with a vernier graticule; C, point adjustable in focus plane of the microscope; D, height adjustment of the microscope; E, flat sample of low energy surface; F, spreading liquid, G, glass vessel.

RESULTS OF NEGATIVE RUPTURE PRESSURES

It is believed that the phenomenon described measures a negative rupture pressure. In table 1, the critical rupture thickness h of a number of liquids on Teflon measured by this method are given. Even pure hydrocarbons and ethylene glycol ruptured spontaneously.

TABLE 1.—CRITICAL RUPTURE THICKNESS OF PURE LIQUIDS ON TEFLON

liquid	ρ g/cm^2	γ_L dyn/cm	h cm
water	1.0	72	-0.051
benzene	0.88	29	-0.016
ethylene glycol	1.12	49	-0.056
1-decene	0.74	23	-0.026
1-tetradecene	0.775	27	-0.027
1-octadecene	0.79	28.4	-0.027

That the process of rupture was not some artifact due to irregularities of the solid surface was shown by comparing the critical disjoining thickness of a flat and curved wax surface. The curved surface was formed by coating a watch-glass with paraffin wax. These results (table 2) indicate that the effect of curvature on the critical rupture thickness is relatively small, and that therefore the effect of gross unevenness of a solid surface cannot account for the large rupture thickness. The results of water rupturing on a Teflon and on a wax surface (tables 1 and 2) show that the rupture thickness is critically dependent on the nature of the solid surface.

TABLE 2.—CRITICAL RUPTURE THICKNESS OF WATER FILMS ON PARAFFIN WAX SURFACES

	h (cm)
flat surfaces	-0.031
curved surface (radius 6 cm)	-0.032

The effect of spreading aqueous solutions of KNO_3, tetradecyltrimethyammonium bromide (t.t.a.b.) and the sodium salt of disec-octylsulphosuccinate (d.o.s.) on Teflon was also investigated.   It was found that the critical rupture thicknesses of solution of KNO_3 followed closely those of solutions of t.t.a.b. as seen from fig. 6.   At

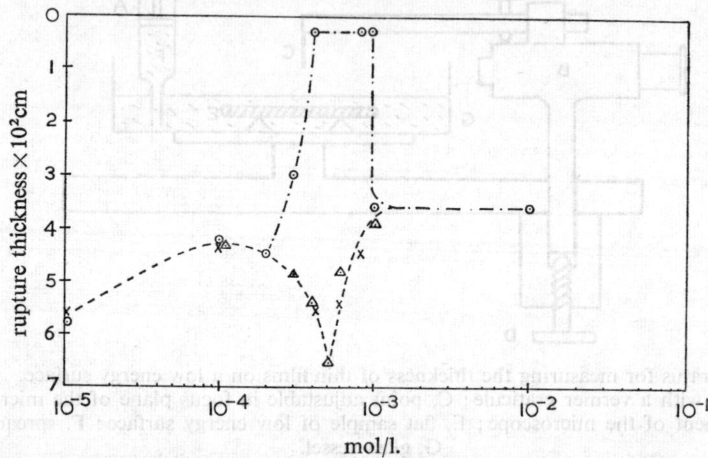

FIG. 6.—Critical rupture thickness of aqueous solution on teflon.   ×, tetradecyltrimethylammonium bromide ; ⊙, sodium dioctyl sulphosuccinate ; △, KNO_3.

concentration below 3×10^{-4} M only small changes in thickness were observed, but between 3×10^{-4} and 10^{-3} M the rupture pressure decreased suddenly and then increased to a value slightly greater than that of pure water. D.o.s. behaved differently.   In the same concentration range the anionic surfactant d.o.s. produced a film that was stable at all measurable thicknesses and would not spontaneously disjoin (see fig. 6).   In the concentration range 3×10^{-4} to 10^{-3} M, both d.o.s. and t.t.a.b. adsorb at both the liquid-air and the low energy solid-liquid interfaces,[30, 31] thus it is likely that the main contribution to these forces is the electric term arising from monolayers of adsorbed ionic surfactants.

The comments of Dr. J. A. Kitchener are gratefully acknowledged.

APPENDIX (ADDED IN PROOF)

The following evidence was obtained from further experiments, performed since writing the paper, and was presented in the Discussion.

SURFACE DEFORMATION PRIOR TO RUPTURE

In a further series of experiments, the rupture of a thin film of water at a curved wax surface was followed using a high-speed cine camera recording at approximately 1000 frames s^{-1}, while the liquid layer was continuously thinned.   These sequences showed clearly that a

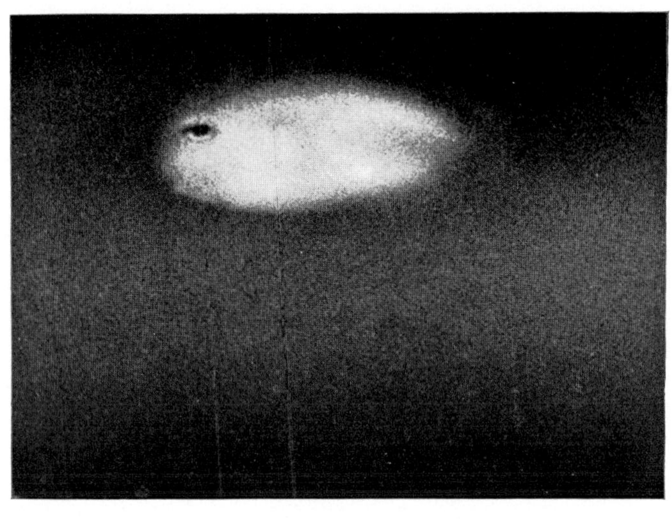

(a)

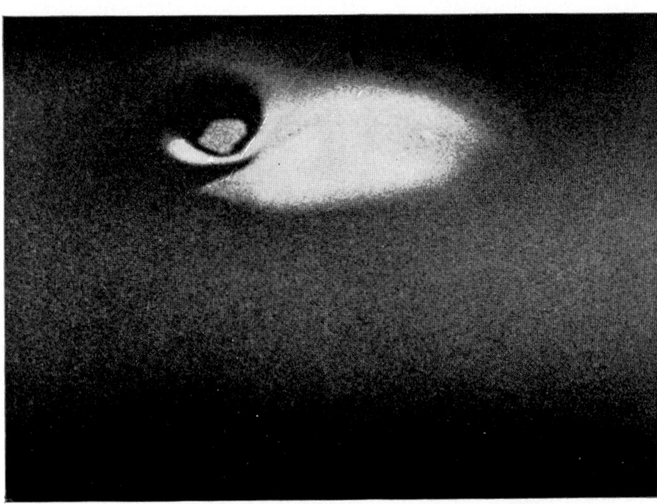

(b)

FIG. 7.—Formation of dimple and rupture of a thin film of water on wax.   Approx mag. ×2.
Radius of wax surface 7 cm.

(*a*) Formation of dimple (towards left-hand edge of white patch) ; (*b*) rupture revealing wax surface
below.

To face page 73.]

local indentation of the water/air interface occurred at the position where the convex wax surface was nearest the liquid-air surface. In fig. 7, prints of two frames from this cine sequence are shown. The whitish patch is a specular reflection in the liquid/air interface, produced by a powerful lamp. In the upper picture (a), the indentation is seen as a dimple in the surface near the edge of the specular reflection. Some 300 ms later, the dimple broke through to the wax surface as shown in the lower picture (b). Once the wax was revealed, the border expanded rapidly as a receding meniscus. Although these observations were performed on a dynamic system they served to show the formation of the dimple prior to rupture. A cross-section of it is shown in fig. 8(a). The dimple was also produced with

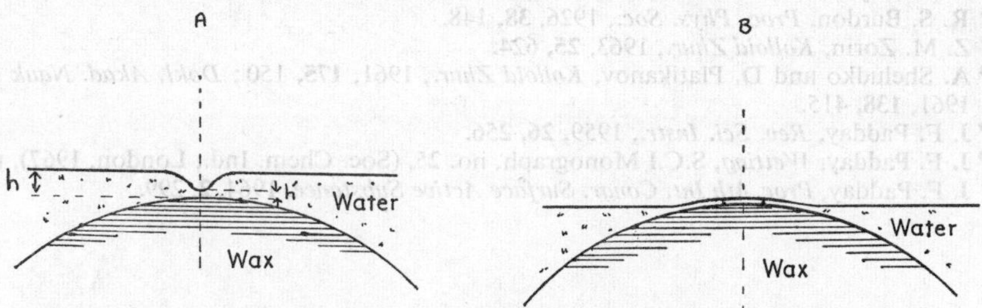

Fig. 8.—A, apparent cross-section profile of the dimple of fig. 7(a); B, apparent cross-section profile of deformation with wax that had been polished with cellulose fibres.

the apparatus of fig. 5 but with the flat Teflon surface replaced by a convex glass surface coated with pure paraffin wax. Fig. 8(b) shows a second type of deformation obtained with a wax surface artificially polished. A dimple formed and appeared to be stable for many minutes while observations were being made.

Vibrations, artificially induced, at the liquid/air interface appeared to be damped out as they reached the dimple and only high energy knocks succeeded in forcing the dimple to break through prematurely. It was not found possible to estimate the depth of the dimple from the cine film but the appearance showed that the dimple penetrates a considerable part of the thickness of liquid layers. Thus, the thickness of the liquid film between the bottom of the dimple and the top of the convex wax-liquid interface will be considerably less than the critical disjoining thickness of tables 1 and 2. Furthermore, the apparent negative disjoining forces at the bottom of the dimple will be balanced by a further restoring pressure P_1 given by

$$P_1 = \rho g t - (2\gamma/\gamma_2), \tag{8}$$

where t is the depth of the dimple and r_2 the radius of curvature (negative in value) at the bottom of the dimple.

[1] W. Hardy, *Proc. Roy. Soc. A*, 1912, **86**, 610; 1923, **104**, 25; 1925, **108**, 1; 1926, 112, 62.

[2] D. H. Bangham and N. Fakhoury, *J. Chem. Soc.*, 1931, 1324.

[3] D. H. Bangham, N. Fakhoury and A. F. Mohamed, *Proc. Roy. Soc. A*, 1934, **147**, 152 and 175.

[4] D. H. Bangham and S. Mosallam, *Proc. Roy. Soc. A*, 1938, **165**, 552; 1938, **166**, 558.

[5] D. H. Bangham and R. L. Razouk, *Trans. Faraday Soc.*, 1937, **33**, 1459.

[6] D. H. Bangham and Z. Saweris, *Trans. Faraday Soc.*, 1938, **34**, 554.

[7] B. V. Deryaguin, *J. Phys. Chem.*, 1932 ,3, 29; *Z. Phys.*, 1933, **34**, 657; *J. Phys. Chem.*, 1934, **5**, 379; *Sov. Phys.*, 1933, **4**, 431.

[8] B. V. Deryaguin and E. Obuchov, *J. Colloid Chem.*, 1935, **1**, 385; *Acta physicochim.*, 1936, **1**, 5.

[9] B. V. Deryaguin and M. Kussakov, *Bull. Acad. Sci., U.R.S.S.*, 1936, 471.

[10] B. V. Deryaguin and M. Kussakov, *Acta physicochim.*, 1939, **10, 25**.

[11] B. V. Deryaguin and L. M. Shcherbakov, *Colloid J. U.S.S.R.* (*trans*), 1961, **23**, 33.

[12] E. M. Duyvis, *Thesis*, (Utrecht, 1962).

[13] J. A. Kitchener, *Endeavour*, 1962, **22**, 118; *Wetting* S.C.I. Monograph no. 25, 1967; (Soc. Chem. Ind. London, 1967), p. 300.

[14] A. Sheludko, *Adv. Colloid Interface Sci.*, 1967, **1**, 391.

[15] I. Langmuir, *J. Chem. Phys.*, 1938, **6, 873**.

[16] A. N. Frumkin, *Acta physicochim.*, 1938, **9**, 313.
[17] E. J. W. Verwey and J. T. G. Overbeek, *Theory of the Stability of Lyophobic Colloids*, (Elsevier, Amsterdam 1948).
[18] B. V. Deryaguin and L. Laundau, *Zhur. Exsp. Teor. Fig.*, 1941, **11**, 802.
[19] A. D. Read and J. A. Kitchener, *J. Colloid Interface Sci.*, 1969, **30**, 391.
[20] D. H. Bangham, *J. Chem. Phys.*, 1946, **14**, 352.
[21] B. V. Deryaguin and Z. M. Zorin, *Proc. 2nd Int. Congr. Surface Activity*, (London), 1957, **2**, 145.
[22] N. N. Fedyakin, *Colloid J.* (*trans.*), 1962, **24**, 425.
[23] B. V. Deryaguin *et al.*, *Teor. Eksp. Khim.*, 1968, **4**, 527.
[24] Z. M. Zorin and N. V. Churayev, *Colloid J.*, (*trans.*), 1968, **30**, 279.
[25] J. Satterly and R. Turnbull, *Trans. Roy. Soc. Can.*, 1929, **3**, 95.
[26] R. S. Burdon, *Proc. Phys. Soc.*, 1926, **38**, 148.
[27] Z. M. Zorin, *Kolloid Zhur.*, 1963, **25**, 624.
[28] A. Sheludko and D. Platikanov, *Kolloid Zhur.*, 1961, **175**, 150; *Dokl. Akad. Nauk S.S.S.R.*, 1961, **138**, 415.
[29] J. F. Padday, *Rev. Sci. Instr.*, 1959, **26**, 256.
[30] J. F. Padday, *Wetting*, S.C.I Monograph, no. 25, (Soc. Chem. Ind., London, 1967), p. 234.
[31] J. F. Padday, *Proc. 4th Int. Congr. Surface Active Substance*, 1964, **2**, 299.

Mr. D. W. J. Osmond (*I.C.I. Ltd., Slough*) said: My colleagues and I have been very interested in the paper of Andrews *et al.* We have been concerned with the behaviour of sterically stabilized particles in concentrated solutions and " melts ". There are now semi-quantitative theories for the action of polymeric steric barriers in low molecular weight environments. These theories suggest that a large part of the repulsion arises from the non-ideal osmotic pressure generated by the increased concentration of polymer segments in the region in which the barriers have overlapped. Clearly, however, this change in segment concentration and hence excess osmotic pressure, is much reduced in concentrated solutions of unadsorbed stabilizing polymer in the low molecular weight solvent, and disappears altogether in an environment which comprises solely molten polymer. F. A. Waite and I have considered the problem of particle stability in such cases and have concluded, first, that the concentration of segments of stabilizing (adsorbed) polymer chains in the barrier will rise, in the limiting case of the melt, to 100 %, and secondly, that these concentrated barriers will be stiffer and stronger than the more diffuse barrier formed in low molecular weight solvents. We propose that, where the particle surface has free access to more stabilizing molecules, and there are no packing limitations of the anchor groups at the surface, more stabilizer will be adsorbed to maintain the barrier of similar thickness to that of the barrier in low molecular weight solvent. Where these conditions cannot be fully met, the additional rise in concentration must be met by contraction of the barrier back towards the surface. In some preliminary experiments we have demonstrated that stable dispersions of polymethyl methacrylate, stabilized by polymethyl methacrylate/poly(12 OH stearic acid) graft copolymers, can be prepared in molten methyl ester of poly(12 OH stearic acid). In these dispersions, the amount of stabilizer adsorbed per unit area of particle surface appears to be several times that adsorbed from low molecular weight aliphatic hydrocarbon continuous phases. In the latter case, the effective barrier thickness is believed to be about 100 Å (10 nm) and the mean stabilizing segment concentration about 10-15 %. In the melt therefore the segment density may be about 50 % with a barrier still of 100 Å (10 nm) or, more probably, the segment density may be near 100 % with the barrier contracted to about 50 Å (5 nm) thickness.

The experimental techniques of Dr. Haydon and his colleagues are far more refined than ours and they support our ideas closely. It would be of interest if the work could be repeated with higher molecular weight soluble chains. In this case the lower segment densities in low molecular weight solvents would allow a much larger rise in concentration as the solvent molecular weight was raised, and, freed from the limitation of anchor group spacing which may operate with oleyl monoglyceride, the predicted increase in adsorption might also be observed.

Dr. H. E. Ries (*Chicago*) said: Although the interesting studies of Andrews, Manev and Haydon were performed on systems considerably different from those investigated in our laboratories, the overall results and their implications are strikingly different from the stand-point of chain-length compatibility in mixtures of polar and non-polar compounds. In our work we have found that (*a*) mixed films of stearic acid and n-hexadecane at the water/air interface retain considerable hydrocarbon in the monolayer when subjected to elevated surface pressures ; and (*b*) radiotracer adsorption experiments at the solid/liquid interface indicate the formation of mixed films of stearic acid and n-hexadecane.[1]   In the latter experiments, radiostearic acid is adsorbed directly from n-hexadecane solutions on vapour-deposited metal films on the window of a Geiger tube.

Moreover, studies of rust-preventive films clearly demonstrate strong interaction and chain-length compatibility in thin films of fatty acids and hydrocarbons. Fatty acids and hydrocarbons of the same length and geometry apparently form stronger mixed films and thus provide better rust protection,[2] as well as greater resistance to scuffing in four-ball studies by Cameron and others.[3]

Dr. H. Sonntag (*Deutsche Akademie der Wissenschaften, Berlin*) said: With regard to the paper by Haydon *et al.*, did the thickness of the hydrocarbon region change reversibly or irreversibly with applied potential? We obtained with the same method the thickness of oil layers between mercury droplets reversibly up to a certain applied potential, and we could calculate the elasticity of the adsorbed layer which was unaffected by the thickness and had a value for oleic acid layers (thickness 54 Å) of 10^6 dyn/cm^2.[4]   Above this potential the thickness altered irreversibly.   From Haydon's measurements we calculated an elasticity of 10^5-10^6 dyn/cm,[2] which is of the same order as our values.   It it possible that the change of elasticity with applied potential could be explained by an alteration of the film diameter? We measured an increasing diameter of the film with increasing applied potentials in water/oil emulsions.

Prof. J. Th. G. Overbeek (*University of Utrecht*) said: The sharp c.m.c. in the presence of 0.1 M NaCl and the much wider transition region in the presence of saturated NaCl (see fig. 3 of the paper by Haydon *et al.*) point to small aggregates formed with saturated NaCl and larger ones with the more dilute solution. It is worth while to look for a confirmation by light scattering or ultracentrifugation. The difference in aggregate size might be connected with the difference in water activity if water is a necessary partner in at least the bigger aggregates; or the small aggregates might be formed around ions, solubilized in the oil phase. The latter explanation would require an increased conductance of the oil phase in contact with saturated NaCl.

Dr. D. A. Haydon (*Physiological Lab., Cambridge*) said: In reply to Osmond, we have been interested for some time in the possibility of working with black lipid films stabilized by molecules of chain lengths both longer and shorter than that of the oleyl derivatives. A limited investigation has been reported using chain lengths between n-C_{12} and n-C_{22} [5] but owing to the nature of the stabilizing molecules a detailed interpretation of the results was not possible. It is hoped, however, to find satisfactory stabilizers of chain length greater than C_{22}. The problem of steric stabilization in one component liquids such as polymer melts, to which Osmond draws attention, underlines the necessity to develop further the formulation of the configurational free energy along the lines attempted by Mackor and van der Waals and now by Findenegg and Ash.[6]

[1] H. D. Cook and H. E. Ries, Jr., *J. Phys. Chem.*, 1959, **63**, 226.
[2] H. E. Ries, Jr. and J. Gabor, *Chem. and Ind.*, 1967, 1561.
[3] A. Cameron and R. F. Crouch, *Nature*, 1963, **198**, 475.
[4] *Tenside*, 1969, **6**, 61.
[5] J. L. Taylor and D. A. Haydon, *Disc. Faraday Soc.*, 1966, **42**, 51.
[6] this Discussion.

In reply to Ries, the spread films of stearic acid and n-hexadecane at the air/water interface, and the adsorption of stearic acid from n-hexadecane on to metal-coated mica surfaces should perhaps not be compared too closely with the adsorbed monolayers of glyceryl mono-oleate at the n-hexadecane/water interface. In the spread film experiment the n-hexadecane is presumably squeezed out from between the stearic acid molecules as the compression proceeds. Towards the end of the compression (where, from Ries's graph the surface pressure becomes constant) it is not clear where the hexadecane is. Dawson [1] estimated that the maximum adsorption of stearic acid at the n-heptane/water interface (at *ca.* 0.02 mol/l.) corresponded to *ca.* 60 Å²/molecule. The results reported in our paper for glyceryl mono-oleate and adsorption studies on n-alkanols do not indicate any great influence of the solvent on the adsorption, and it seems probable that 60 Å² per molecule would hold also for n-hexadecane. If this is so the space available for n-hexadecane in the monolayer would be substantially greater than in the glyceryl mono-oleate systems, where the area per molecule is *ca.* 30 Å². A similar situation appears to exist in the metal-coated mica systems examined by Ries. The data indicate a maximum area per molecule for stearic acid adsorbed from n-hexadecane of *ca.* 65 Å². Again, there is apparently considerable space available for solvent. The inclusion of n-hexadecane into such stearic acid monolayers does not therefore seem in any way remarkable, and it would be expected that shorter chain solvents would be included to a similar, or even greater, extent. This does not exclude the possibility that in hexadecane the monolayers may have exceptional *mechanical* properties.

In reply to Sonntag, under the conditions described in our paper the thickness of the hydrocarbon region of the film changed reversibly with applied potential. As stated, however, the equilibrium condition was for some systems established only after considerable time. For higher applied potentials the films tended to rupture. In all membranes, the application of a potential caused an increase in the diameter of the film. This effect was allowed for in the calculation of the specific capacitance of the film, and it is not thought that it could account for the change of compressibility with applied potential.

In reply to Overbeek, no attempt was made to examine in detail the aggregation of the glyceryl mono-oleate at high concentrations in the hydrocarbons. Nevertheless, for the oleate in n-heptane at 25°C (equilibrated with saturated NaCl), the vapour pressure curve was measured to concentrations (*ca.* 1.2×10^{-2} M) at which it became roughly linear. In this region the aggregation number was *ca.* 26. The interfacial tension curves certainly suggest that, at lower concentrations at least, the aggregates formed in systems equilibrated with saturated NaCl are smaller than in those equilibrated with 0.1 M NaCl. While no conclusive evidence is available, miscellaneous observations during the course of the experiments indicated that the presence of water in the hydrocarbon phase favoured the formation of larger aggregates.

Dr. G. H. Findenegg (*University of Vienna, Austria*) said: In relation to the stability of thin liquid films I would like to mention some theoretical work by Ash and myself.[2] We have considered a solution of chain molecules between two plane interfaces. The chain molecules have an " active " end-group which is preferentially adsorbed to the interfaces. The calculations are based on a multi-layer lattice model for

[1] Dawson, *Ph.D. Thesis*, (Cambridge, 1963).
[2] S. G. Ash and G. H. Findenegg, *Trans. Faraday Soc.*, to be published.

adsorption from monomer + r-mer solutions.[1] Different chain-conformations are allowed for by considering a number of configurational species of r-mer, according to the sequence of their segments in the lattice layers.[1] We have calculated for r-mers up to $r = 4$, the concentrations of the adsorbed configurational species, the Gibbs adsorption of r-mer, and the configurational Helmholtz free energy F_c, as a function of the following parameters : the thickness of the liquid film ; the concentration of r-mers (in a bulk solution at equilibrium with the liquid film) ; the strength of preferential adsorption of the active end-groups ; and the segment interchange energy of active and non-active segments (monomers are energetically equivalent to non-active segments of r-mers).

Many features of this model are in agreement, qualitatively, with the properties of non-compressed lipid films in aqueous media, as reported by Haydon *et al.* We find that only a fraction of the r-mers are in their fully extended form normal to the interface, and hence the concentration of r-mer chain segments decreases towards the centre of the film. The surface excess of the configurational free energy, F_c^σ, has a (shallow) minimum at a thickness of the film equal to twice the extended form of r-mer (8 layers for $r = 4$). F_c^σ then increases with decreasing thickness, hence there exists a repulsive force. At a given film thickness F_c^σ increases with increasing chain length and concentration of r-mer, as well as by an increase in the preferential adsorption of active segments, and the interchange energy. For films consisting of monomers + homogeneous r-mers (all its segments preferentially adsorbed to the interfaces) we find a decrease in F_c^σ with decreasing film thickness, hence such films should not be stable.

Dr. B. Vincent (*I.C.I. Ltd., Slough*) said: I would suggest some data and calculations that could help resolve the apparent discrepancy between the results presented by Sonntag at this meeting on oil droplets in water, and those of Ottewill and Walker [2] on aqueous polystyrene latex dispersions, both systems containing ethylene oxide (E.O.) stabilizers, but showing different trends in stability as the stabilizer concentration is increased.

I have recently determined some Hamaker constant values by the method of Gregory [3] using refractive index data. Values relevant to this work are :

material	$A \times 10^{20}$ J
octane	5.3
polystyrene	7.8
poly(ethylene oxide)	6.9
water	3.8

Appropriate models were set up (flat-plates for octane ; spheres for the polystyrene latices) in which the thickness of, and ethylene oxide concentration in, the adsorbed layer, together with the surface separation of the particles, were allowed to vary. The relevant equations for the attraction energy V_A^p (flat plates) and V_A^s (spheres) were derived from eqn (2) of Sonntag's paper, and the equation of Vold,[4] respectively. Retardation effects were neglected for the purposes of these calculations. Fig. 1 indicates how the Hamaker constant A_3 of the adsorbed layer varies as a function of the wt fraction w of E.O. in the layer. Fig. 2 and 3 show how V_A^s varies as a function

[1] S. G. Ash, D. H. Everett and G. H. Findenegg, (*a*) *Trans. Faraday Soc.*, 1968, **64**, 2645 ; (*b*) 1970, **66**, 708.
[2] R. H. Ottewill and T. Walker, *Kolloid-Z. Polymere*, 1968, **227**, 108.
[3] J. Gregory, *Adv. Colloid Interface Sci.*, 1969, **2**, 396.
[4] M. J. Vold, *J. Colloid Sci.*, 1961, **16**, 1.

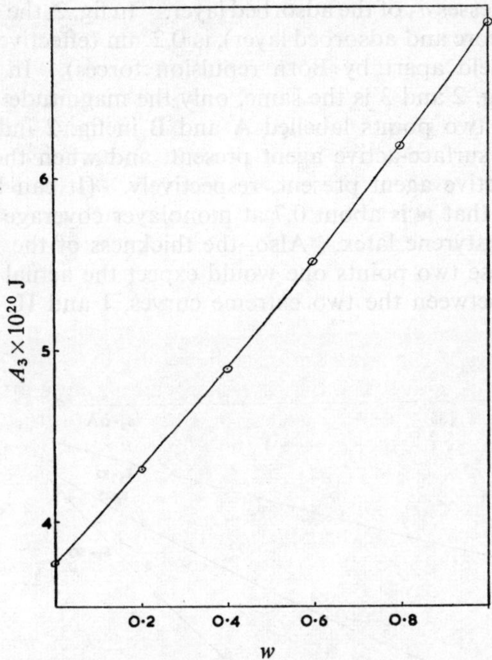

FIG. 1.—Hamaker constant of adsorbed layer A_3 as a function of wt fraction w of ethylene oxide in the adsorbed layer.

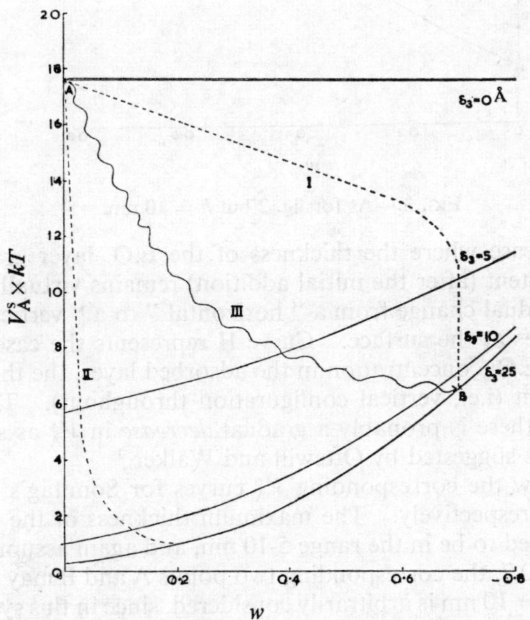

FIG. 2.—Attraction energy between two aqueous polystyrene latex particles (radius 25 nm), covered with adsorbed layer, thickness δ_3, as a function of ethylene oxide wt fraction w; surface separation, $h = 0.2$ nm.

of w, for various thicknesses δ_3 of the adsorbed layer.   In fig. 2, the surface separation h between the particles (core and adsorbed layer), is 0.2 nm (effectively the two particles are in contact, i.e., held apart by Born repulsion forces).   In fig. 3, $h = 10$ nm. The general form of fig. 2 and 3 is the same, only the magnitude of V_A^s being significantly different.   The two points labelled A and B in fig. 2 indicate the values of V_A^s when there is no surface-active agent present, and when there is a monolayer of adsorbed surface-active agent present, respectively.   (It can be shown from the adsorption isotherm [1] that w is about 0.7 at monolayer coverage of n-dodecyl hexaethylene oxide on polystyrene latex.   Also, the thickness of the E.O. layer is about 2.5 nm).   Between these two points one would expect the actual locus of points for V_A^s, curve III, to be between the two extreme curves, I and II, in fig. 2.   Curve I

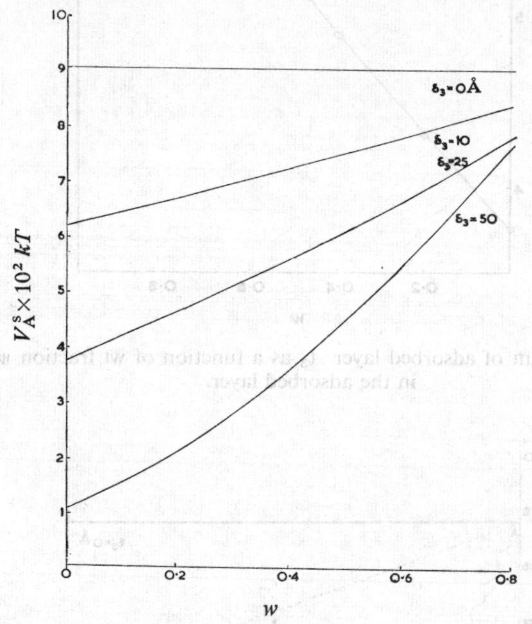

Fig. 3.—As for fig. 2 but $h = 10$ nm.

corresponds to the case where the thickness of the E.O. layer gradually increases, but that its E.O. content (after the initial addition) remains virtually constant.   This corresponds to a gradual change from a " horizontal " to a " vertical " configuration of the E.O. molecules at the surface.   Curve II represents the case where there is a gradual increase in E.O. concentration in the adsorbed layer, the thickness remaining more or less constant (i.e., vertical configuration throughout).   The real curve III, would indicate that there is probably a gradual *decrease* in V_A^s as stabilizer is added to the system, as was suggested by Ottewill and Walker.[1]

Fig. 4 and 5 show the corresponding V_A^p curves for Sonntag's system, again for $h = 0.2$ and 10 nm, respectively.   The maximum thickness of the E.O. layer in this case would be expected to be in the range 5-10 nm, and again assuming the maximum value for w is about 0.7, the corresponding two points A and B may be plotted (fig. 5). This time the case $h = 10$ nm is arbitrarily considered, since in this system the electrical double layer repulsion will contribute to the equilibrium separation.   However, again, the general form of the curves in fig. 4 and 5 is similar, although for $h = 0.2$ nm, the curves for δ_3 are virtually coincidental.   (This results from the r^{-6} dependence

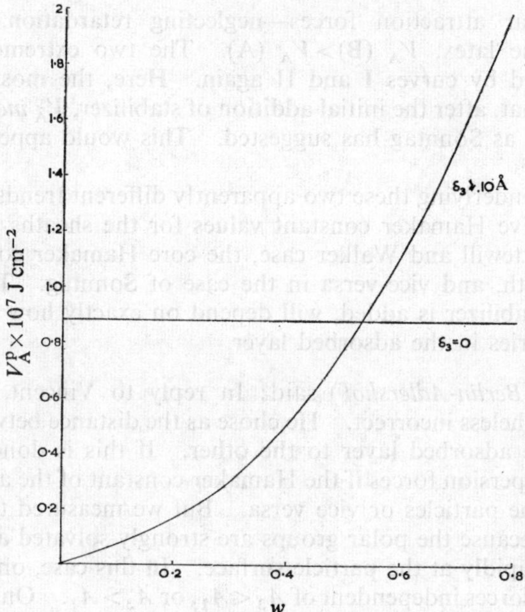

FIG. 4.—Attraction energy between two octane droplets in water, having an adsorbed layer thickness δ_3 as a function of ethylene oxide wt fraction, w; surface separation $h = 0.2$ nm.

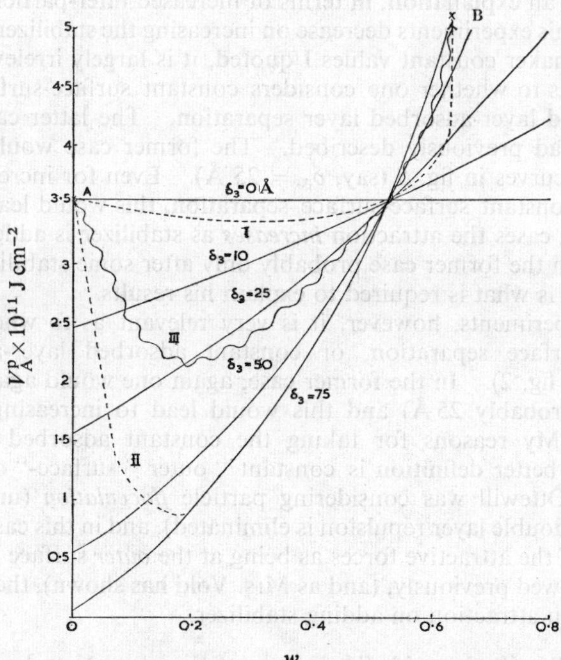

FIG. 5.—As for fig. 4, but $h = 10$ nm.

of the inter-molecular attraction forces—neglecting retardation.) In this case, unlike the polystyrene latex, V_A (B)$> V_A$ (A). The two extreme loci, discussed above, are represented by curves I and II again. Here, the most probable locus (curve III) indicates that, after the initial addition of stabilizer, V_A^p *increases* on further addition of stabilizer, as Sonntag has suggested. This would appear to resolve the apparent discrepancy.

The main reason underlying these two apparently different trends in the attractive force lies in the relative Hamaker constant values for the sheaths and cores in the two cases. In the Ottewill and Walker case, the core Hamaker constant is greater than that of the sheath, and vice-versa in the case of Sonntag. The *exact* way in which V_A varies as stabilizer is added, will depend on exactly how its configuration and concentration varies in the adsorbed layer.

Dr. H. Sonntag (*Berlin-Adlershof*) said: In reply to Vincent, his calculations are correct, but nevertheless incorrect. He chose as the distance between the particles the spacing from one adsorbed layer to the other. If this is done, he will indeed find a reduction of dispersion forces if the Hamaker constant of the adsorbed material is less than that of the particles or vice versa. But we measured the distance from particle to particle, because the polar groups are strongly solvated and the refractive index changes most rapidly at the particle surface. In this case, one always obtains increasing dispersion forces independent of $A_3 < A_1$, or $A_3 > A_1$. Only if the Hamaker constant of the adsorbed material is less than that of the water phase can one calculate decreasing dispersion forces. Unfortunately the discrepancy between Ottewill's and our paper cannot be explained so easily.

Dr. B. Vincent (*University of Bristol*) said: I agree partly with Sonntag, but I am also wondering if he has missed the main point of my argument. We are both looking, I think, for an explanation, in terms of increased inter-particle attraction, of why the spacings in his experiments decrease on increasing the stabilizer concentration. In terms of the Hamaker constant values I quoted, it is largely irrelevant in the case of his experiments as to whether one considers constant surface-surface separation, or constant adsorbed layer-adsorbed layer separation. The latter case corresponds to the situation I had previously described. The former case would correspnd to one of the fixed δ_3 curves in fig. 5 (say, $\delta_3 = 25$ Å). Even for increasing adsorbed layer thickness at constant surface-surface separation, this would lead to increasing attraction. In both cases the attraction *increases* as stabilizer is added (in the latter case continuously, in the former case probably only after some stabilizer has initially been added). That is what is required to explain his results.

In Ottewill's experiments, however, it is very relevant as to whether one takes constant surface-surface separation, or constant adsorbed layer-adsorbed layer separation (see e.g., fig. 2). In the former case, again one would again have to take a fixed δ_3 (again probably 25 Å) and this would lead to increasing attraction on adding stabilizer. My reasons for taking the constant adsorbed layer-adsorbed layer (or perhaps a better definition is constant " outer " surface-" outer " surface) separation is that Ottewill was considering particle *flocculation* (under conditions where the electrical double layer repulsion is eliminated), and in this case it is necessary to take the origin of the attractive forces as being at the *outer* surface of the particles. In this case, as I showed previously, (and as Mrs. Vold has shown), the most probable result is a *decrease* in attraction on adding stabilizer.

Prof. A. Scheludko (*Sofia*) said: The paper of Sonntag, Netzel and Unterberger is a further step in the investigation of emulsion films. First, the use of a porous plate [1] as a modification of Mysels' porous ring method [2] is of considerable interest. In this particular case the role of the porous walls is to fix the film to the holder, something which until now has been difficult to achieve in a smooth-walled glass cell and which represented a major obstacle. This method makes it possible to vary over a wide range the capillary pressure which although not made use of in their work in fact was its original purpose.[2] Also this is the first application, in real emulsion films, of the microscopic method of measuring the contact angle film/bulk liquid.[3] As with foam films the contact angle reaches significant values after transition to the Perrin films, obtained in this case at sufficiently high electrolyte concentration (5×10^{-1} M $MgSO_4$). The reported values of the contact angle at low concentration (a few hundredths of a degree) are greater than the sensitivity of the method.

The observed influence of the surfactant concentration on the equilibrium thickness is somewhat unexpected. According to these observations the equilibrium thickness decreases with increase of the surfactant concentration. This corresponds to the negative component of the disjoining pressure prevailing over the positive component. In fact, the effect is even greater than can be seen in fig. 2, because the surface tension and therefore the outer capillary pressure decrease, when the surfactant concentration is increased.

We, however, cannot agree with their generalization of the effect to any dispersion system since in other cases, as e.g., with some foam films, the opposite effect of the surfactant has been observed. Thus, in ref. (3), the deviations from the DVO theory were in the direction of the positive disjoining pressure prevailing over the negative component for sufficiently thin films. The influence of the surfactant on foam films was directly demonstrated in the investigation of the critical thickness of rupture.[4] In the latter, a decrease of the critical thickness at high surfactant concentration was obtained which meant an increase of the positive disjoining pressure i.e., the reverse of the effect described in their paper.

Unfortunately, the curves of equilibrium thickness against electrolyte concentration in the paper are not interpreted according to the DVO theory with the simplest assumption that Π_{vw} is inversely proportional to the third power of the film thickness. It is hardly appropriate to take into account the thickness of the adsorption layers (in fact, merely the thickness of the polar " heads ") when considering films with such a large total thickness (200-550 Å). The thickness of the polar groups is of the order of magnitude of the experimental error. That is why their interpretation does not appear convincing.

Dr. H. Sonntag (*Deutsche Akademie der Wissenschaften, Berlin* (*communicated*)): In reply to Scheludko, one finds an alteration of the critical thickness of rupture with increasing surfactant concentration until only C_{bl} is reached. Above this concentration the rupture thickness remains constant. We always measured the equilibrium thicknesses above C_{bl} and therefore a comparison of his results with ours is impossible. On the other side, his coworker Exerowa also found a decrease of equilibrium thickness with increasing concentration of non-ionic surfactant.

Prof. J. Lyklema (*Wageningen, Netherlands*) said: In looking for an explanation for the effect of surfactant concentration on the equilibrium thickness, as shown in

[1] D. Exerowa and S. Scheludko, *Compt. Rend. Acad. Bulg. Sci.*, in press.
[2] K. Mysels, *J. Phys. Chem.*, 1964, **68**, 3741.
[3] A. Scheludko, B. Radoev and T. Kolarov, *Trans. Faraday Soc.*, 1968, **64**, 2213; T. Kolarov, A. Scheludko and D. Exerowa, *Trans. Faraday Soc.*, 1968, **64**, 2864.
[4] I. Ivanov, B. Radoev, E. Manev and A. Scheludko, *Trans. Faraday Soc.*, 1970, **66**, 1262.

fig. 2 of Sonntag *et al.*'s paper one could also think of a kind of " tele-entropic " stabilization mechanism according to the following principle. If two emulsion droplets, each covered with surfactant molecules and bearing at least some charge, approach each other, the diffuse double layers start to overlap and counterions are transferred from the Gouy-layer to the Stern-layer. When the degree of occupancy of the liquid/liquid interface with surfactant molecules is low, it is possible that these counterions are pressed in between these molecules, which could reduce their configurational entropy, and hence constitute a repulsion. The more compact the surfactant layer becomes, the less this effect and, at sufficiently high degrees of occupancy this repulsive entropic term is completely absent. Thus, the model can account for a repulsion that is only operative at low surfactant concentrations.

Dr. H. Sonntag (*Deutsche Akademie der Wissenschaften, Berlin*) said: With regard to the paper by Padday, the rupture of liquid films on solid surfaces is a very complicated example of hetero-coalescence. I think his measuring device is not sufficiently sensitive to decide what happens at the rupture process, because the thickness of the film is inhomogeneous before breaking. We investigated the rupture process of aqueous films on cyclohexane with the apparatus described in our paper. The thickness of rupture decreased with increasing concentration of surface-active agent from several 1000 Å to 300 Å. Above a certain concentration the thickness of rupture remained unaltered and we obtained a stable film whose thickness depended only on the electrolyte concentration.

Dr. J. F. Padday (*Kodak Ltd., Harrow*) In reply to Sonntag, in subsequent experiments described in the appendix of my paper I have pointed out that high-speed cine photographs revealed an indentation or dimple in the surface of my experiments. Thus, the critical thickness between the liquid-air and solid-liquid interfaces may well be of the same order as Sonntag's rupture thickness because the thickness I have measured is that of h in the fig. 8 whereas Sonntag measures h^1.

Dr. R. G. Picknett (*C.D.E., Porton Down, Wilts*) said: Padday has given a modified form of the Kelvin equation which relates capillary pressure, disjoining pressure and vapour pressure :

$$\gamma\left(\frac{1}{r}+\frac{1}{r^1}\right)-\Pi = \frac{RT}{v_m}\ln\frac{P}{P_s}. \tag{1}$$

Deryaguin [1] made it clear that for thin liquid layers the curvature is not constant, but varies with film thickness in accordance with the disjoining pressure Thus, in applying eqn (1) to liquid in a capillary or at the point of contact of sphere and plane, it is incorrect to calculate Π from the total liquid thickness; instead, it must be evaluated for each point on the liquid surface This has been done for water between a 720 μm diam. glass sphere and plane, the work being part of an adhesion investigation by Cross and myself. The surfaces were assumed to be smooth, and the vertical profile of the water/air interface was calculated using Π values derived from the work of Deryaguin and Zorin. [2] The adsorbed water film far from the point of contact was so thin that it had only a negligible effect. The maximum thickness obtained for the water annulus is shown in fig. 1(*a*) as a function of P/P_s : it is larger than the thickness calculated from the simple Kelvin equation by a factor of 1.2-1.3. Knowing the surface profile, the adhesion due to the surface tension of the water

[1] B. V. Deryaguin, *Proc. 2nd Int. Congr. Surface Activity*, (London, 1957), **2**, 153.
[2] B. V. Deryaguin and Z. M. Zorin, *Proc. 2nd Int. Congr. Surface Activity*, (London), 1957, **2**, 145.

can be obtained as a function of P/P_s. These adhesions are shown in fig. 1(b) and 1(c) as solid lines, the diagrams referring to two different separations h of sphere and plate corresponding to different degrees of surface roughness. Experimental adhesions were measured using plates with artificially-induced roughness, and are represented in fig. 1(b) and 1(c) by crosses. The experimental values lie well below the solid curves derived from eqn (1), and in fact lie more closely on the broken curves which were derived from the simple Kelvin equation. The discrepancy is thought to be significant, and indicates that eqn (1) does not apply to the sphere+plane system. The explanation may well lie in the assumption used in eqn (1) that the

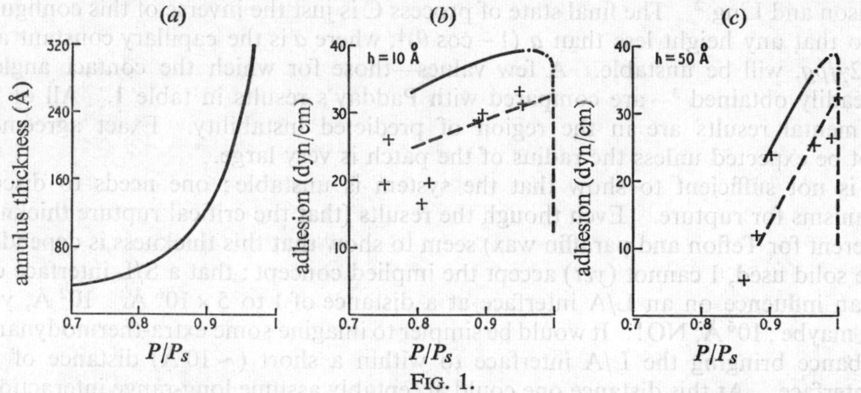

Fig. 1.

capillary pressure and the disjoining pressure are additive. If, instead of this assumption, allowance is made for change in surface tension as well as for disjoining pressure, then better agreement with experiment may well be found.

Dr. J. F. Padday (*Kodak Ltd., Harrow*) said: I am grateful to Picknett for pointing out to me that Deryaguin, in 1957, proposed a modification to the Kelvin equation that took into account the effects of disjoining forces on the condensation of liquid within a pore. This equation is general in principle but is difficult to apply particularly to the system of liquid condensing between the ball and plate of Picknett's experiments. In applying either form of the Kelvin equation modified for thin film effect, Picknett has made two assumptions. The first is that the surface tension is a constant value at both the thin and thick parts of the meniscus and that therefore the angle of contact between thin and thicker film is zero. This assumption is reasonably well justified because the force of adhesion is predominantly a surface pressure term and in the results quoted in the comments depends on the " bulk " surface tension and the radii of curvature at the neck of the liquid bridge. The second assumption concerns the effect of geometry of the axisymmetric liquid bridge on the disjoining pressure at any element of liquid between the solid/liquid and liquid/air interfaces. I understand that Picknett calculated the data of fig. 1 by first deriving the shape of the liquid bridge using a graphic integration method with eqn (6) of my paper. The sum of the Π terms were derived from the values given by Deryaguin and Zorin, and the integration process produced a shape from which shape parameters such as the two principle radii of curvature at the narrowest point at the neck of the liquid-air interface were obtained. The sum of these curvatures multiplied by the " bulk " surface tension was then, it appears, used to calculate both the adhesion force and the vapour pressure in equilibrium with it. In applying these calculations one must assume that the disjoining pressure measurements of Deryaguin and Zorin for disjoining of a thin layer of liquid between two flat parallel surfaces are equally suitable when the two surfaces deviate appreciably from parallelism. Both van der Waals' and electric-double layer forces appear to be sufficiently sensitive to the angle between the two surfaces bounding the thin film to suggest that at the neck where the angle is nearly 90°, the value of the disjoining pressure would be greatly reduced and thereby explain Pickett's results.

Dr. L. M. Dormant (*Bristol University*) said: Process C of Padday's paper can be examined in a more quantitative manner. The equilibrium shape of a very large drop in a gravity field leading to the Young and Dupré equation was solved by Adamson and Ling.[2] The final state of process C is just the inverse of this configuration so that any height less than $a(1-\cos\theta)^{\frac{1}{2}}$, where a is the capillary constant and $a^2 \equiv 2\gamma/\rho g$, will be unstable. A few values—those for which the contact angle θ was readily obtained [3]—are compared with Padday's results in table 1. All of the experimental results are in the region of predicted instability. Exact agreement cannot be expected unless the radius of the patch is very large.

It is not sufficient to show that the system is unstable; one needs to discuss mechanisms for rupture. Even though the results (that the critical rupture thickness is different for Teflon and paraffin wax) seem to show that this thickness is dependent on the solid used, I cannot (*yet*) accept the implied concept: that a S/L interface can exert an influence on an L/A interface at a distance of 1 to 5×10^6 Å. 10^2 Å, yes; 10^3 Å, maybe; 10^6 Å, NO! It would be simpler to imagine some extra-thermodynamic disturbance bringing the L/A interface to within a short (~ 10 Å) distance of the S/A interface. At this distance one could acceptably assume long-range interactions, thereby forcing the L/A interface to remain there and thus forming the *dimple* that Padday described.

TABLE 1.—THERMODYNAMIC AND EXPERIMENTAL RUPTURE THICKNESSES

	a^2 [a] (mm²)	θ [b] (deg.)	h° [c] (mm)	h_{crt} [d] (mm)
benzene/Teflon	14.7	46	1.4	0.16
decane/Teflon	6.73	24	0.7	0.26
water/Teflon	6.34	108	4.4	0.21
water/paraffin	6.34	106	4.3	0.31

(a) $a^2 = 2\gamma/\rho g$; (b) from ref. (3); (c) from $h^\circ = a(1-\cos\theta)^{\frac{1}{2}}$; (d) from table 1 and 2 of Padday.

The mechanism which might help to explain this postulated disturbance is the formation of capillary gravity waves and the subsequent amplification of certain wavelengths due to the geometry of the system. One could make a somewhat dubious analogy to the placing of the more dense of two immiscible liquids on top of the less dense liquid. The primary mode of rupture for such *bulk* phases is that some of the wavelengths are selectively amplified until rupture occurs.[4] This system is only stable to wavelengths less than $\sqrt{2}\pi a$. The analogy is useful in so far as it poses three other questions: (i) what would happen if instead of a large piece of Teflon, a sample a few mm² or less was used, and (ii) what would happen if his trough was only a few cm wide? These experiments could show whether Padday

[1] W. A. Zisman, *Adv. Chem.*, no. 43, (Amer. Chem. Soc., Washington, D.C., 1964), p. 11.
[2] A. W. Adamson and I. Ling, *Adv. Chem.*, no. 43, (Amer. Chem. Soc., Washington, D.C., 1964), p.72..
[3] A. W. Adamson, *Physical Chemistry of Surfaces*, (Intersci., N.Y., 1967), p. 364, 389.
[4] R. Bellman and R. H. Bennington, *Quat. Appl. Math.*, 1954, **12**, 151.

is indeed measuring a fundamental property or only some property connected with the geometry of the system. The amplification step may take a considerable time (the time factor was not mentioned) and leads to the remaining questions: what would happen if the system was kept slightly above the critical rupture thickness but held there a factor of 10 times longer than the remainder of the experiments? Also what would happen if vibrations were deliberately induced?

The question of impurity has already been brought up. Judging from fig. 6, it seems that a small concentration of impurity would have a negligible effect. Impurities should be negligible if the mechanism is a long-range effect. Yet ,if a wave phenomenon was involved, a surface tension gradient of 0.01 dyn/cm (N/km) could have pronounced effects [1] perhaps explaining the difference between Teflon and paraffin. Similar conceptual difficulties occur with the salt and surfactant solutions. The double layer thickness of a 10^{-3} M solution is approximately 10^2 Å so that it would be difficult to postulate effects at distances of 10^4 times the double-layer thickness regardless of the charge on the Teflon surface. On the other hand, the damping of waves by the addition of a surface-active agent leading to greater stability is a common phenomenon. [2] I do not believe that a negative damping coefficient has ever been previously observed.

Dr. J. F. Padday (*Kodak Ltd., Harrow*) said: I disagree with the first comment of Dormant by which it is implied that any thin film at a solid surface with a thickness less than [3] $(2\gamma_L (1-\cos\theta)/\rho g)^{\frac{1}{2}}$ is unstable. For water on wax this thickness is about 4.5 mm and all experiments show clearly that layers between 1 and 4.5 mm in thickness are stable over many days. Column 3 of the table in his remarks is thus not relevant as there is *no* instability. Layers in this region of thickness may have a degree of metastability but this will depend much more on the forces that give rise to dimple formation than to the contact angle properties of a wetting meniscus. As pointed out in the appendix, the dimple, once formed, is resistant to low energy vibrations or ripples at the liquid-air interface. The process by which the dimple first forms could possibly be the trapping of a wave form at the point where the liquid is thinnest but the high speed cine films do not show this.

The geometry of the vessel and to some extent, the solid surface, has been altered and this does not appear to vary the process described. As the phenomenon is also observed on a vibration-free mounting surface effects arising from very small surface tension gradients induced by waves may be discounted. The apparent electrical effects do not necessarily operate over large distances because the dimple reduces the interaction distance. Negative damping coefficients do not have to be invoked although they are well known, [4] because it is believed that the dimple is in an asymmetric force field.

Dr. G. Frens (*Philips Res. Lab., Eindhoven*) said: Padday's work may explain why layers of (aggregated) particles form at the surface of perfectly stable gold and silver sols (and of many other colloidal dispersions). Let us suppose that Brownian motion brings a particle with a hydrophobic surface near the surface of the dispersion. Padday shows there is a chance for the water layer between the particle and the surface to break. Then the particle will rise, and despite its larger density, it will be kept afloat by surface tension forces. Now let us consider the coagulation

[1] M. van den Tempel, private communication.
[2] E. H. Lucassen-Reynders and J. Lucassen, *Adv. Colloid Int. Sci.*, 1970, **2**, 347.
[3] J. F. Padday, *Proc. 2nd Int. Congr. Surface Activity*, 1957, **3**, 136, 187.
[4] J. T. Davies and E. K. Rideal, *Interfacial Phenomena*, (Academic Press, London, 1961), p. 360.

of two such hydrophobic particles. Instead of meeting the surface they approach each other, in a Brownian collision. The water layer between the two particles will break, if only because the two particles come in close contact. If the water layer is broken at some point the capillary forces will push the water back from the narrow pores of the aggregate until the radius of curvature of the newly created interface becomes large enough to be in equilibrium with the hydrostatic pressure outside the aggregate. There is a close analogy between this situation inside a floc in a hydrophobic sol and that in mercury porosimetry [1,2] and the work of Kruyt and van Selms [3] on dispersions in mixed solvents (one wetting, the other non-wetting) comes to mind, as well as that of Vanderhoff *et al.* on the coalescence of latex particles.[4]

In general, it can be concluded that an aggregate of hydrophobic particles should be a three-phase instead of a two-phase system. This will have consequences for the properties of such aggregates, e.g., for their repeptization, since the situation inside these aggregates will be greatly influenced by the addition of surface-active agents to the dispersion medium,[5] and to a far lesser extent by agents which might influence the electrical properties of the interface between the original particles and the solution.

Dr. J. F. Padday (*Kodak Ltd., Harrow*) said: In reply to Frens, in principle, the process of rupture described in part 2 of my paper could explain the coagulation of two or more hydrophobic particles which approach each other at some critical rupturing distance when embedded in a non-wetting liquid. A similar explanation [6] was used to account for the driving force that gives rise to the aggregation of two cyanine dye ions to form dimers against electrostatic repulsion forces. However, by invoking such an explanation, the very small particles may well give rise to rupture and cohesion forces that are very different from those of my experiments because the reciprocal of the radius of curvatures in the two cases differ by more than an order of magnitude. The reverse explanation, i.e., that hydrophobic particles near the surface of the liquid gives rise to the dimple, does not appear to be possible because the strong disjoining action of a " pillar " of small hydrophobic particles would lead to immediate rupture and not the intermediate formation of the dimple.

[1] R. P. Iczkowski, *Ind. Eng. Chem. Fund.*, 1966, 516.
[2] R. P. Mayer and R. A. Stowe, *J. Phys. Chem.*, 1966, **70**, 3867.
[3] H. R. Kruyt and F. G. van Selms, *Rec. Trav. Chim.*, 1943, **62**, 407.
[4] J. W. Vanderhoff, H. L. Tarkowski, M. C. Jenkins and E. B. Bradford, *J. Macromol Chem.*, 1966, **1**, 361.
[5] E. J. Clayfield and E. C. Lumb, *Disc. Faraday Soc.*, 1966, **42**, 285.
[6] J. F. Padday, *J. Phys. Chem.*, 1968, **72**, 1259.

Thermodynamic Properties of Thin Films of Some Dipolar Liquids Adjacent to Fused Silica Surfaces

By K. H. Adlfinger and G. Peschel

Institut für Physikalische Chemie der Universität Würzburg, 87 Würzburg, Markus-straße 9-11, Germany

Received 31st March, 1970

The disjoining pressure of some dipolar organic liquids ($PhNO_2$, $PhCN$, $PhCH_2CN$, $PhCF_3$, $MeCN$) forming thin layers between two fused silica surfaces shows maximum values in temperature ranges which refer to thermodynamic transitions of higher order. Here the molecules in the bulk liquid seem to gain an additional rotational degree of freedom about one axis, whereas the molecules in the oriented surface zone still show a rotational restriction about this axis. Thermodynamic quantities for the liquid boundary layers are introduced and discussed.

Since the experiments of Deryaguin [1,2] evidence has been accumulated that a solid surface can alter the structure of an adjacent liquid layer [3-5] up to a thickness of about 10^{-5} cm. This phenomenon is assumed to be caused by the solid surface inducing a molecular long-range orientation in the vicinal liquid. This orientation implies a decrease of the thermodynamic potential of the surface zone compared with that of the bulk liquid; it gives rise to the disjoining pressure. Thus, if two solid plates immersed in a liquid showing long-range orientation approach to distances smaller than about 10^{-5} cm under the action of an external force an oppositely directed disjoining force arises holding the plates apart by a distance which is dependent on the external force. Knowing the geometry of the plates the disjoining pressure can be evaluated. Hitherto, only few details are known about the mechanism of the long-range orientation which also alters the viscosity,[6,7] the thermal conduction,[8] and the dielectric constant of liquid surface zones.[9]

The most frequently investigated liquid in this respect is water where the disjoining pressure is of the order of magnitude of 10^5 dyn/cm^2.[10,11] The most interesting effect, however, is represented by the temperature dependence of the disjoining pressure. The present authors found significant maxima of the disjoining pressure of water at temperatures at which structural transitions of higher order occur.[12,13] These temperatures at about 14, 32, 45, and 61°C are repeatedly quoted, and are discussed by Drost-Hansen.[14] By a special experimental device [7,10,11] we succeeded also in determining the viscosity of aqueous boundary layers which show likewise maxima at the characteristic temperatures.[13]

Our interpretation agrees with that of Deryaguin,[15] viz., that small molecular aggregates determine the structure of aqueous boundary layers. The four maxima suggest the existence of four different molecular aggregates being stabilized in different temperature ranges. Applying the theory of viscous flow of Eyring [16] and determining approximate values for the activation energy of viscous flow, it can be assumed that the aqueous boundary layers above 14°C consist of small aggregates mainly built up of three or four water molecules. Above 32°C the aggregates are larger, and above

45°C the size of the aggregates seems to be comparable with that of the molecular entities existing above 14°C. Results for the temperature range about 61°C are not yet available because of experimental difficulties. All these effects get smaller with increasing plate distance and vanish at about 1.6×10^{-5} cm. This value was also found by Deryaguin and coworkers [1,4] by measuring the shear modulus of the oriented aqueous surface zone. The results for the non-polar liquids benzene and cyclohexane indicate a steep decrease of the disjoining pressure with rising temperature starting from the melting point.[17]

THERMODYNAMIC RELATIONS FOR ORIENTED SURFACE ZONES

Regarding two surface areas, each of 1 cm^2, and separated by an intermediate oriented liquid layer, the free excess energy for the approaching of the surfaces from a distance h^+ to a distance $h < h^+$ is given by

$$(\Delta F^E)_h = -\int_{h^+}^{h} \Pi \, dh'. \tag{1}$$

Π is the disjoining pressure which is dependent on h. h^+ is the maximum distance at which the disjoining pressure can no longer be detected; it is dependent on the sensitivity of the experimental apparatus. During the approach of the surfaces, an oriented liquid layer with the thickness $h^+ - h$ gets disordered and is pressed into the outer bulk liquid.

A formal case is now considered. In order to press a liquid layer having infinitesimal thickness and lying in the medium plane between the surfaces into the bulk liquid the free excess energy

$$d(\Delta F^E)_h = -\frac{\partial (\Delta F^E)_h}{\partial h} dh \tag{2}$$

has to be brought into the system. This case, however, is not a practicable one and it is convenient to replace dh by Δh, the last quantity being the thickness of a monomolecular layer. Presuming that $\Delta h \ll h$ is always valid the free excess energy to remove the monomolecular oriented liquid layer from the medium plane between the surface areas is

$$\Delta(\Delta F^E)_h = -\frac{\partial (\Delta F^E)_h}{\partial h} \Delta h. \tag{3}$$

The total excess energy to remove this layer can be evaluated from the Gibbs-Helmholtz relation using $\Delta(\Delta F^E)_h$ and its temperature derivative.

Since the density of the oriented surface zone is—except for water—unknown, conversion into molar quantities is only possible if the surface zone density is replaced by the density ρ_e of the bulk liquid.[17] This, however, may imply large errors; for the extreme case of aqueous surface zones the densities might exceed the bulk value by 20 or even 30 %.[18]

The temperature dependence of Π was assumed to have the form

$$\Pi = C \exp(-nh), \tag{4}$$

with C and n being constants.

Using eqn (4) and taking molar quantities the total excess energy is obtained in the form:

$$\Delta(\Delta U_m^E)_h = \frac{M}{\rho_l}\left[C \exp\left(-nh\right) + \right.$$

$$\left. T\left(\frac{C}{\rho_l}\exp\left(-nh\right)\frac{d\rho_l}{dT} - \exp\left(-nh\right)\frac{dC}{dT} + hC \exp\left(-nh\right)\frac{dn}{dT}\right)\right]. \qquad (5)$$

If the temperature dependence of $\Delta(\Delta U_m^E)_h$ is accessible, the corresponding excess molar heat $\Delta(\Delta C_{v,m}^E)_h$ can be evaluated.

EXPERIMENTAL

For the determination of the disjoining pressure a spherically formed and a planar fused silica surface both highly polished, totally covered with hydroxyl groups, and immersed in the liquid to be investigated were pressed against each other. This system was chosen to avoid the problem that dust particles in the region between the plates appreciably falsify the measurements. For two plane-parallel surfaces this problem seems to be a serious one as Hayward and Isdale [19] have pointed out.

In our experimental device, the planar fused-silica plate is mounted to the bottom of a container which can be filled with the liquid in question. The spherically formed plate facing the planar one is fastened to one end of a balance which can be operated electro-magnetically. Above the spherically formed plate the measuring pin of a displacement transducer is pressed upon the balance with a definite force. The displacement transducer is connected with an electronic strain gauge measuring bridge connected to a xy-recorder. All deflections of the balance can thus be registered with high accuracy.[7, 10, 11] The distance between the plates can be determined by a rather complicated method which takes sufficient account of the surface roughness derived from the surface profilograms and interferograms of the polished fused silica plates.[11, 20]

The whole apparatus is placed in a big container which can be evacuated. In this way, contact of the liquid with the humidity of the surrounding air can be avoided. At lower temperatures this precaution becomes important. The distilled and carefully degassed liquid can be sucked into the evacuated container so that the formation of air bubbles can be excluded. The temperature of the liquid in the container can be registered by a special recorder. Mechanical vibrations of the apparatus due to external causes and which disturb the measurements can be reduced by a special suspension. From a theory given by the present authors [11] the parameters n and C in eqn (4) can be evaluated.

RESULTS

The liquids chosen for investigation are nitrobenzene, benzonitrile, benzyl cyanide, benzotrifluoride, and acetonitrile. By applying eqn (4) for different plate distances ranging between 10^{-6} and 5×10^{-6} cm and plotting Π against the temperature fig. 1-3 were obtained. It is still uncertain if for all liquids investigated in this work the oriented surface zone really extends beyond the maximum distances chosen in fig. 1-3 or abruptly turns into the bulk phase at some shorter distance.

This problem could not be elucidated previously, since our apparatus lacks the sensitivity to yield sufficiently precise results.

For the aromatic compounds, benzotrifluoride shows the largest disjoining pressure, acetonitrile, an example of an aliphatic dipolar liquid, exhibits a still larger effect. The most striking evidence, however, is the fact that the disjoining pressure seems to assume large values only in distinct temperature ranges lying about 30° above the respective melting point T_m of the compound under consideration.

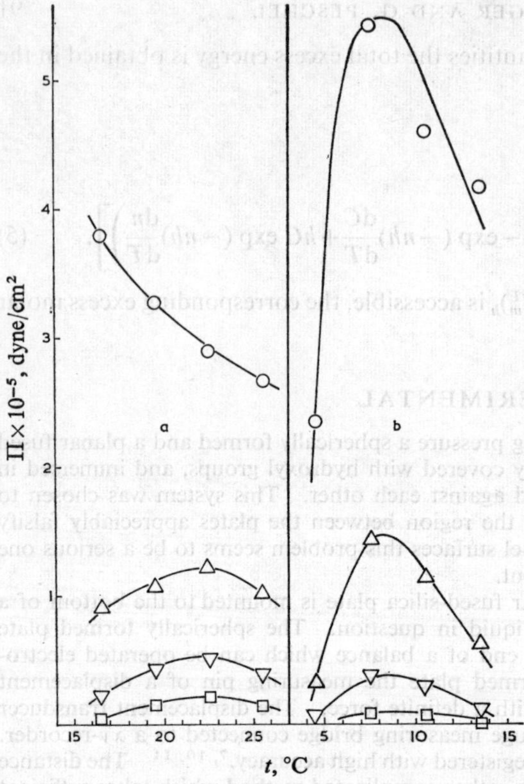

FIG. 1.—Temperature dependence of the disjoining pressure Π for PhCN (*a*) and PhCH₂CN (*b*) at different plate distances h: 10^{-6} cm, ○; 2×10^{-6} cm, △; 3×10^{-6} cm, ▽; 4×10^{-6} cm, □.

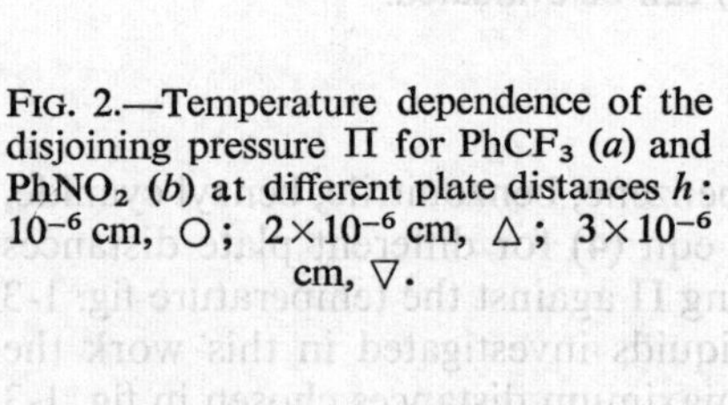

FIG. 2.—Temperature dependence of the disjoining pressure Π for PhCF₃ (*a*) and PhNO₂ (*b*) at different plate distances h: 10^{-6} cm, ○; 2×10^{-6} cm, △; 3×10^{-6} cm, ▽.

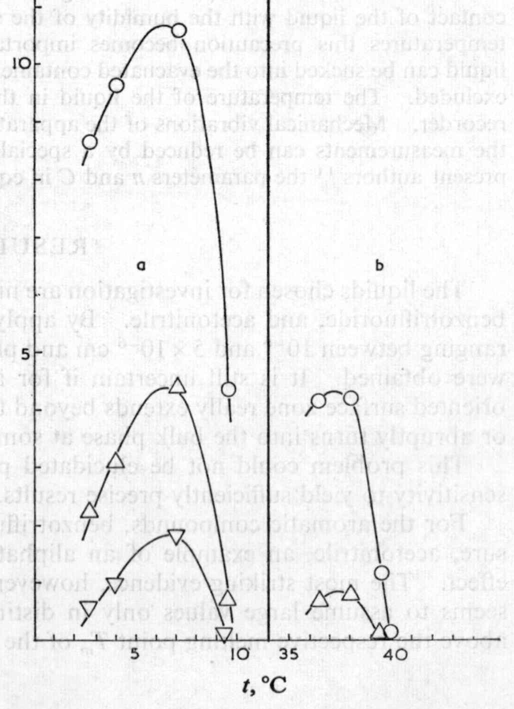

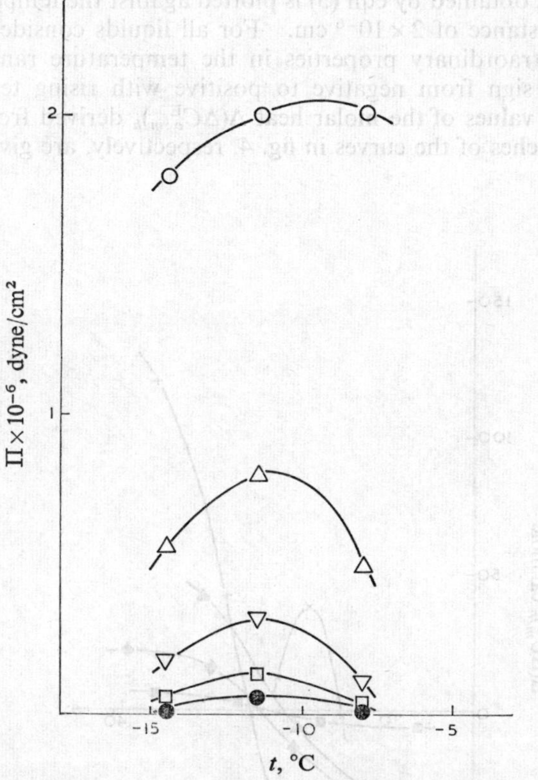

FIG. 3.—Temperature dependence of the disjoining pressure Π for MeCN at different plate distances h: 10^{-6} cm, $\bigcirc$; 2×10^{-6} cm, $\triangle$; 3×10^{-6} cm, $\triangledown$; 4×10^{-6} cm, $\square$; 5×10^{-6} cm, $\bullet$.

TABLE 1.—PARAMETERS n AND C

liquid	temp. (°C)	$n\times10^{-6}$ (cm^{-1})	$C\times10^{-6}$ (dyn/cm²)
nitrobenzene	36.5	1.78	2.47
	38	1.66	2.21
	39.5	2.74	1.81
	40	4.79	1.22
benzonitrile	16	1.43	1.59
	19.5	1.01	0.90
	22.5	0.86	0.69
	25.5	0.96	0.70
benzyl cyanide	4.5	2.00	1.75
	7.5	1.34	2.09
	10.5	1.39	1.85
	13.5	1.66	1.57
benzotrifluoride	2.7	1.36	3.38
	4	1.15	3.10
	7	0.87	2.54
	9.5	1.87	2.83
acetonitrile	−14.5	1.16	5.73
	−11.5	0.91	4.99
	−8	1.41	8.26

In fig. 4, $\Delta(\Delta U_m^E)_h$ obtained by eqn (5) is plotted against the temperature difference $T-T_m$ for a plate distance of 2×10^{-6} cm. For all liquids considered in this paper $\Delta(\Delta U_m^E)_h$ exhibits extraordinary properties in the temperature ranges in question; thus, it changes its sign from negative to positive with rising temperature. The corresponding mean values of the molar heat $\Delta(\Delta C_{v,m}^E)_h$ derived from the ascending and descending branches of the curves in fig. 4, respectively, are given in table 2.

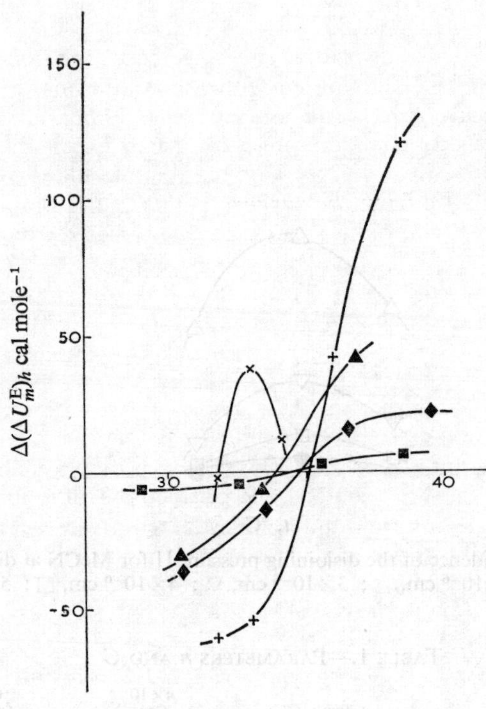

FIG. 4.—Behaviour of the total excess energy $\Delta(\Delta U_m^E)_h$ in the reduced temperature range: PhNO$_2$, $\times$; PhCN, $\blacksquare$; PhCH$_2$CN, $\blacklozenge$; PhCF$_3$, $+$; MeCN, $\blacktriangle$.

TABLE 2.—$\Delta(\Delta C_{v,m}^E)_h$ FOR $h = 2\times 10^{-6}$ cm

liquid	for the ascending branch	for the descending branch
	cal mol^{-1} K^{-1}	
nitrobenzene	42	−41
benzonitrile	2	
benzyl cyanide	9	
benzotrifluoride	41	
acetonitrile	11	

The values for $\Delta(\Delta U_m^E)_h$ and those for $\Delta(\Delta C_{v,m}^E)_h$ imply large errors which may exceed 50 %.

DISCUSSION

There is still no satisfactory theory to explain the magnitude of the disjoining pressure found by us and other authors.[5] The disjoining pressure calculated on the basis of long-range dispersion forces extending from the solid surfaces is thought to have much smaller values.[10, 21]

In a former paper [17] the present authors introduced the concept of molecular rotation restriction in oriented surface zones. For benzene, e.g., one may imagine the molecules to be strongly adsorbed to the surface hydroxyl groups and oriented with the C_6-axis perpendicular to the surface, rotational movement being only possible about this axis.[22] This adsorption effect decreases the intermolecular distance to any vicinal molecular layer compared with the conditions in the bulk liquid, in which a benzene molecule can rotate almost freely about all three axes.[23] The decreased distance, on the other hand, gives rise to an excess dispersion interaction causing the disjoining pressure. In this way, many molecular layers can be built up, but, with increasing distance from the surface, rotational vibrations about the other two axes will become important and diminish the stabilizing excess interaction.

If this concept is correct, a disjoining pressure should exhibit a maximum value in a temperature range, where the liquid in question is known to have a rotational transition; because for rotational freedom in the bulk liquid and rotational restriction of one molecular axis, e.g., in the surface zone, relatively large differences in the chemical potential, i.e., $\Delta(\Delta F_m^E)_h$, are expected. By raising the temperature, the oriented surface zone suffers destruction and the disjoining pressure vanishes. Below the transition range, rotational restriction prevails in the surface zone as well as in the bulk liquid and $\Delta(\Delta F_m^E)_h$ and Π, respectively, should become small.

Indeed, our results seem to confirm the correctness of this concept, since the temperature ranges found for the liquids investigated are in accordance with those revealing anomalous bulk properties, which can be regarded as evidence for a rotational transition in liquid phase.[23] Regarding the viscosity of nitrobenzene and benzonitrile, respectively, the plot of $\log \eta$ against $1/T$ (fig. 5) shows deviations in the temperature range in question. Moreover, the activation energy of flow seems to show an abrupt change in this range which likewise points to an alteration of molecular rotational behaviour. The viscosities were determined by a capillary viscometer over only small temperature steps. Our data for nitrobenzene agree with those found in literature,[24] which show the same deviations as cited above. For acetonitrile, values of the molar heat C_p are available.[25] The temperature derivative of C_p against T (fig. 6), indicates an anomaly in the temperature range as found in fig. 3. Work is in progress to obtain accurate evidence for the abrupt change in molecular rotational behaviour of these liquids. For halogenobenzenes we succeeded in showing that in the corresponding transition ranges, likewise lying about 30° above the respective melting points, the change in rotational behaviour refers to the molecular C_2-axis. For these considerations we applied the method of Davies and Matheson [23] and calculated the rotational volume of a molecule in the gas phase and the space available for the same molecule in the condensed liquid phase at the characteristic temperature.[26]

It is difficult to explain why the disjoining pressure differs in its magnitude and its temperature dependence for the four aromatic liquids quoted in this paper. However, the disjoining pressure is dependent on the structure of the oriented surface zone as well as on the structure of the bulk liquid, which is usually unknown. Investigations of other dipolar aromatic liquids, to be published later, revealed a tendency for molecules with large functional groups (e.g., $PhCH_2CN$) which give rise to steric hindrance in the molecular movement to show marked maxima of the disjoining

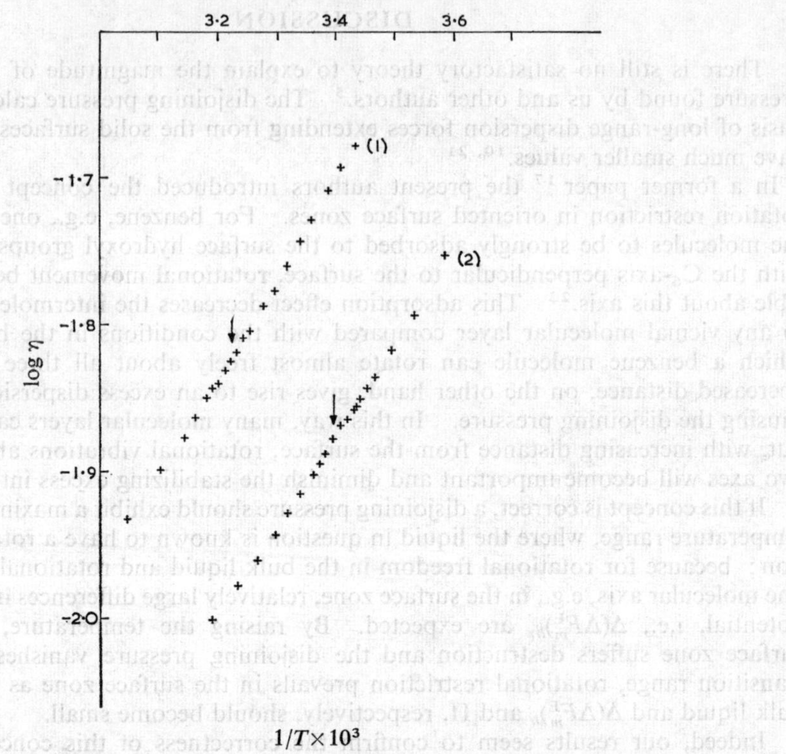

FIG. 5.—Log η against $1/T$ for PhNO$_2$ (1) and PhCN (2). The arrows indicate the centres of the transition ranges found by the disjoining pressure.

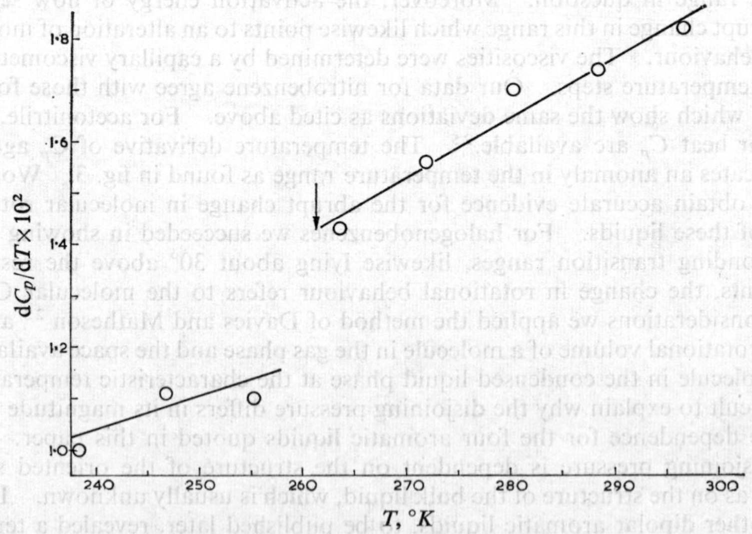

FIG. 6.—Temperature dependence of $\mathrm{d}C_p/\mathrm{d}T$ for MeCN. The arrow indicates the centre of the transition range found by the disjoining pressure.

pressure, whereas molecules with small functional groups (e.g., benzonitrile) exhibit less marked maxima, especially at small plate distances.

Regarding the excess energies $\Delta(\Delta U_m^E)_h \ll RT$, the orientational molecular packing effect in the surface zone cannot be pronounced except perhaps for those molecules with great rotational restriction. The values of $\Delta(\Delta C_{v,m}^E)_h$ are rather high because of the presence of a transition range.

We are indebted to Prof. Dr. G. Briegleb and the *Deutsche Forschungsgemeinschaft* for support of this work. Thanks are due to Mrs. G. Nöll for performing the precise viscosity measurements.

[1] B. V. Deryaguin, *Z. Phys.*, 1933, **84**, 657.

[2] B. V. Deryaguin and E. Obuchov, *Acta physicochim.*, 1936, **5**, 1.

[3] J. C. Henniker, *Rev. Mod. Phys.*, 1949, **21**, 322.

[4] B. V. Deryaguin, *Disc. Faraday Soc.*, 1966, **42**, 109.

[5] A. Sheludko, *Colloid Chemistry* (Elsevier Publ. Co., Amsterdam, 1966).

[6] B. A. Kholodnitskii, *Vestn. Leningrad Univ., Fiz. Khim.*, 1968, **23**, 153.

[7] G. Peschel and K. H. Adlfinger, *Z. Naturforsch.*, 1969, **24a**, 1113; *Ber. Bunsenges. phys. Chem.* 1970, **74**, 351.

[8] M. S. Metsik and O. S. Aidanova, *Research in Surface Forces*, ed. B. V. Deryaguin (Consultants Bureau, New York, 1966), vol. 2, p. 169.

[9] G. Peschel and R. Schnorrer, in preparation.

[10] G. Peschel, *Z. phys. Chem. (N.F.)*, 1968, **59**, 27.

[11] K. H. Adlfinger and G. Peschel, *Z. phys. Chem. (N.F.)*, in press.

[12] G. Peschel and K. H. Adlfinger, *Naturwiss.*, 1967, **54**, 614; *Chem. Labor Betrieb*, 1970, **21**, 193.

[13] G. Peschel and K. H. Adlfinger, *Naturwiss.*, 1969, **56**, 558.

[14] W. Drost-Hansen, *Ind. Eng. Chem.*, 1969, **61**, (11), 10.

[15] B. V. Deryaguin, I. G. Ershova, V. K. Simonova and N. V. Churayev, *Teor. Eksp. Khim.*, 1968, **4**, 527.

[16] S. Glasstone, K. J. Laidler and H. Eyring, *The Theory of Rate Processes* (McGraw-Hill, New York, 1961).

[17] G. Peschel and K. H. Adlfinger, *Z. phys. Chem. (N.F.)*, 1969, **63**, 150.

[18] B. V. Deryaguin, Z. M. Zorin and N. V. Churayev, *Dokl. Akad. Nauk.*, *S.S.S.R.*, 1968, **182**, 811.

[19] A. T. J. Hayward and J. D. Isdale, *Brit. J. Appl. Phys. (J. Phys. D)*, 1969, **2**, 251.

[20] K. H. Adlfinger, R. Schnorrer and G. Peschel, *Z. angew. Phys.*, 1970, **29**, 136.

[21] J. Frenkel, *Kinetic Theory of Liquids* (Dover Publ. Inc., New York, 1955).

[22] A. V. Kiselev and D. P. Pashkus, *Dokl. Akad. Nauk.*, *S.S.S.R.*, 1958, **120**, 843; D. Michel, *Z. Naturforsch.*, 1968, **23a**, 339.

[23] D. B. Davies and A. J. Matheson, *Disc. Faraday Soc.*, 1967, **43**, 216.

[24] B. P. Nikolski, *Handbuch des Chemikers* (VEB Verlag Technik, Berlin, 1956), vol. 1.

[25] W. E. Putnam, D. M. McEachern, Jr., and J. E. Kilpatrick, *J. Chem. Phys.*, 1965, **42**, 749.

[26] G. Peschel and R. Schnorrer, *Ber. Bunsenges. phys. Chem.*, 1969, **73**, 917.

Boundary Viscosity of Polydimethylsiloxane Liquids and their Binary Mixtures

By B. V. Deryaguin, V. V. Karasev, I. A. Lavygin, I. I. Skorokhodov and
E. N. Khromova

Institute of Physical Chemistry, Academy of Sciences of U.S.S.R.
31 Lenin Prospect, Moscow, U.S.S.R.

Received 30th April, 1970

A systematic study is made of the boundary viscosity of a number of polydimethylsiloxane liquids (PMS) of molecular weights ranging from 1,000 to 40,000 by the blow-off method, film thicknesses being measured by a modulation-polarimetric (elliposmetric) precedure. It is established experimentally that the viscosity of the liquids studied is not the same throughout the thickness of the boundary later.

PMS applied to glass and steel substrates retain their bulk viscosity values down to a layer thickness of *ca.* 150-200 Å, below which the viscosity increases slightly. On further decrease of the distance to 10-15 Å from the substrate, the viscosity becomes anomalously low, amounting to about 10-20 % of the bulk value. The existence of a layer of anomalously low viscosity may be attributed to orientation of the PMS molecules in the plane of the substrate, and hence, to the ease with which these oriented layers slip relative to one another.

When studying the boundary viscosity of binary mixtures of PMS of different molecular weights, boundary phases of elevated viscosity were found to exist at a distance of 15-30 Å from the substrate. This phenomenon may be attributed to a change in concentration of the mixture at the solid surface i.e., to an increase in content of the component with the higher molecular weight in the wall-adjacent layer.

The importance of organosilicon liquids in present-day engineering can hardly be overestimated, since they possess many outstanding properties that have found wide application as bases for greases, instrument oils, and other lubricants. However, the further development of new lubricants based on organosilicon liquids is retarded, in particular, by the fact that the properties and molecular structure of these polymers in the region bounding on the solid surface have hardly been studied. We have studied the viscosity of boundary layers of polydimethyl siloxane liquids (PMS) with molecules of linear structure,[1, 2] using the blow-off method,[3, 4] the essentials of which are as follows. A thin layer of liquid (2) (see fig. 1) is applied to one of the walls of a slit formed by two plane-parallel plates (1). If a uniform current of air is then blown through this slit, the tangential stress caused by the air current gives rise to a flow in the liquid film which changes its profile. If certain conditions, enumerated in ref. (4) are observed, this flow will be of a single-dimensional layer-by-layer nature, i.e., the velocity of the particles in each layer parallel to the substrate will be strictly the same and will be a function only of the distance of the layer from the wall surface. In this case, the viscosity of the liquid η can be expressed as a function of the distance l from the solid wall as [3]

$$\eta = (H/2)\mathrm{grad}\, p(\mathrm{d}l/\mathrm{d}x)t,$$

where H is the slit height, x is the distance from the part of the film where its thickness is l to the wetting boundary, ρ is the gradient of pressure of the air along the slit, and t is the duration of blowing.

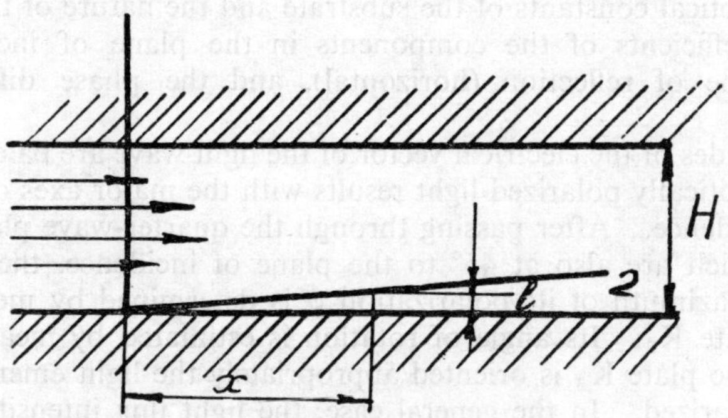

FIG. 1.—Blowing-off a liquid film.

Thus, the viscosity of the liquid is directly related to the steepness of film profile. At constant viscosity, if all the other parameters are constant, the film profile should be strictly linear, and if the viscosity changes, the slope of the line should vary so that the higher the viscosity of the liquid in any given elementary layer, the steeper will be the line at layers level, and vice versa.   Investigations of viscosity as a function of the distance to the solid surface make it possible to elucidate not only a number of structural features in boundary layers of PMS, but also their variation from the solid surface into the bulk of the liquid, since viscosity is a property very sensitive to the molecular structure of a liquid.   Measurements of film thickness, in order to plot its profile, were carried out on a modulation-polarimetric unit (fig. 2) using a LG-55 gas laser as the light source,[5] which considerably improves the accuracy of measurement.

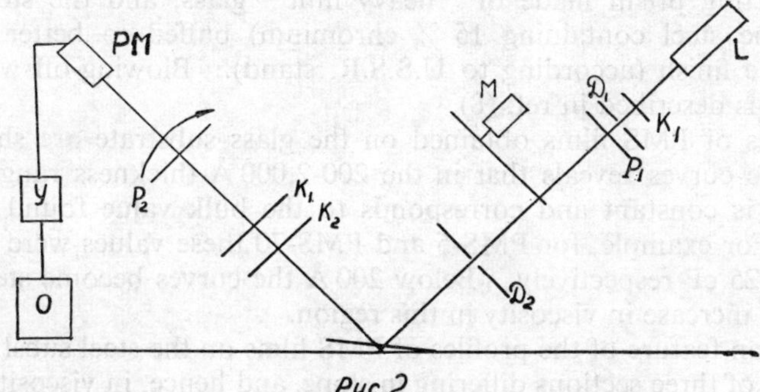

FIG. 2.—Unit for measuring film thickness.

From the small-size laser L the light ray passes through a quarter-wave ($\lambda/4$) plate K_1, diaphragms D_1 and D_2, polarizer P_1, and interrupter M on to the object (substrate with film).   On passing through plate K_1, the linearly polarized light of the laser acquires circular polarization.   This is necessary to avoid a change in intensity of the light on rotation of polarizer P_1.   Diaphragm D_1 restricts the size of the light beam, D_2 decreases the aperture angle, and interrupter M modulates the light.

After reflection from the substrate with the film, the light passes through two plates K_2 and K_3 and a rotating polaroid filter and is incident on the cathode of a photoelectric multiplier (PM) connected to an electron oscillograph O via a narrow-band amplifier Y with a RC filter.   The film thickness can be calculated by formulae containing the optical constants of the substrate and the nature of the film, the ratio of reflection coefficients of the components in the plane of incidence (vertical) and in the plane of reflection (horizontal), and the phase difference between them.[6, 7]

If the amplitudes of the electrical vector of the light wave are balanced by rotating polarizer P_1, elliptically polarized light results with the major axes oriented at 45° to the plane of incidence.   After passing through the quarter-wave plate K_2, the main directions of which are also at 45° to the plane of incidence, the light is linearly polarized.   The azimuth of its polarization ψ is determined by means of a second quarter-wave plate K_3.   Its angle of rotation is estimated by measuring the phase difference.   If the plate K_3 is oriented appropriately the light emerging from it will be circularly polarized.   In the general case, the light flux intensity varies in time, so that the ray on the oscillograph screen forms a straight line of varying length. At the moment of compensation circularly polarized light arises, which is not affected by the rotating polarizer.   Then the length of the line on the oscillograph screen stops varying.   At this moment, readings are taken on analyzer and polarizer.

Film thicknesses l were calculated from the change in polarization azimuth ψ of the light reflected from the substrate with the film, by means of formulae which take account of interference on ray reflection from the thin film.[6, 7]   The course of the $\psi = f \mid l \mid$ curve depends on the coefficient of refraction of the film material, and therefore we calculated the dependence of ψ on l for each liquid studied on BESM-1 and BESM-4 electronic computers.   The calculated $\psi = f \mid l \mid$ curves were then used as calibration curves for passing from the experimentally-found polarization azimuth to the corresponding film thickness.

The viscosity of a PMS series with molecular weights ranging from 1,000 to 40,000 was studied in the thin layers on glass and metal surfaces.   The glass substrate was a totally reflecting prism made of " heavy flint " glass, and the steel plate (ball-bearing chrome steel containing 15 % chromium) buffed to better than the 14b class of surface finish (according to U.S.S.R. stand).   Blowing-off was carried out in the apparatus described in ref. (8).

The profiles of PMS films obtained on the glass substrate are shown in fig. 3. Analysis of the curves reveals that in the 200-2,000 Å thickness range the viscosity of the liquids is constant and corresponds to the bulk value found in a capillary viscometer.   For example, for PMS-5 and PMS-70 these values were 4.92 and 4.70, 69, 68 and 68.25 cP respectively.   Below 200 Å the curves become steeper, which is evidence of an increase in viscosity in this region.

The common feature of the profiles of PMS films on the steel substrate (see fig. 4) is the presence of three sections differing in slope, and hence, in viscosity of the corre-sponding liquid interlayers.   The table lists average data on measurements of the relative viscosity of these interlayers and their thickness.   At a definite distance from the substrate of *ca.* 20-30 Å all the PMS studied display a sharp decrease in viscosity. Above this boundary the viscosity of the liquids is constant, and at 150-200 Å it decreases by approximately 1.2-1.4 times, sharply in most cases, subsequently remaining constant up to distances of 2,200-2,500 Å.

The change in viscosity of a liquid at a definite distance from a solid surface had been observed earlier as was pointed out in ref. (9)-(12).   The increase and decrease of viscosity in the boundary layers was attributed to phenomena related to orientation of the liquid molecules under the influence of the solid surface.   Indeed, the change in viscosity of a liquid as a result of orientation of its molecules in one way or another has been confirmed in studies carried out by other methods.   For instance, orientation of molecules in the direction of flow of a liquid reduces its viscosity.[13]

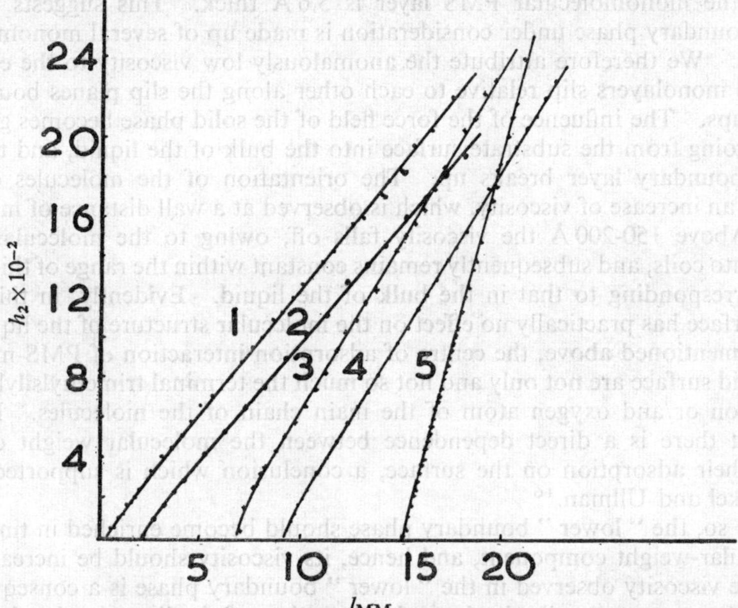

FIG. 3.—Profiles of PMS films on a glass substrate.   1, PMS-5 ; 2, PMS-15 ; 3, PMS-70 ; 4, PMS-400 ; 5, PMS-2 000.

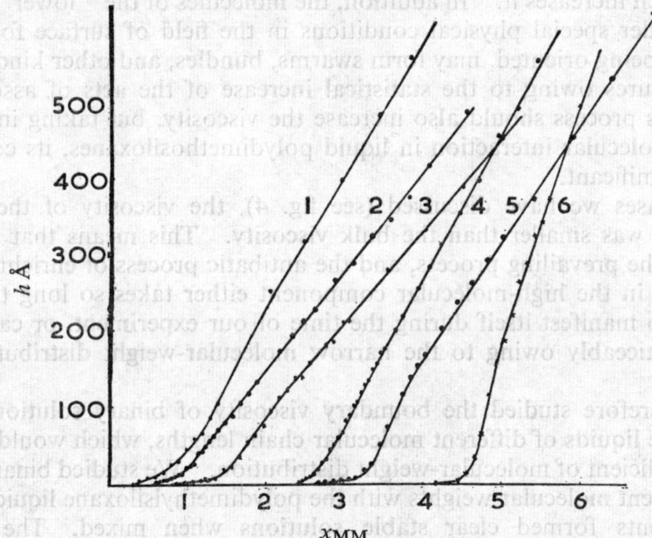

FIG. 4.—Profiles of PMS films on a steel substrate.   1, PMS-5 ; 2, PMS-15 ; 3, PMS-25 ; 4, PMS-70 ; 5, PMS-400 ; 6, PMS-2 000.

The change in viscosity of PMS observed in the layer up to 20 Å is most probably related to a sharp disturbance of conformational equilibrium towards straightening of molecular coils and subsequent orientation of the straightened chains in the plane of the substrate.   Owing to the high polarizatility of the Si—O bond,[14] each of the silicon and oxygen atoms may theoretically be a centre of attraction interaction with the solid phase, which is the reason why these molecules orient themselves horizontally when adsorbed. According to published data [15] for horizontal orientation of molecules, the monomolecular PMS layer is 5.6 Å thick. This suggests that the " lower " boundary phase under consideration is made up of several monomolecular PMS layers. We therefore attribute the anomalously low viscosity to the ease with which these monolayers slip relative to each other along the slip planes bounded by methyl groups. The influence of the force field of the solid phase becomes gradually weaker in going from the substrate surface into the bulk of the liquid, and the poly-molecular boundary layer breaks up. The orientation of the molecules changes, resulting in an increase of viscosity, which is observed at a wall distance of more than 20-30 Å. Above 150-200 Å the viscosity falls off, owing to the molecular chains rolling up into coils, and subsequently remains constant within the range of thicknesses studied, corresponding to that in the bulk of the liquid. Evidently, in this region the solid surface has practically no effect on the molecular structure of the liquid.

As was mentioned above, the centre of adsorption interaction of PMS molecules with the solid surface are not only and not so much the terminal trimethylsilyl groups, as any silicon or/and oxygen atom of the main chain of the molecules. Hence it follows that there is a direct dependence between the molecular weight of linear PMS and their adsorption on the surface, a conclusion which is supported by the data of Perkel and Ullman.[16]

If this is so, the " lower " boundary phase should become enriched in time in the high-molecular-weight component, and hence, its viscosity should be increased. In this case the viscosity observed in the " lower " boundary phase is a consequence of two competing processes, viz., horizontal orientation of the linear molecules, which decreases the viscosity, and increasing concentration of the high-molecular-weight component, which increases it. In addition, the molecules of the " lower " boundary phase, being under special physical conditions in the field of surface forces of the solid phase and being oriented, may form swarms, bundles, and other kinds of super-molecular structures owing to the statistical increase of the acts of association of molecules. This process should also increase the viscosity, but taking into account the weak intermolecular interaction in liquid polydimethosiloxanes, its contribution is evidently insignificant.

In all the cases we have discussed (see fig. 4), the viscosity of the " lower " boundary phase was smaller than the bulk viscosity. This means that orientation of molecules is the prevailing process, and the antibatic process of enrichment of the boundary phase in the high-molecular component either takes so long that it does not have time to manifest itself during the time of our experiment, or cannot affect the viscosity noticeably owing to the narrow molecular-weight distribution of the samples studied.

We have therefore studied the boundary viscosity of binary solutions of poly-dimethylsiloxane liquids of different molecular chain lengths, which would artificially increase the coefficient of molecular-weight distribution. We studied binary mixtures of PMS of different molecular weights with the polydimethylsiloxane liquids PMS-10. These components formed clear stable solutions when mixed. The boundary viscosity was studied on a metallic substrate, since the most interesting range of thicknesses in which viscosity anomalies could be expected was up to 400 Å.

Fig. 5 shows the film profiles obtained after blowing-off PMS-10 as such, and of solutions of other PMS liquids in it, on a metallic substrate. The concentration of the solutions was 10 % by weight in all cases.

It is evident from fig. 5 (curve 1) that the profiles of the PNS-10 film is similar to those of the other PMS (see fig. 4), observed on a metallic substrate, and that they exhibit a boundary phase of low viscosity, the thickness of which is, as in the previous cases, 15-20 Å.

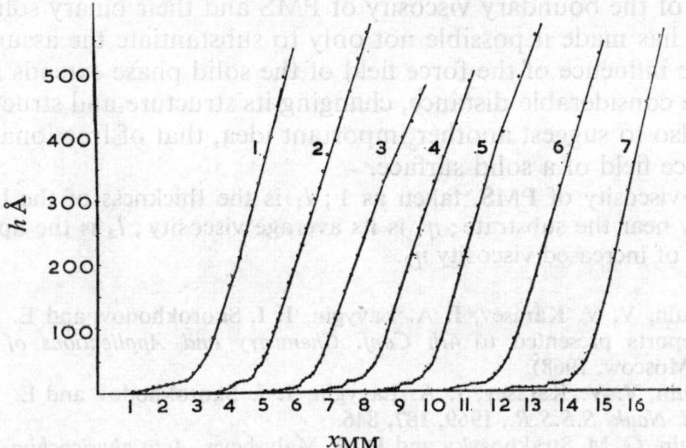

FIG. 5.—Film profiles of PMS solutions on a steel stubstrate. 1, PMS-10; 2, PMS-10+PMS-70; 3, PMS-10+PMS-400; 4, PMS-10+PMS-2,000; 5, PMS-10+PMS-2×10; 6, PMS-10+PMS-5× 10^4; 7, PMS-10+PMS-10^6.

The film profiles of PMS solutions (see fig. 5, curves 2-7) in the 10-30 Å range are of a distinctly stepped nature, the height of the steps, as well as the general shape of the curve, being readily reproduceable in duplicate experiments.   This step height is independent of the molecular weight of the dissolved PMS.   The curve is of a stepped nature when blowing is carried out after an interval of $\frac{1}{2}$ h or more from the moment of applying the liquid sample to the substrate.   If the layer is blown-off immediately the steps on the film profile curve are either indistinct or absent.

TABLE 1.—CHARACTERISTICS OF PMS AND THICKNESSES AND RELATIVE VISCOSITIES OF THEIR BOUNDARY LAYERS ON A METALLIC SUBSTRATE

substance	molecular weight	refractive index η_D^{20}	$\frac{\eta_1}{\eta_0}$	l_1 Å	$\frac{\eta_2}{\eta_0}$	l_2 Å
PMS-5	1,070	1.3992	0.15	25	1.35	235
PMS-15	1,580	1.4042	0.25	15	—	—
PMS-25	2,080	1.4050	0.11	20	1.40	140
PMS-70	2,860	1.4058	0.38	30	1.11	160
PMS-400	6,700	1.4062	0.09	10	1.35	130
PMS-2,000	36,900	1.4080	0.04	6	1.40	155

The step profile of the boundary phase is not a consequence of the instability of the film profile, since on maturing the layer after blowing, for 1.0-1.5 h, several intermediate measurements of its thickness showed that the film profile retains its shape satisfactorily throughout this time interval.   We conclude from this that the stepped profile of the film reflects phenomena at the liquid-substrate interface and is evidence of changes in viscosity and structure of the liquid, localized in the thickness range up to 30 Å.   The nature of the film profile observed in this case is proof of the existence of liquid boundary phases of changed viscosity similar to the cases discussed earlier.[10]   The steps formed on the curves is a result of slip of liquid interlayers of different structure along one another.   The fact that horizontal orientation of molecular chains occurs in this case too, is confirmed by the absence of any relation between molecular weight and step height of the film profile for different solutions. The step-film profile characteristic of the samples held on the substrate for some time before blowing off, is evidence of a definite kinetics of formation of the boundary layer, the study of which is beyond the scope of this paper.

Investigation of the boundary viscosity of PMS and their binary solutions by the blow-off method has made it possible not only to substantiate the assumption made earlier [17] that the influence of the force field of the solid phase extends into the bulk of the liquid to a considerable distance, changing its structure and structure-sensitive properties, but also to suggest another important idea, that of fractionating polymer liquids in the force field of a solid surface.

η_0 is the bulk viscosity of PMS, taken as 1 ; l_1 is the thickness of the liquid layer of decreased viscosity near the substrate; η_1 is its average viscosity; l_2 is the upper boundary of the liquid layer of increased viscosity η_2.

[1] B. V. Deryaguin, V. V. Karasev, I. A. Lavygin, I. I. Skorokhodov, and E. N. Khromova, *Theses of Reports* presented to 4*th Conf. Chemistry and Applications of Organo-silicon Compounds*, (Moscow, 1968).

[2] B. V. Deryaguin, V. V. Karasev, I. A. Lavygin, I. I. Skorokhodov and E. N. Khromova, *Doklady Akad. Nauk. S.S.S.R.*, 1969, **187**, 846.

[3] B. V. Deryaguin, G. M. Strakhosvky and D. S. Malysheva, *Acta physicochim.*, 1944, **19**, 541.

[5] V. V. Karasev, Yu. M. Luzhnov and N. V. Churayev, *Zhur. Fiz. Khim.*, 1968, **42**, 558.

[6] B. V. Deryaguin, V. I. Gol'dansky and V. V. Karasev, *Doklady Akad. Nauk. S.S.S.R.*, 1947, **57**, 697.

[7] A. A. Vlasov, *Lens Coating*, ed. Acad. I. V. Grebenshchikov, Moscow, (Leningrad, 1946).

[8] B. V. Deryaguin and V. F. Pichugin, *Proc. 2nd U.S.S.R. Conf. Friction and Wear*, 1949, **1**, 103.

[9] V. V. Karasev and B. V. Deryaguin, *Doklady Akad. Nauk. S.S.S.R.*, 1948, **62**, 762.

[10] V. V. Karasev and B. V. Deryaguin, *Kolloid. Zhur.*, 1953, **15**, 366.

[11] V. V. Karasev and B. V. Deryaguin, *Zhur. Fiz. Khim.*, 1959, **33**, 100.

[12] B. V. Deryaguin, V. V. Karasev, N. N. Zakhavayeva and V. P. Lazarev, *Zhur. Tekhn., Fiz*, 1957, **27**, 1076.

[13] A. Bondy, *Appl. Phys.*, 1945, **16**, 539.

[14] C. Eaborn, *Organosilicon Compounds*, (Butterworth and Co., London, 1960), chap. 8, 9.

[15] H. W. Fox, P. M. Taylor and W. A. Zisman, *Ind. Eng. Chem.*, 1947, **39**, 1401.

[16] R. Perkel and R. Ullman, *J. Polymer Sci.*, 1961, **54**, 127.

[17] B. V. Deryaguin, *Mineral'noye Syryo*, 1934, no. 2, 33.

Boundary Layers of Pure Liquids at the Graphon Surface

By S. G. Ash * and G. H. Findenegg

Institut für Physikalische Chemie, Universität Wien, 1090 Wien, Austria.

Received 14th April, 1970

Measurements of the volume changes which accompany the wetting of Graphon by n-alkanes, benzene and water, and the heats of wetting for the same systems, are analyzed. For the long-chain alkanes there is evidence for a pre-freezing phenomenon near the Graphon/liquid interface. For liquids composed of smaller molecules this effect is not found.

In this paper, the results of thermodynamic measurements on solid/liquid interface systems are explained in terms of the structure of the boundary layer of the liquids. Robert [1] found that the heat of wetting of Graphon by a series of n-alkanes increases with increasing chain length. This result was confirmed by Clint et al.[2] and by Everett and Findenegg.[3] At 25°C the heat of wetting per unit surface area of Graphon increases by a factor of 2 from hexane to hexadecane. For tetradecane and hexadecane the heat of wetting decreases with increasing temperature.[2] To explain these results for the long-chain alkanes it was postulated that, close to the freezing point of the liquid, the adsorbed layer has properties intermediate between those of the liquid and the crystal.[3] The heat of wetting would therefore include a contribution related to the heat of fusion of these alkanes.

If the above concept is correct the adsorbed layer should have a higher density than the bulk liquid. Therefore, the alkane + Graphon systems were further investigated by measuring the density of the boundary layer of the liquid. Details of these measurements will be reported elsewhere, but some results are summarized here. The emphasis of the present paper is on the correlation of the volumetric results with the heats of wetting for the same systems.

EXPERIMENTAL

The volumetric behaviour of the liquid boundary layers was studied as a function of temperature by making precision density measurements. Known weights of Graphon were introduced into calibrated pyknometers, which were then evacuated. The liquids were introduced into the evacuated pyknometers and the total weights and volumes measured at several temperatures. The results are expressed in terms of the surface excess mass of liquid,

$$m^\sigma = m_l - \rho_l^\circ v_l, \tag{1}$$

where m_l is the mass and v_l the volume of the liquid in the pyknometer, equal to the total volume of the pyknometer minus the volume of Graphon; ρ_l° is the bulk density of the liquid at the given temperature. The values of ρ_l° are determined in separate experiments. The surface concentration is the excess mass of liquid per unit surface area of solid,

$$\Gamma = m^\sigma / A^\sigma. \tag{2}$$

* present address : Shell Research Ltd., Thornton Research Centre, P.O. Box 1, Chester, CH1 3SH, England.

The volume of Graphon in the pyknometer can be calculated from the weight of Graphon and its density. However, the absolute density of Graphon is not precisely known, and instead a " reference density " is used. This is determined by making the density measurements with a reference liquid, and by equating the surface excess, for this system, to zero. The surface concentration is thus a relative quantity based on a chosen reference liquid for which, by definition, $\Gamma = 0$. In the present work the reference liquid is cyclohexane. Density measurements for the Graphon + cyclohexane system have been made at temperatures from the melting point of cyclohexane (6.5°C) to 50°C. Within the accuracy of the experiment, a constant coefficient of expansion of Graphon was obtained.

The surface area of Graphon was 87 m^2 g^{-1} based on the B.E.T. method and a molecular area of N$_2$ = 16.2 Å^2. Details of the experiment will be given elsewhere.[4]

RESULTS AND DISCUSSION
n-ALKANES C$_{14}$, C$_{16}$, C$_{18}$

The surface concentrations of tetradecane, hexadecane, and octadecane, plotted against the experimental temperature minus the melting point temperature of the alkane, $t - t_f$, are shown in fig. 1. From these results we note : (i) at a given value of $t - t_f$ the surface concentration increases with increasing chain length of the alkane ; (ii) Γ is strongly temperature dependent at temperatures close to the freezing point of the alkane, and this effect increases with increasing chain length of the alkane.

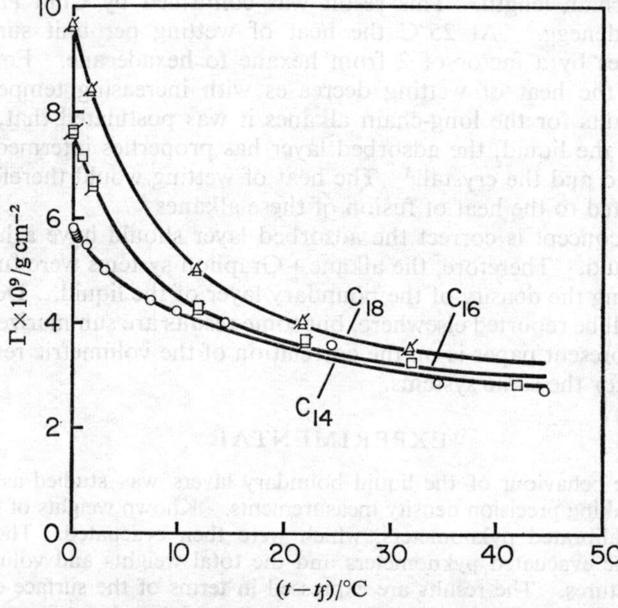

Fig. 1.—The surface concentrations of tetradecane, hexadecane and octadecane as a function of the experimental temperature minus the freezing point temperature of the alkane.

Qualitatively, the dependence of Γ on the number of carbon atoms per molecule, and on temperature, is similar to the variation of the heat of wetting of Graphon by these liquids.[2, 3] For a quantitative correlation of the two effects we consider the definition of the surface concentration. On locating the Gibbs dividing surface at the Graphon/liquid boundary,

$$\Gamma = \int(\rho(x) - \rho_i^\circ)\mathrm{d}x, \tag{3}$$

where $\rho(x)$ is the actual density of the liquid as a function of the distance x normal to the boundary. The exact form of the function $\rho(x)$ is not known. It is convenient to adopt a simple model of the boundary layer by making the following assumptions: (i) the excess mass of liquid is distributed uniformly over a surface zone, of density ρ_s and thickness h_s; (ii) the properties of this surface zone are the same as the properties of the solid alkane (at the melting point). The thickness of the surface zone, obtained from eqn (3), is

$$h_s = \Gamma/(\rho_s - \rho_l^\circ), \tag{4}$$

and the number of moles of alkane contained in the surface zone is

$$n_s = A^\sigma h_s \rho_s / M, \tag{5}$$

where M is the molecular weight of the alkane. On wetting the Graphon surface, n_s moles of the alkane liquid form a surface zone with properties of the solid alkane. This process is accompanied by the evolution of an energy $n_s H_f$ (H_f, molar heat of fusion of the alkane), in addition to the Graphon-alkane interaction energy. To calculate h_s and n_s from eqn (4) and (5), the density of the solid alkanes has been calculated from the density of the corresponding liquid alkane by assuming that, at the melting point, $(\rho_s - \rho_l)/\rho_l = 0.1100$. This relationship approximately holds for a large number of n-alkanes.[5] At $t - t_f = 0.05°$ the solid-like surface zones of tetradecane, hexadecane, and octadecane have thicknesses: $h_s = 6.8$, 9.0 and 11.3 Å, respectively. The quantity $n_s H_f$ is plotted against $t - t_f$ in fig. 2 for tetradecane and in fig. 3 for hexadecane. The molar heats of fusion of the alkanes have been taken from the literature.[6]

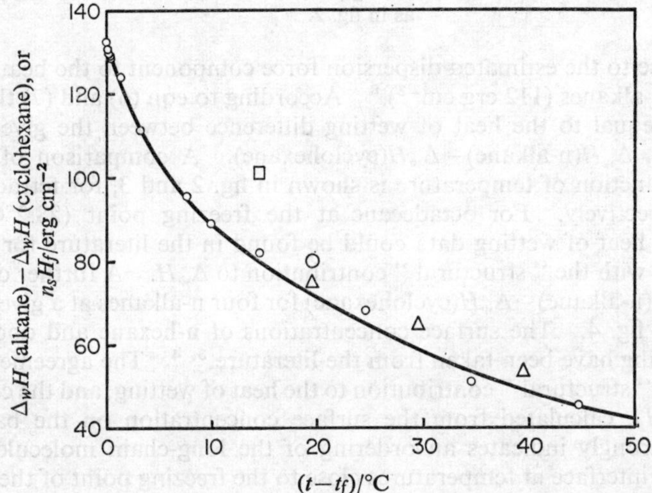

FIG. 2.—The structural contribution to the heat of wetting of Graphon by n-tetradecane as a function of the experimental temperature minus the freezing point temperature of n-tetradecane: (a) $\circ$, $n_s H_f$, calculated from the surface concentration on the basis of the proposed model, (b) experimental values of $\Delta_w H$ (n-tetradecane) $-\Delta_w H$ (cyclohexane): $\square$, Robert[1]; $\triangle$, Clint[2]; $\circ$, Everett.[3]

The observed heat of wetting, $\Delta_w H$, can be considered to be the sum of two terms: (i) the heat H_w resulting only from Graphon-alkane interactions; (ii) a heat connected with structural changes in the surface layer of the alkane. In terms of the present model,

$$\Delta_w H = H_w + n_s H_f. \tag{6}$$

For the wetting of Graphon by alkane liquids, H_w is considered to be nearly constant, i.e., independent of the nature of the molecules of the liquid.[7]   Since cyclohexane is our reference liquid for which $\Gamma = 0$, we set

$$H_w = \Delta_w H(\text{cyclohexane}) = 107 \text{ erg cm}^{-2}. \tag{7}$$

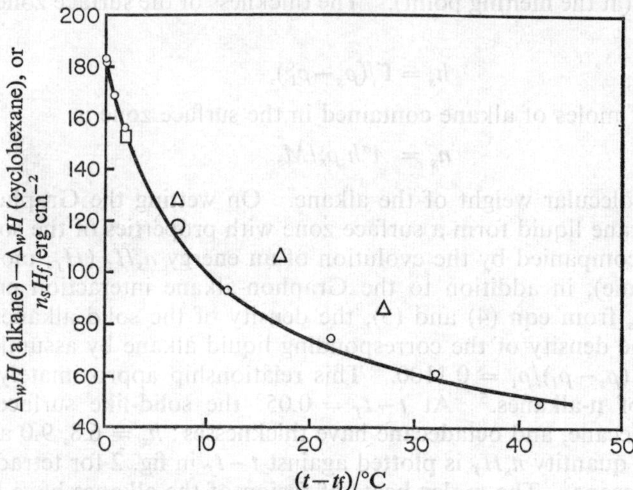

FIG. 3.—The structural contribution to the heat of wetting of Graphon by n-hexadecane as a function of the experimental temperature minus the freezing point temperature of n-hexadecane; conventions as in fig. 2.

This value is close to the estimated dispersion force component to the heat of wetting of Graphon by n-alkanes (112 erg cm^{-2}).[8]   According to eqn (6) and (7) the quantity $n_s H_f$ should be equal to the heat of wetting difference between the given n-alkane and cyclohexane, $\Delta_w H(\text{n-alkane}) - \Delta_w H(\text{cyclohexane})$.   A comparison of these two quantities as a function of temperature is shown in fig. 2 and 3, for tetradecane and hexadecane, respectively.   For octadecane at the freezing point (28.2°C) $n_s H_f = 235$ erg/cm^2; no heat of wetting data could be found in the literature for this liquid to compare $n_s H_f$ with the " structural " contribution to $\Delta_w H$.   A further comparison of $n_s H_f$ and $\Delta_w H(\text{n-alkane}) - \Delta_w H(\text{cyclohexane})$ for four n-alkanes at a given temperature is shown in fig. 4.   The surface concentrations of n-hexane and dodecane and the heats of wetting have been taken from the literature.[3, 4]   The agreement between the experimental " structural " contribution to the heat of wetting, and the corresponding quantity $n_s H_f$, calculated from the surface concentration on the basis of our simple model, strongly indicates an ordering of the long-chain molecules near the Graphon/alkane interface at temperatures close to the freezing point of these alkanes.

From measurements of adsorption from solution [9, 10] long-chain alkanes are strongly preferentially adsorbed onto Graphon.   For dilute solutions of n-C$_{32}$ in n-heptane, Groszek concluded that the adsorption is confined entirely to the basal planes of the graphite structure.   The specific interaction between the graphite basal plane and the n-alkane molecules is attributed to the good geometrical fit of the molecules on to the hexagons of the graphite basal plane.[10]   A consideration of the thickness of the surface zone of pure n-alkanes at the Graphon/alkane interface (of h_s above) shows that in pure long-chain alkane liquids the surface affects a much wider zone than a monolayer of molecules oriented parallel to the surface. The strong temperature dependence of the quantities $\Delta_w H$ and Γ at temperatures close to the freezing point of the substances is evidence for a highly co-operative character of the forces which are responsible for the ordered structure of the zone. It is possible that the ordered structure consists of several layers of molecules adsorbed parallel to the surface. But the possibility of a different structure (e.g., a closely packed monolayer of molecules oriented normal to the surface) cannot be ruled out on the basis of the present results.

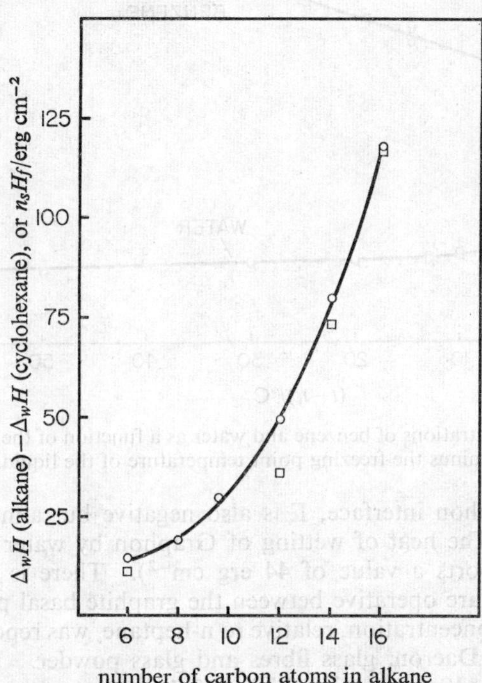

Fig. 4.—The structural contributions to the heats of wetting of Graphon by n-alkanes at 25°C: (a) ○, experimental values of $\Delta_w H$(alkane)$-\Delta_w H$(cyclohexane), Everett [3]; (b) □, $n_s H_f$ calculated from the surface concentration on the basis of the proposed model.

BENZENE AND WATER

The surface concentrations of benzene and water at the Graphon/liquid interface, plotted against $t-t_f$, are shown in fig. 5. The situation is different from that with the chain alkanes : for both liquids Γ is negative over the temperature range investigated. For the n-alkanes we have found that a positive value of Γ is correlated with a positive value of $\Delta_w H$(n-alkane)$-\Delta_w H$(cyclohexane). For benzene, however, Γ is negative while $\Delta_w H$ is somewhat larger than the value for cyclohexane (benzene : 112 erg cm^{-2}, cyclohexane 107 erg cm^{-2}, both values at 20°C [1]). This result could be the consequence of a weak specific interaction between the graphite basal plane and the benzene molecule at certain configurations of the molecules on the surface. If the surface zone is assumed to consist of a monolayer of benzene molecules oriented with the plane of the aromatic ring parallel to the solid surface the thickness of this layer is about 3.4 Å.[11] The density of the zone (assuming $\Gamma = 0$ for the cyclohexane/Graphon interface) is then found, from eqn (4), to be about 5 % less than the density of bulk benzene (at the freezing point). With increasing temperature the density of the surface zone approaches the density of the bulk liquid. At about

50°C above the freezing point $\Gamma = 0$. There are no data for the heat of wetting of Graphon by benzene as a function of temperature, with which to correlate these density measurements.

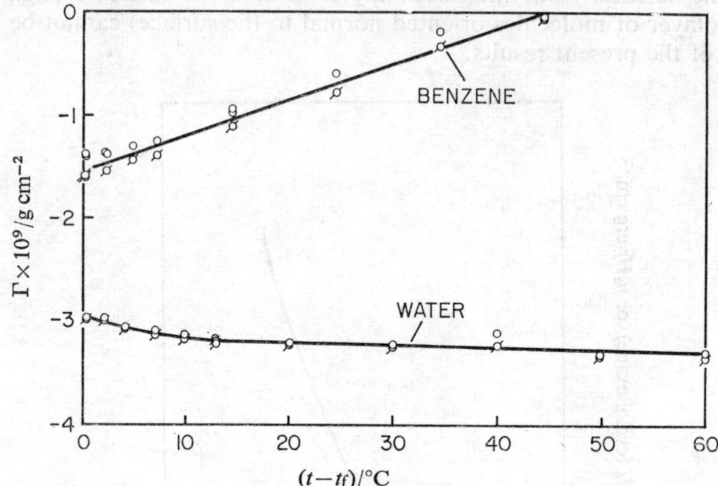

Fig. 5.—The surface concentrations of benzene and water as a function of the experimental temperature minus the freezing point temperature of the liquid.

For the water/Graphon interface, Γ is also negative but almost independent of temperature (fig. 5). The heat of wetting of Graphon by water is low; $\Delta_w H = 32$ erg/cm^2 [7a] (Robert reports a value of 44 erg cm^{-2}). There is evidence that only dispersion interactions are operative between the graphite basal plane and water.[8]

A positive surface concentration, relative to n-heptane, was reported by van Gils [12] for water near Nylon, Dacron, glass fibres and glass powder. His Γ values range from 0.7×10^{-5} to 3.0×10^{-5} g cm^{-2}, which is 3-4 powers of ten larger in absolute value than the (negative) value obtained by us for the Graphon/water interface. Drost-Hansen [13] has explained the large positive surface concentration of water observed at these polar surfaces by postulating that close to the polar surface there exists an extensive zone of disordered water molecules which are more closely packed than in bulk water. Conversely, the negative surface concentration of water at the non-polar surface of Graphon may be caused by the stabilization of bulky aggregates of water molecules. The distance over which significant structuring may occur is guessed by Drost-Hansen " at somewhere between tens and hundreds (or more) of molecular diameters ". According to our surface concentration, a density of the surface zone, e.g. 1 % less than that of the bulk liquid, corresponds to a thickness of this zone of approximately 30 Å (to be compared with the nearest neighbour distance in water of less than 3 Å).

CONCLUSION

The most significant result of this work is the different behaviour, near the freezing point temperature, of the n-alkanes tetradecane, hexadecane, and octadecane on the one hand, and cyclohexane, benzene, and water, on the other hand, at the Graphon/liquid interface. For the chain alkanes there is sharp increase in the surface excess mass and in the heat of wetting when the temperature approaches the freezing point. These effects can be understood as " pre-freezing " phenomena, i.e., as a cooperative ordering of the chain molecules with their long axes parallel to one another, which is induced by the surface forces. For the smaller molecules of cyclohexane and benzene such an effect is not observed. The behaviour of water is similar to the behaviour of these non-polar liquids, but a marked negative surface concentration is observed.

We thank the Royal Society for an award to S. G. A.

[1] L. Robert, *Compt. rend.*, 1963, **256**, 655; *Bull. Soc. Chim.*, 1967, 2309.
[2] J. H. Clint, J. S. Clunie, J. F. Goodman and J. R. Tate, *Nature*, 1969, **223**, 51.
[3] D. H. Everett and G. H. Findenegg, *Nature*, 1969, **223**, 52; *J. Chem. Thermodynamics*, 1969, **1**, 573.
[4] G. H. Findenegg, *J. Colloid Interface Sci.*, 1971, **35**, 249.
[5] H. Sackmann and F. Sauerwald, *Z. phys. Chem.*, 1950, **195**, 295. H. Sackmann and P. Venker. *Z. phys. Chem.*, 1952, **199**, 100.
[6] J. Timmermans, *Physico-Chemical Constants of Pure Organic Liquids* (Elsevier, 1950 and 1965).
[7] (a) F. H. Healey, J. J. Chessick, A. C. Zettlemoyer and G. J. Young, *J. Phys. Chem.*, 1954, **58**, 887.
(b) F. E. Bartell and R. M. Suggitt, *J. Phys. Chem.*, 1954, **58**, 36.
(c) A. C. Zettlemoyer and K. S. Narayan, *The Solid-Gas Interface* ed. E. A. Flood, (Dekker/Arnold, New York/London, 1967), chap. 6.
[8] J. A. Lavelle and A. C. Zettlemoyer, *J. Phys. Chem.*, 1967, **71**, 414.
[9] R. Aveyard, *Trans. Faraday Soc.*, 1967, **63**, 2778.
[10] A. J. Groszek, *Proc. Roy. Soc. A*, 1970, **314**, 473.
[11] L. Pauling, *The Nature of the Chemical Bond* (Cornell University Press, New York, 1960), p. 260.
[12] G. E. van Gils, *J. Colloid Interface Sci.*, 1969, **30**, 272.
[13] W. Drost-Hansen, *Ind. Eng. Chem.*, 1969, **61**, No. 11, 10.

Contact Between a Gas Bubble and a Solid Surface and Froth Flotation

By A. Scheludko, Sl. Tschaljowska and A. Fabrikant

University of Sofia, Dept. of Chemistry, Institute of Physical Chemistry,
Institute of Mining and Geology, Sofia, Bulgaria

Received 9th April, 1970

The process of formation of the contact silica surface/air when pressing a liquid meniscus on to the solid/liquid interface is investigated. The kinetics of expansion of the contact area and the final states characterized by contact angles in both directions (by spreading and withdrawing of the contact) are examined. Preliminary data about the effect of surfactants on the velocity of expansion of the contact and on the contact angle and its hysteresis are presented. The new methods applied in the investigations are described. The results are compared with flotation experiment data.

The purpose of this paper is to investigate the contact formed by pressing an air bubble on to a solid surface, employing the method and approach used in the investigation of microscopic free black films.[1,2] The data obtained are compared to flotation measurements on the basis that flotation is only possible when the film separating the particle from the gas/liquid interface breaks.[3]

ANGLES OF CONTACT

A liquid meniscus is forced through a thin-walled circular silica tube of internal radius $R = 0.102$ cm, towards a polished plate of vitreous silica until Newton's rings appear around the tip of the meniscus, i.e., at a distance of several 1000 Å from the plate. The pressure upon the meniscus, equal to

$$2\gamma \cos \theta_R/R + h\rho_0 g, \tag{1}$$

is maintained strictly constant. Its value is registered photo-electrically on a precise aqueous manometer. In eqn (1), γ is the surface tension of the solution, ρ_0 its density, g gravitational acceleration and h the depth of immersion of the tube in the solution.

The separating film formed from appropriate solution of dodecylamine hydrochloride (1.35×10^{-5} mol/l) in these conditions breaks, forming an exposed contact whose radius r increases to r_0. This process is observed and recorded in reflected light on a metallographic microscope with a camera. As the silica tube is much narrower than the vessel containing the solution, the depth of immersion remains constant, so that in all the following treatments, the hydrostatic component of the pressure $h\rho_0 g$ is eliminated.

The profile of the meniscus after formation of contact is shown in fig. 1. After solving the variational problem for the minimum surface at constant volume of a body of rotation about the axis OZ, as described in ref. (1), (4), we obtain for the capillary pressure

$$P_\gamma = 2\gamma(R \cos \theta_R - r \sin \theta_r)/(R^2 - r^2). \tag{2}$$

In the process treated here both of the angles θ_R and θ_r form as a result of the recession of the liquid on identical surfaces of vitreous silica. Thus, at $r = r_0$, we obtain $\theta_R = \theta_r = \theta_0$. At constant external and hydrostatic pressure according to (1) and (2) we obtain

$$\tan \theta_0 = r_0/R. \tag{3}$$

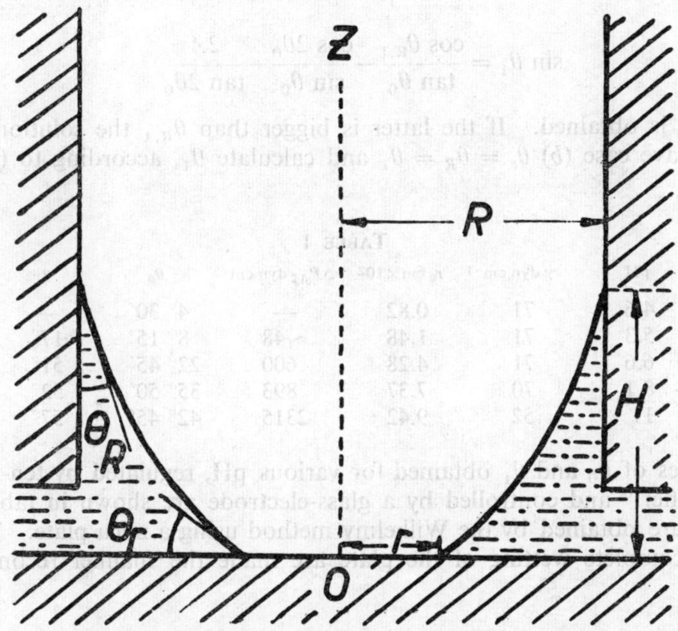

FIG. 1.—Profile of a meniscus contacting a plane surface.

After establishment of r_0 we gradually reduce the external pressure. At first, the circumference remains fixed. It begins to narrow only after a specific lowering of the pressure ΔP_M. This irreversibility (hysteresis) of the process is due to the fact that in order to start moving in the direction of wetting the angle θ_r has to increase from θ_0 to θ_1. The angle θ_R also becomes bigger than θ_0 and two states are now possible.

(a) θ_R increases more slowly than θ_r and the shift of the circumference at $\theta_r = \theta_1$ takes place at fixed position in the region $\theta_R = \theta_{R,1}$ i.e., at constant H, equal to H_0.

(b) θ_R reaches θ_1 before θ_r, and the shift of θ_r corresponds to the condition $\theta_r = \theta_R = \theta_1$.

The solution of the problem for the shape of the body shown in fig. 1 for $\theta_R = \theta_r = \theta_0$ is

$$H_0 = R(1 - \tan \theta_0), \tag{4}$$

and for the case (a) $H = H_0 = \text{const.}$,

$$\frac{H_0}{R} = \frac{x_1}{2y_0}\Delta E - \frac{2y_1}{x_1}\Delta F, \tag{5}$$

where $\Delta E = E(\lambda_r,q^2) - E(\lambda_R,q^2)$ and $\Delta F = F(\lambda_r,q^2) - F(\lambda_R,q^2)$ (E and F are elliptic integrals of the second and first kind, tabulated, e.g., in ref. (5)),

$$y_0 = A + \cos\theta_0, \quad y_1 = y_0 - \cos\theta_{R,1}, \quad A = \Delta P_M R/2\gamma, \quad x_{1,2} = 1 \pm \sqrt{1 + 4y_0 y_1},$$

$$q^2 = 1 - (x_2/x_1)^2, \quad \lambda_r = \arcsin(1/q)\sqrt{1 - (2y_0 \tan\theta_0/x_1)^2}, \quad \text{and}$$

$$\lambda_R = \arcsin(1/q)\sqrt{1 - (2y_0/x_1)^2}.$$

From these equations the intermediate value $\theta_{R,1}$ is calculated, and from the expression

$$\sin\theta_1 = \frac{\cos\theta_{R,1}}{\tan\theta_0} - \frac{\cos 2\theta_0}{\sin\theta_0} - \frac{2A}{\tan 2\theta_0} \tag{6}$$

the angle θ_1 is obtained. If the latter is bigger than $\theta_{R,1}$ the solution is correct. If not, we have case (b) $\theta_r = \theta_R = \theta_1$ and calculate θ_1, according to (6) but with $\theta_{R,1} = \theta_1$.

TABLE 1

pH	γ dyn cm^{-1}	r_0 cm $\times 10^2$	ΔP_M dyn cm^{-2}	θ_0	θ_1
4.6	71	0.82	—	4° 30′	—
5.3	71	1.48	~48	8° 15′	~17°
6.6	71	4.28	600	22° 45′	51°
8.1	70	7.37	893	35° 50′	52°
11	52	9.42	2315	42° 45′	57°

The values of θ_0 and θ_1 obtained for various pH, regulated by ten-fold diluted universal buffer [6] and controlled by a glass-electrode are shown in table 1.* The values of γ are obtained by the Wilhelmy method using a mica plate. The corrections for incomplete wetting of the plate are made the manner recommended in ref. (7).

EXPANSION OF THE CONTACT

The process of expansion of the contact $r(t)$ is either recorded photographically when it is fast, or visually read on an eye-piece micrometer when slow. Following ref. (2), the velocity of expansion is represented as

$$dr/dt = a\bar{\Delta}_t. \tag{7}$$

Here

$$\bar{\Delta}_t = \gamma(\cos\theta_t - \cos\theta_0) \tag{8}$$

is the tangent to the solid-surface force acting at time t from the beginning of the process and a is the instantaneous mobility (the reciprocal coefficient of friction). Both co-factors correspond to a unit length of the circumference of wetting. As $\bar{\Delta}_t$ is a function of r through the non-equilibrium angle $\theta_r = \theta_t$, in order to obtain the values of a for different r, the slopes dr/dt (obtained by graphical differentiation of the curves $r(t)$) ought to be divided by $\bar{\Delta}_t$. The latter is easy to calculated with the aid of eqn (2),

$$\bar{\Delta}_t = \gamma[\sqrt{1 - (r\cos\theta_0/R)^2} - \cos\theta_0], \tag{9}$$

if we assume that the angle $\theta_R = \theta_0$ remains constant during expansion at constant pressure.

---

* The details of the method, its theory and the additional measurements of γ and floatability will be published separately.

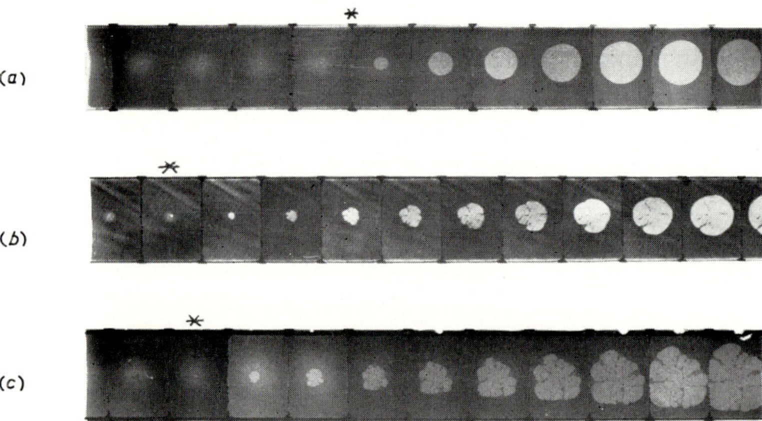

FIG. 2.—Photographs of the contact expansion frequency 12 frames/s: (*a*) pH 5.3 at 100× magnification; (*b*) and (*c*) pH 8.1 at 40× and 100×.   The appearance of the contact is marked by an asterisk.

At low pH the expansion is very rapid, as is seen from the photographs in fig. 2a, and the final state r_0 is established roughly within a minute.   On fig. 3, curve 1, the corresponding values of $a(r)$, are given.   Within the limits of scatter, they are the

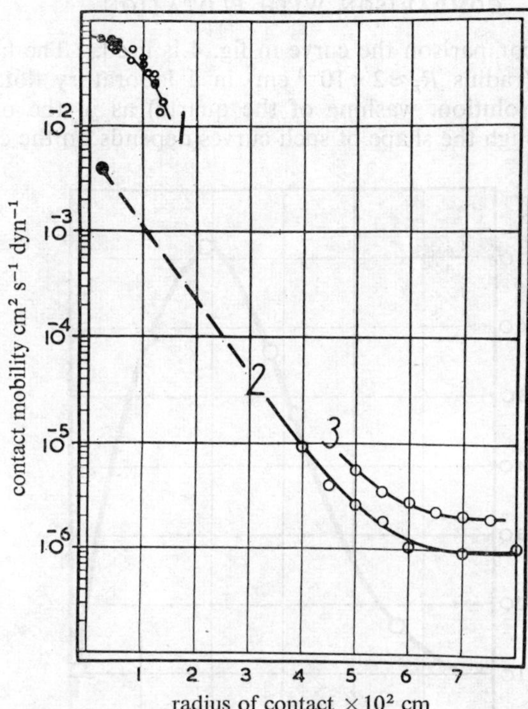

FIG. 3.—Contact mobility as a function of the contact radius.   Curve 1, for low pH (●, pH 4.6 and ○, pH 5.3).   Curve 2 for pH 11 (the first point ● is obtained from τ_m); curve 3 for pH 8.1.

same at pH = 4.6 and 5.3 and of the order of 10^{-2} cm² s⁻¹ dyn⁻¹.   Apparently, the friction here is determined by the viscosity in the volume of the liquid which is almost independent of pH.   The volume mechanism of the dissipation of energy in this case is corroborated by the fact that a here coincides with $a = 5 \times 10^{-2}$ obtained in ref. (2) for expansion of a black film of radius 7×10^{-3} cm.   In that paper the volume mechanism of friction was proved by the independence of a on the film tension at given r over a wide range of values of the film tension.

At high pH the situation is completely different.   The time of establishment of r_0 is of the order of several hours.   The dependence $a(r)$ at pH = 11 (fig. 3, curve 2) shows that a in this case first reaches values which are four orders lower, i.e., the friction is 10 thousand times greater than at low pH.   As the volume viscosity cannot rise significantly with increase of pH, the velocity of expansion is determined by surface or peripheral friction.   It is interesting as well that a now decreases extremely steeply with the expansion so that the data are represented on a log scale. This effect is completely rheological—the friction increases dramatically when the deforming force diminish.

The process of surmounting the rheological obstructions by big initial forces is clearly demonstrated at intermediate values of pH≈8.   On the photographs in fig. 2b and c it is seen that the circumference of wetting first tears and a starlike expansion takes place, its velocity being of the same order as at low pH.   With decreasing

$\bar{\Delta}_r$, a regular circular circumference forms gradually (fig. 2b) which expands very slowly, the values of a being close to those at pH 11 (fig. 3 curve 3).

COMPARISON WITH FLOTATION

As a basis of this comparison the curve in fig. 4 is used. The latter is obtained with quartz particles (radius $R_p \approx 2 \times 10^{-2}$ cm) in a laboratory flotation cell under the same conditions (solution, washing of the quartz) as in the measurements of contact angles. Although the shape of such curves depends on the construction and

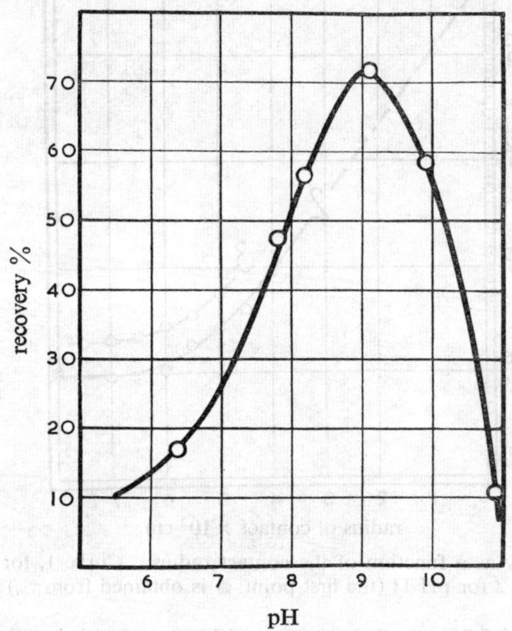

FIG. 4.—Flotation recovery of quartz as a function of pH.

regime of the flotation cell, the shape of the curve in fig. 4 is sufficiently typical (see also ref (8)) to permit a comparison with the contact investigation.

The decrease of floatability at high pH, although the contact angles retain their high values, is due to the sharp deceleration of the contact expansion. Hence our interpretation is based on the kinetics of contact expansion.

According to the theory of attachment of particles to bubbles,[9, 10] a spherical particle under the action of a detaching force G, breaks away from the bubble when the chord of the contacted part becomes equal to or smaller than $2r_m$ at *

$$r_m = \sqrt{GR_p/2\pi\gamma} \tag{10}$$

Therefore, it is necessary that the contact shall succeed in expanding at least to this size, in order to retain the attachment, or flotation. We simplify the problem, by neglecting the inertial forces, and take

$$G = G_p = \tfrac{4}{3}\pi R_p^3(\rho - \rho_0)g, \tag{11}$$

* In order to simplify the calculation, the hysteresis of the contact angle [10] is not taken into account here and later.

where ρ is the density of the particles, we obtain $r_m \approx 2 \times 10^{-3}$ cm at $\rho - \rho_0 = 2$ and γ (see table 1) in the range 71 dyn cm^{-1} ($r_m = 1.7 \times 10^{-3}$ cm) to 52 dyn cm^{-1} ($r_m = 2 \times 10^{-3}$ cm).

At such small r_m the chord and the diameter of contact are almost identical and the time τ_m for development of the contact is very short. In these conditions, the integration of (7) is simple, as the initial ($r \to 0$) values $\gamma(1 - \cos \theta_0)$ for $\bar{\Delta}_t$, and a_0 for a can be used, so that

$$\tau_m = \frac{R_p^2 \sqrt{2(\rho - \rho_0)g/3}}{a_0 \gamma^{\frac{3}{2}} (1 - \cos \theta_0)} \tag{12}$$

From fig. 3, curve 1 we obtain at low pH $a_0 \approx 7 \times 10^{-2}$ and from this $\tau_m = 3.3 \times 10^{-2}$ s for pH = 5.3 ($\gamma = 71$, $\theta_0 = 8°$ 15'). Since at pH = 5.3 the recovery is some 10 % (fig. 4), then if we assume that it is roughly inversely proportional to the time of contact, we find the latter to be of the order of a few milliseconds in good agreement with other data.[11] This result confirms such an intrepretation and allows a possible estimate of a for $r_m = 2 \times 10^{-3}$ at the other end of the flotation curve, at pH 11. Here the recovery is of the same order of magnitude but the direct determination of the initial value of a is very difficult, because of the steep change of a with r. Substituting in (12) $\tau_m = 3.3 \times 10^{-2}$, $\gamma = 52$ and $\theta_0 = 42°$ 45', according to table 1, we obtain $a = 4.4 \times 10^{-3}$ for $r_m = 2 \times 10^{-3}$ at pH = 11, the first point of curve 2, fig. 3. This result indicates that friction in the periphery of the contact, which hampers the recovery at high pH, is for small $r = r_m$ much less than that measured for larger contacts. This first point confirms the sharp decrease of a and the rheological character of the peripheral friction. The latter is apparently closely connected with the hysteresis of the contact angle.

In conclusion, the present paper shows that the main part of the time of contact in this investigation of froth flotation is not the time of thinning of the separating film to its rupture,[3] but the time of expansion of the contact after rupture to the size necessary for attachment.

In other conditions, e.g., for very small particles, the time τ_m strongly decreases (eqn. (12)) and it is possible that the thinning of the separating film becomes the controlling process. Possibly, this is the cause of the decrease in floatability of small particles.

[1] A. Scheludko, B. Radoev and T. Kolarov, *Trans. Faraday Soc.*, 1968, **64**, 2213.

[2] T. Kolarov, A. Scheludko and D. Exerowa, *Trans. Faraday Soc.*, 1968, **64**, 2864.

[3] A. Scheludko, *Kolloid-Z.*, 1963, **191**, 52; A. Scheludko, Sl. Tschaljowska, A. Fabrikant, H. Schulze and B. Radoev, *Freiberger Forschung.*, in press.

[4] D. Exerowa, I. Ivanov and A. Scheludko, *Ann. Université, Sofia*, 1961/62, **56**, 157.

[5] V. Beliakov, R. Kravtzova and M. Rappoport, *Tables of Elliptic Integrals*, (Russ.), (Moscow 1962), vol. 1.

[6] H. Britton and R. Robinson, *J. Chem. Soc.*, 1931, 1456.

[7] A. W. Neumann and W. Tanner, *Tenside*, 1967, **4**, 220.

[8] R. Dean and P. Ambrose, *U.S. Bur. Mines., Bull.*, 1944, 449; A. Gaudin, *Flotation*, (New York, 1957), chap. 10, §6; A. Gaudin and D. Fuerstenau, *Min. Eng.*, 1955, 66; M. Eigeles and M. Volova, *8th Int. Mineral Processing Congr.*, (Leningrad, 1968), S 12, p. 1.

[9] C. W. Nutt, *Chem. Eng. Sci.*, 1960, **12**, 133.

[10] A. Scheludko, B. Radoev and A. Fabrikant, *Ann. Université Sofia*, in press.

[11] V. Klassen and V. Mokrousov, *An Introduction to the Theory of Flotation*, (Russ.), (Moscow, 1959), pp. 139-144; H. Kirchberg and E. Töpfer, *7th Int. Min. Process. Cong.*, (New York), p. 157.

Interfacial Energies of Clean Mica and of Monomolecular Films of Fatty Acids Deposited on Mica, in Aqueous and Non-Aqueous Media

By Anita I. Bailey,* Andrea G. Price

Medical Research Council, Biophysics Research Unit, King's College,
London, England

and Susan M. Kay

Dept. of Physics, Royal Holloway College, Egham, Surrey, England

Received 3rd June, 1970

The influence of a variety of media on solid/fluid interfacial energies has been measured by a cleavage technique. The solid used was mica, chosen because of its near perfect cleavage and ideal bulk properties. Solid/vapour and solid/liquid interfacial energies, γ_{SV} and γ_{SL}, were measured by cleaving specimens in the form of strips, first, in an atmosphere of the vapour and then with the specimens completely immersed in the corresponding liquid. Samples of mica coated with mono-molecular films of fatty acids were constructed in such a way that separation took place between the oppositely oriented films. Polar and non-polar liquids and vapours were used in these experiments. The results allow an investigation into the validity of Young's equation for contact angles which are zero or positive. For well-behaved systems in which no adsorption takes place during the cleavage, the relation is valid. For fatty acid films in an aqueous environment, it is necessary to introduce an additional term into the relation to preserve the equality, thus

$$\gamma_{SL} = \gamma_{SV} - \gamma_{LV} \cos\theta + \gamma_H.$$

The energy γ_H has been associated with entropic effects occurring in the liquid phase. Its existence shows that there are measurable anisotropies in the water. The results can be explained if a single layer of the water in the immediate vicinity of the hydrophobic interface, is assumed to become preferentially oriented.

Model-building experiments have been used to estimate the number of water molecules involved in the interaction and a value of 213 cal/mol of oriented water was obtained for the energy involved. This allows one to predict values for the free energy of solution of small hydrophobic molecules in water, which are in good agreement with those of Frank and Evans. The effect disappears in a concentrated solution of urea indicating that the structure of the liquid near the interface is, in this case, effectively the same as it is in bulk, i.e., the urea has caused a breakdown in the ordered structure in the neighbourhood of the interface. This is consistent with current views of the denaturant action of urea on proteins and other bimolecules.

Young's equation describes the equilibrium of a three-phase solid-liquid-vapour system in terms of the interfacial energies between the different phases. Thus, the equilibrium of a drop of liquid resting on a solid surface is described by the equation

$$\gamma_{SV} = \gamma_{SL} + \gamma_{LV} \cos\theta,$$

where γ_{SV} is the interfacial energy of a solid in contact with the saturated vapour of the liquid; γ_{SL} is the interfacial energy between the solid and liquid phases; γ_{LV} is the interfacial energy of the liquid in equilibrium with its vapour, and θ is contact angle between solid and liquid phases.

Johnson [1] has made a comprehensive theoretical study of the validity of this equation, including in his analysis the effects of gravity, adsorption and curvature of the liquid vapour interface. He has shown that it is in general valid provided

* present address: Institut für Physik und Chemie der Grenzflächen, Fraunhofer-Gesellschaft, 7000 Stuttgart 1, Römerstr. 32.

the phases remain homogeneous right up to the interface, and adsorption effects are taken into account.

We have carried out some experiments to test the validity of the equation experimentally. This has been done by measuring separately the energies of the solid-liquid and solid-vapour interfaces using a cleavage technique.[2-4] For simple solid-liquid-vapour systems the equation holds. The experiments were then extended to include systems in which effects due to reorientation of molecules take place.[5]

EXPERIMENTAL

Muscovite mica was used as the solid phase. This material has several important properties which enable its surface energy to be measured accurately by cleavage. Cleavage takes place along single planes in the crystal over large areas. It is thus possible to obtain specimens which are of constant thickness. Such cleavage faces are molecularly smooth and the real surface area is equal to the geometrical value. Another important property of muscovite mica is that it deforms elastically. Energy is therefore stored reversibly in the portions of the sheet which become bent during the cleavage, and this quantity can be calculated. Hence, the amount of energy to be associated with the formation of new surfaces can be determined.

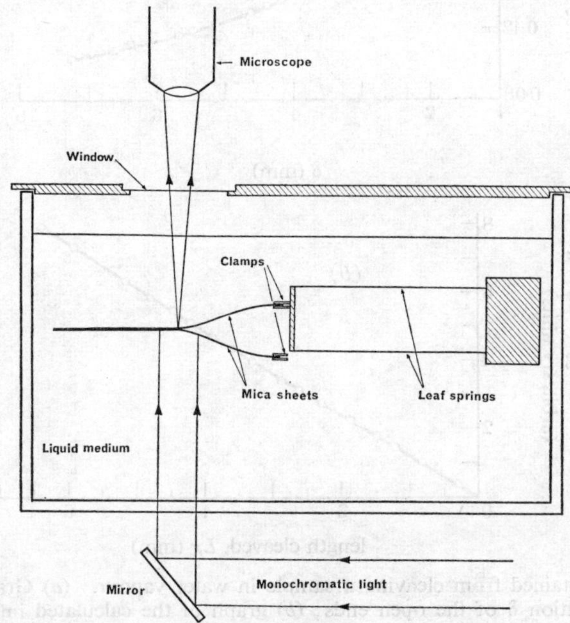

Fig. 1.—Schematic diagram of the apparatus. Cleavage of the sample is produced by lowering the bottom clamp a known distance δ. The deflection of the top clamp gives a measure of the force applied to the specimen and the position of the cleavage line is located by observing the Fizeau fringes produced by the sample.

A schematic diagram of the apparatus is shown in fig. 1. The specimens are in the form of narrow strips. Symmetrical cleavage is initiated at one end, and vertical forces F are applied to the open ends so that the strip assumes the equilibrium shape shown with a maximum separation δ at the point of application of the force. The total work done by the applied forces to increase the separation from δ_i to δ_f is $\int_{\delta_i}^{\delta_f} F d\delta$ and the amount of energy stored as elastic energy in the bent sheets during the process is increased by $\frac{1}{2}\Delta[F\delta]_i^f$. The area of crystal cleaved is determined by examination of the high-resolution multiple-beam

Fizeau fringes formed by the sheets.   Cleavage was carried out slowly so that the amount of kinetic energy dissipated in the system was small.   The whole cleavage mechanism was contained in a trough with optically worked faces, so that the sample could be completely surrounded by the liquid or vapour under examination.

Mica is also a suitable substance to act as a substrate on which to deposit monomolecular films of low-energy substances such as the fatty acids.   By this means we could extend observations to include low-energy surfaces.   These specimens were prepared by depositing films on the mica immediately after cleavage in a clean atmosphere.   This was done by retraction from non-polar solvent, a technique due to Zisman and coworkers,[6] who showed that monolayers so formed consist of an oriented layer of close-packed molecules.   Saturated solutions in n-decane and n-hexadecane, were prepared at about 30-60°C and subsequently allowed to cool.   Before use, they were filtered through the finest-grade Millipore filter to remove any dust particles or small crystals which had come out of solution.   After immersion

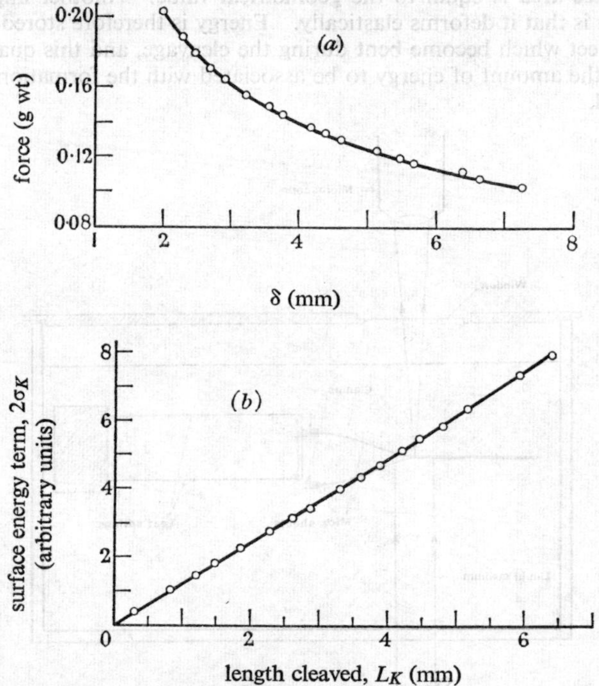

FIG. 2.—Results obtained from cleaving a sample in water vapour.   (a) Graph of the measured force against separation δ of the open ends; (b) graph of the calculated interfacial energy term (expressed in arbitrary units) against the length of sample cleaved.   The value of γ_{SV} is obtained from the slope of this line.

in the saturated solution, the sheets of mica emerged dry and two such sheets were then immediately placed together and allowed to seal up uniformly.   The quality of the deposition was judged by examining the " sandwich " using high-dispersion Fizeau interference fringes produced between the outer surfaces by a mixture of monochromatic radiations.   In this method small changes in the overall thickness of a sample can produce large variations in the intensity and hue in the interference pattern.[7]   Samples showing deviations from uniformity of the transmitted light were discarded.

Fig. 2 shows typical results obtained by this method for mica cleaved in water vapour at near saturation pressure.   The curve (a) shows the variation of the cleavage force with increasing separation δ of the ends of the sample.   The lower graph (b) shows the total interfacial energy σ (expressed in arbitrary units) plotted against the area of the sample which has been cleaved. The value of γ is obtained from the slope of this graph. Similar curves were obtained for the other cases quoted in table 1 and further examples are shown in fig. 3 and 4.

cleavage of stearic acid monolayer in water

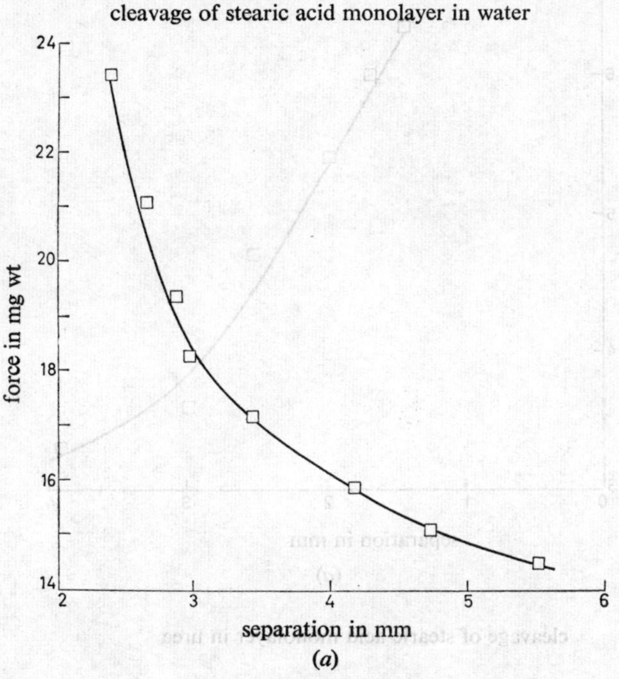

cleavage of stearic acid monolayer in water

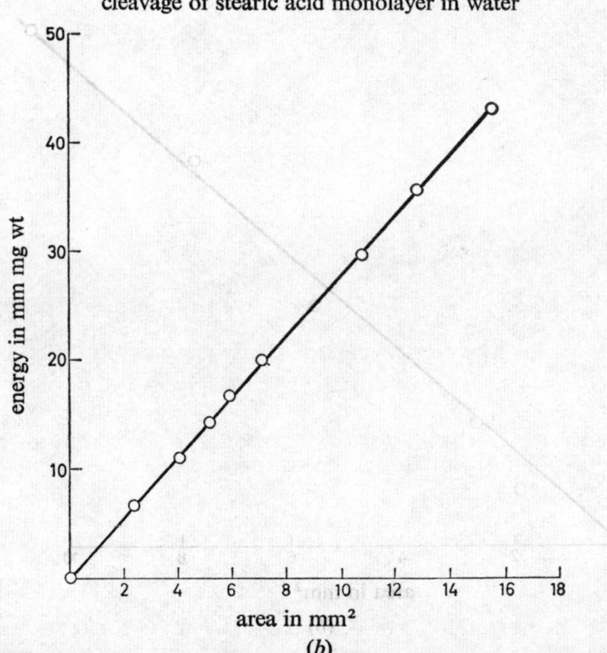

FIG. 3.—Results obtained for cleaving hydrophobic samples in water.

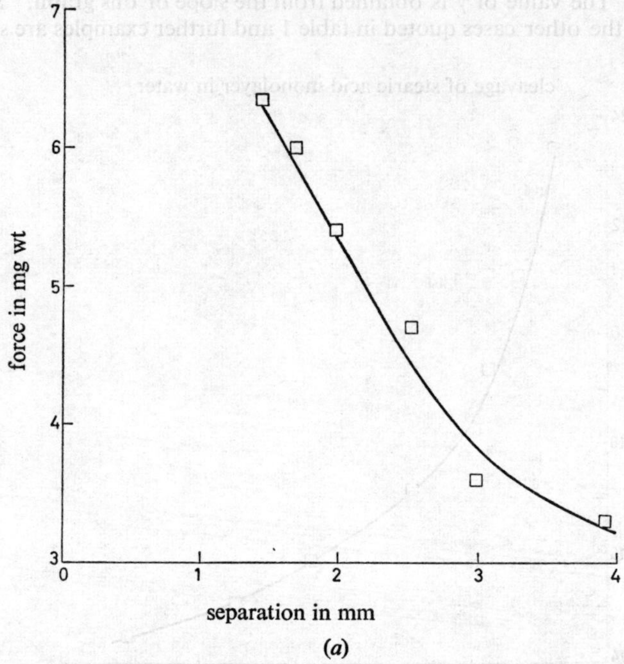

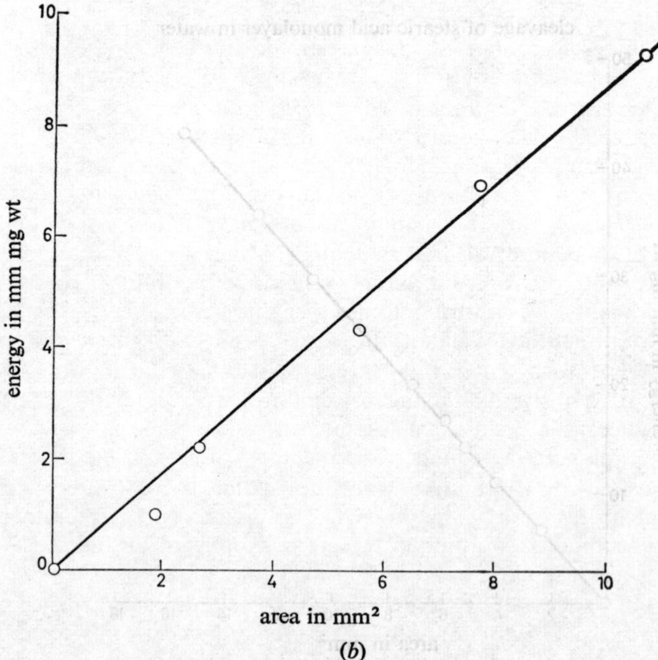

Fig. 4.—Similar results obtained for cleavage in a concentrated solution of urea.

TABLE 1

solid	medium	contact angle θ_A	interfacial energy ergs cm^{-2}
new mica	dry air (r.h. 1 %)	—	308
	moist air (r.h. 50-60 %)	—	220
	water vapour (r.h. 90 %)	—	183
	water	0	107
	hexane vapour	—	271
	hexane liquid	0	255
resealed mica	dry air matching sheets		260
	dry air non-matching sheets		120
lauric acid	room air	—	37
stearic acid	room air	—	25
perfluorodecanoic acid	room air	—	12
lauric acid	n-decane	14°	15
stearic acid	n-decane	32°	15
perfluorodecanoic acid	n-decane	74°	5
lauric acid	water	60°	8
stearic acid	water	65.5°	13

RESULTS

VERIFICATION OF YOUNG'S EQUATION

Water and n-hexane both wet a clean mica surface, but a finite contact angle is formed between n-decane and a mica surface coated with a monomolecular film of a fatty acid. The results give values for γ_{SL} of 107 ± 2 ergs cm^{-2} for water and 255 ± 2 ergs/cm^{-2} for n-hexane, while those calculated from Young's equation were 110 and 252.6 ergs/cm^{-2} respectively.

The contact angles formed by n-decane on the three monolayers studied were well-behaved and showed no hysteresis. Measured values of γ_{SL} for stearic acid, lauric acid and perfluorodecanoic acid were 15,15 and 5 ergs/cm^{-2} respectively and the corresponding values calculated from Young's equation were 14.5, 14.7 and 5.4 ergs/cm^{-2}. That Young's relation holds shows that it can be valid for quite complicated interfaces provided the interfaces and surrounding media do not alter during the experiments. Hence, although solids do not have equilibrium surfaces,[8] they so nearly approach this condition that the equation may be usefully applied to them.

With water we obtained quite different values for the contact angles depending on whether sessile drops or captive bubbles were used to make the measurements; with n-decane both methods yielded the same results. Again, hysteresis effects were always observed with water but were absent when the liquid was n-decane. We are indebted to Mr. B. A. Levine for making independent measurements on these systems to check the results. These are shown in table 2.

Results for monolayers deposited from the melt were not significantly different from those deposited by retraction from non-polar solvent. We attribute the different angles given by the two methods to the greater ease with which the small sessile drops can become contaminated. This is supported by the fact that on first contact the angles are large and comparable with the values obtained using bubbles, but they rapidly fall to the values recorded above. The monolayers are not removed into the water since if a stream of clean water was allowed to flow over the surface before the measurement of contact angle was made the results were identical. The values used in the calculation of interfacial energies are shown in table 1.

The hysteresis of contact angle suggests that some rearrangement of the molecules constituting the monolayers must take place. Monomolecular layers formed by retraction from n-alkane solutions contain a high proportion of solvent molecules unless the samples remain immersed for a long time.[9,10] Immersion times in these experiments did not usually exceed 15-30 s and one would expect under these circumstances that only about one-third of the absorbed molecules are acid molecules. Since the films are stable, these polar molecules must be uniformly distributed throughout the film allowing the solvent molecules to be anchored and oriented. A mixed film of this kind would present a surface consisting primarily of CH_3 groups in air or in a non-polar environment. When, on the other hand, the environment is a polar liquid, some of the acid molecules reorient so that the head group is exposed to the liquid. That such overturning and mobility between adjacent monolayers can take place has been adequately demonstrated by many workers, e.g., Gaines.[11]

TABLE 2.—CONTACT ANGLE MEASUREMENTS ON MONOMOLECULAR FILMS DEPOSITED ON MICA

method	monolayer	water		decane		urea
		θ_A	θ_R	θ_A	θ_R	θ_A
sessile drop	stearic acid	36°	6°	31°		36°
	lauric acid	10°	2°	14°		10°
	perfluorodecanoic acid	27°	1°	74°		27°
captive bubble	stearic acid	$65\frac{1}{2}$°	35°-42°	31°		$65\frac{1}{2}$°
	lauric acid	60°	42°	14°		60°
	perfluorodecanoic acid	79°	42°	74°		79°

Since the film does not become detached, only a proportion of the molecules can be in this state at any time. The surface of mica which has been exposed to moist air is covered with a strongly adsorbed monolayer of water, so it is likely that, on the average, half of the acid molecules in a film immersed in water will be oriented towards the bulk liquid. For a gaseous environment the proportion will be lower. Advancing contact angles were reproducible to ± 1° for measurements made with several bubbles on different parts of the same solid sample; for successive measurements on the same region of a sample, and for different samples. A spread of 5-7° was observed on measurements of the receding angle.

The measured values of the monolayer/liquid interfacial energies are lower than we, at first, expected. If the monolayers presented a surface composed entirely of methyl groups to the liquid phase one might expect that the interfacial energy would have a value intermediate between that of the solid and the liquid.[12-15] The overturning of some of the polar molecules is playing a part in the actual value of the interfacial energy with water. Similar effects are observed in the measurements of liquid/liquid interfacial energies. Table 3 shows results for such systems taken from the *Int. Crit. Tables*, from which it may be seen that the polar liquids have an interfacial energy value with water which lies below that of either pure liquid with air, while the nonpolar liquids produce an intermediate value.

By measuring the energy of formation of hydrocarbon/water interfaces it should be possible to obtain information about hydrophobic interactions. Non-polar molecules dispersed in water show a tendency to associate. The thermodynamics of transfer of simple hydrophobic molecules from a non-polar to an aqueous medium have been discussed by several authors. Such a change is invariably accompanied by a large decrease in the entropy of the system and a decrease in the specific molar volume.[16-18] These effects are now interpreted in terms of structural changes occurring in the water in the immediate neighbourhood of the non-polar molecules.   Each non-polar chain is thought to be surrounded by a thin layer of water in a quasi-crystalline state and the increase in entropy, consequent on the destruction of this interfacial region, favours aggregation.

TABLE 3.—INTERFACIAL ENERGIES FOR LIQUID IN CONTACT WITH WATER TAKEN FROM INTER-NATIONAL CRITICAL TABLES

liquid	liquid/liquid interfacial energy ergs cm^{-2}	liquid/air interfacial energy ergs cm^{-2}
benzene	35	28.9
carbon tetrachloride	45	26.9
n-hexane	51.1	18.5
n-octane	50.8	21.8
n-octylalcohol	8.5	27.5
diethylether	10.7	17.0

When the experiments using monolayers were carried out in water the measured values of γ_{SL} were, in general, larger than those predicted by Young's equation.   The average discrepancy was 7.5 ergs/cm^{-2}.   The equilibrium of the three phases may then be expressed by adding a term γ_H to account for these hydrophobic effects, thus

$$\gamma_{SL} = \gamma_{SV} - \gamma_{SL} \cos\theta + \gamma_H,$$

where γ_H then has the value 7.5 ergs/cm^{-2}.   In order to compare our results with the measurements of Frank and Evans,[17] models of the interfacial region were constructed.   C.P.K. space-filling construction models were used since the aim of the model building was to see how the clustering and orientation of water molecules in the region affects the availability of space for a molecule of water at the interface.

Models of the monolayers consisted of a planar array of close-packed hydrocarbon chains, one-third of the molecules having carboxyl end-groups.   Half of the end groups oriented upwards and we allowed them to take a number of random positions in the plane, so that sometimes the polar heads occurred in groups and at other times they were more uniformly distributed.   The remainder of the molecules were n-alkane molecules aligned symmetrically along the length of the fatty acid chains.

Clusters and chains of hydrogen-bonded water molecules were also assembled, paying as much attention as possible to the average number of hydrogen bonds per molecule as suggested by the work of Senior and Vand.[19-21]   These were then placed in contact with the model interface in such a way that, in the neighbourhood of a non-polar molecule, the water molecules were all oriented with their hydrogen bonds towards the bulk water while near a polar end-group they were allowed to assume any orientation.   This constraint limits the possible shape of the clusters. It provides hollows into which parts of the somewhat bumpy hydrocarbon surface can fit and excludes other parts from participation.   These constructions were carried out many times, and each may be regarded as representing a possible array at any particular instant, since the process must be a dynamic one.   The number of water molecules in the first layer which were oriented with their hydrogen bonds away from the monolayer was counted in each case and an average taken.   One may then estimate the number of oriented water molecules per unit area of interface and associate with them the excess energy measured in the cleavage experiments.   This has the value 213 cal/(mol of oriented water).

In order to see what this predicts for the hydrophobic contribution in experiments such as those of Frank and Evans, it was necessary to construct similar models of oriented clusters to simulate the conditions in their experiments. They used n-alkanes having 1-4 carbon atoms in the chain. For C_4H_{10} the construction was carried out with the molecule either fully extended or folded as compactly as possible so as to present the least interfacial area. From these constructions we may estimate the number of moles of water which orients when one mole of hydrocarbon is dissolved in water and hence predict a value for the free energy ΔF involved. This assumes that the solution is so dilute that the hydrocarbon molecules exist in solution as single isolated entities.

Table 4 shows our predicted values together with the results of Frank and Evans; the agreement is remarkably good. For C_4H_{10} the prediction indicates that most of the molecules are fully extended. This is unlikely to be the case, but the presence of a small proportion of aggregated molecules would also lead to this result.

TABLE 4.—PREDICTED FREE ENERGY CHANGES COMPARED WITH RESULTS OF MEASUREMENTS OF FRANK AND EVANS

system (Frank and Evans)	ΔF cal/mol	n	ΔF predicted cal/mol
CH_4 (a) benzene to water	2,600	15.0	3,200
(b) ether to water	3,300		
(c) carbontetrachloride to water	2,900		
C_2H_6 in (a) benzene to water	3,800	17.0	3,620
(b) carbontetrachloride to water	3,700		
C_3H_8 liquid to water	5,050	23.6	5,030
C_4H_{10} liquid to water	5,850	26.3-28.0	5,600-5,960

n = average number of oriented water molecules/hydrocarbon molecule.

Hydrophobic interactions are thought to be important in determining the tertiary structure of proteins. Urea is effective in causing denaturation of these molecules although the mechanism by which it does so is not clear. Fig. 4(a) and (b) shows curves obtained when mica covered with a monomolecular film of stearic acid was cleaved in a concentrated solution of urea. The results give a value of 4.5 ergs/cm^{-2} for γ_{SL}, while that predicted from Young's equation is 5 ergs/cm^{-2}. It seems therefore that $\gamma_H = 0$ in this case.

DISCUSSION

The results show that Young's equation holds for a number of solid-liquid-vapour systems, provided that the interfaces and the phases do not undergo any fundamental changes. With the hydrophobic monolayers in water the monolayers themselves become partially reoriented and cause a lowering of the contact angle and the interfacial energy when compared with values to be expected for a hydrocarbon/water interface. Young's equation did not hold in this case, and it was found necessary to add a term to restore the balance.

This additional energy has been associated with entropic effects taking place in the immediate neighbourhood of the hydrophobic interface. By means of model-building, the energy has been estimated as 213 cal/mol of oriented water, and the results are in good agreement with those of other workers.

The results also shed some light on processes occurring in the water when urea is added to it. Water has a fairly open structure and many authors have discussed the solubility of short-chain hydrocarbons in terms of their abilty to be accommodated in the cavities in the structure. The increased solubility of these substances in the presence of urea is thought to result from the participation of urea in the formation of clusters of molecules with even larger cavities between them so that more or larger solute molecules may be contained within them.

In these experiments the non-polar material is in a special form.   It is anchored at a solid-liquid interface and presents an extended sheet to the solution.   The surface is nevertheless bumpy on a molecular scale and cavities in the water structure certainly facilitate the formation of a layer which is in good contact with the methyl groups.   The disappearance of γ_H for urea is difficult to explain simply on the basis of increased cavity size, since it is difficult to see how this can affect the ease with which the sheet can be accommodated in the structure.   While this may well be a factor which determines the increased solubility of hydrocarbons in urea solutions, these observations indicate that the medium is now homogeneous and implies a breakdown of the ordered structure of the water consistent with current concepts of denaturant action.

We thank our colleagues at the P.C.S. Laboratory, Cambridge, and the M.R.C. Biophysics Research Unit, King's College, London, for many helpful discussions, in particular, with Dr. D. Tabor and Dr. W Gratzer.   We also thank the Medical Research Council for financial support.

[1] R. E. Johnson, *J. Phys. Chem.*, 1959, **63**, 1655.
[2] J. W. Obreimoff, *Proc. Roy. Soc. A*, 1930, **127**, 290.
[3] A. I. Bailey, *Proc. 2nd Int. Congr. Surface Activity*, 1957, **3**, 406.
[4] A. I. Bailey and S. M. Kay, *Proc. Roy. Soc. A*, 1967, **301**, 47.
[5] A. I. Bailey and A. G. Price, *J. Chem. Phys.*, 1970, in press.
[6] W. C. Bigelow, D. L. Pickett and W. A. Zisman, *J. Colloid Sci.*, 1946, **1**, 513.
[7] S. Tolansky, *Multiple-beam Interferometry of Surfaces and Films*, (Oxford Univ. Press, 1948).
[8] A. W. Adamson and I. Ling, *Adv. Chem. Ser.*, 1964, **43**, 57.
[9] R. T. Mathieson, *Nature*, 1959, **183**, 1803.
[10] L. O. Brockway and R. L. Jones, *Adv. Chem. Ser.*, 1964, **43**, 275.
[11] G. L. Gaines, Jr., *J. Colloid Sci.*, 1960, **15**, 321.
[12] F. M. Fowkes, *Adv. Chem. Ser.*, 1964, **43**, 99.
[13] L. A. Girifalco and R. J. Good, *J. Phys. Chem.*, 1957, **61**, 904.
[14] R. J. Good, L. A. Girifalco and G. Kraus, *J. Phys. Chem.*, 1958, **62**, 1418.
[15] R. J. Good and L. A. Girifalco, *J. Phys. Chem.*, 1960, **64**, 561.
[16] J. A. V. Butler, *Trans. Faraday Soc.*, 1937, **33**, 235.
[17] H. S. Frank and M. W. Evans, *J. Chem. Phys.*, 1945, **13**, 507.
[18] W. L. Masterson, *J. Chem. Phys.*, 1954, **22**, 1830.
[19] W. A. Senior and V. Vand, *J. Chem. Phys.*, 1965, **43**, 1869.
[20] V. Vand and W. A. Senior, *J. Chem. Phys.*, 1965, **43**, 1873.
[21] V. Vand and W. A. Senior, *J. Chem. Phys.*, 1965, **43**, 1878.

ANITA L. BAILEY, ANDREA O. PRICE AND SUSAN M. KAY   127

GENERAL DISCUSSION

Prof. J. Th. G. Overbeek (*University of Utrecht*) said: What is the radius of curvature of the " spherically formed plate " used by Peschel? Is the plate distance h the shortest distance between the planar and the spherical plate?

Dr. G. Peschel (*Universität Würzburg*) said: In reply to Overbeek, the curvature radius of the spherically formed plate was $R = 100$ cm. The plate distance h is the shortest distance between the plates, i.e., it is the distance between the tops of the surface asperities which come into contact when the plates come into contact. Our theory takes account of this fact.

Dr. K. J. Padday (*Kodak Ltd., Harrow*) said: The data of fig. 1b, 2a and 2b of Aldfinger and Peschel seem to indicate that at some temperature either above or below that at which maxima appear, the disjoining pressure decreases to very small values. Although the scale of the ordinates of these graphs do not permit accurate interpolation or extrapolation, the data suggest that in the low pressure region the disjoining pressure is nearly independent of thickness of the layer. Is this a correct interpretation and if so does it mean that the function of the thickness which gives the disjoining pressure changes with temperature?

Prof. L. E. Scriven (*University of Minnesota, Minneapolis*) said: With regard to the paper by Deryaguin *et al.*, from optical measurements by the modulation-polarimetric method, profiles of thin liquid films can be calculated, as the authors report here in ref. (1). One wonders if the profiles after blowing should be interpreted in terms of a simple hydrodynamic model from which the effects of van der Waals' and electrostatic forces have been excluded. It may be asked whether the experimental and electrostatic forces have been excluded. It may be asked whether the experimental measurements are of phenomena peculiar to a thin liquid film bounded by roughly parallel gas/liquid and liquid/solid interfaces, rather than to the boundary region of bulk liquid in contact with solid. Have the authors considered alternative models in which the forces are appropriately accounted for (and in which viscosity could accordingly vary with film thickness without depending on distance from the solid surface)? What are the r.m.s. asperity height and aspect ratio of the " heavy flint " glass prism and of the USSR-14b-surface-finish steel plate used in the experiments?

In previous publications the authors have presented film profiles but seem not to have calculated viscosities of film liquids, one reason evidently being difficulty in locating precisely the " wetting boundary ", i.e., the three-phase contact line at $x = 0$ in their fig. 1. It would be of some interest, therefore, to have available the analysis of the curves in their fig. 3, and the capillary viscometer measurement, both mentioned in the paper. Of special interest would be observations of the behaviour of the wetting boundary during the blowing process. (Are the coordinate axes of fig. 3 labelled correctly?)

Subtle volatility effects have been implicated for anomalous profiles of spreading films and draining films of polymethylsiloxane liquids as well as many other liquids

[1] *Proc. 2nd Int. Congr. Surface Activity*, 1957, **3**, 531.

of low volatility.[1]   One therefore wonders if the authors have considered the possibility that differential depletion of slightly more volatile film fractions by the air-stream in the " blow-off " method might, through surface tension variation or other effects, contribute to anomalies in film profiles, particularly when binary solutions are investigated.   If surface velocity of the film does indeed increase with distance down-stream in the air-stream, the gas/liquid interface suffers a little dilatation that might influence vaporization and other interfacial phenomena.

Prof. B. V. Deryaguin and **Prof. V. V. Karasev** (*Inst. Phys. Chem., U.S.S.R. Acad. Sci., Moscow, U.S.S.R.*) (*communicated*) : In reply to Scriven, for non-polar liquids the linear velocity profile is observed, hence electrostatic and van der Waals forces alone cannot effect the deviation from normal viscosity.   Forces of attraction to the substrate do not influence directly the film flow as they are normal to the flow velocity. Indirect effect of pressure via viscosity is vanishingly small due to smallness of pressure. A curved film profile, being caused by microroughness of the substrate, the linear profile would not be observed also with clean liquid petrolatum.   Thus, microroughness does not influence the film profile.

The dependence of the relative viscosity on the distance from the solid substrate is the main interest of this work.   In several measurements presented in the report and in some previous papers we compared bulk viscosity obtained by using the capillary viscometer and by the blow-off method.   Coincidence of values is obtained with accuracy to $\sim 3\%$.   According to the formula $\eta = A \tan \phi$, viscosity depends only on the slope of the profile and does not depend on the wetting boundary position. Moreover, in all cases the wetting boundary was immobile before, during and after the experiment.   It is not difficult to fix the wetting boundary position.   For complete wetting this boundary does not change its position neither during the blow-off process nor during measurements.   The plot of l is equivalent to an X plot.

The last question concerning volatility and surface tension effects upon film profile can be answered in the following way : all polydimethylsiloxanes were treated in vacuum to distil off all volatile components ; these polydimethylsiloxanes under normal conditions ($t = 20°C$; $p = 760$ mm) are non-volatile liquids.   In addition, evaporation being supposed to take place, the steps length greatly exceeds the wetting boundary shift (which was demonstrated in Bascom's work).

In our experiments, the wetting boundary retained its position even during the blow-off process.   In Bascom's work, in the preliminary vacuum treatment, the spreading also stopped.   The viscosity anomaly and profile distortion connected with it cannot be explained by a surface tension effect as the liquids are non-volatile and for the binary mixture, the surface tensions of both components are almost the same : PMS–10–19.5 dyn/cm ; PMS–2000–20.8 dyn/cm.   The last remark about surface evaporation being influenced by extension is not clear.

Dr. A. J. Smith and **Dr. A. Cameron** (*Dept. Mech. Eng., Imperial College*) said: We have used a 100 cm^3 silicone at 70°C in the apparatus described by us, and investigated the thinning round a steel ball of American specification 52100 or British EN31. It is interesting to note that we find the same type of behaviour as Deryaguin *et al.* do but it occurs at a very much bigger film thickness.   In view of the findings in our own paper, it is interesting to note that the thicknesses at which this effect occurs are in the same order as the reactivities of the substrates.   Glass gives the thinnest film.   The authors' steel, whose composition we do not know, but is presumably a

[1] W. D. Bascom, R. L. Cottington and C. R. Singleterry, *Adv. Chem. Series*, no. 43, pp. 355-9, (Amer. Chem. Soc., Washington, D.C., 1964).

martensitic stainless steel, comes next, and finally the steel we use is the most reactive. We would like the authors' comments on the possibility of the dimethylsiloxane fluids containing some kind of reactive constituent which could have formed the gel-like layer described in our paper.

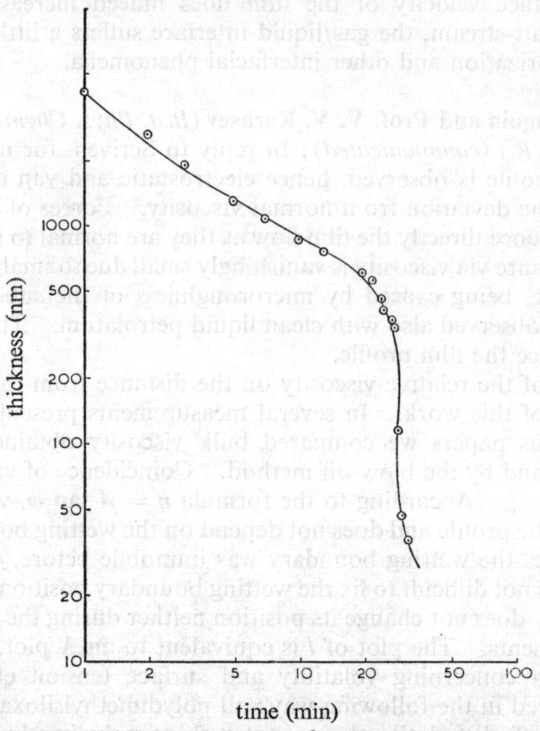

Fig. 1.—Thinning curve of 100 cm³ polydimethylsiloxane on steel.

Prof. S. G. Mason (*McGill University, Canada*) said: As I understand Deryaguin's blow-off method, the tangential stress is taken to be the same on both sides of the air/liquid interface, with the implied assumption that the surface tension σ is independent of the film thickness down to the edge $x = 0$, $h = 0$ (fig. 1). However, if $(d\sigma/dh) \neq 0$, a portion of the tangential stress developed by the air can be borne by the interface itself; this would presumably give rise to an *apparent* change in viscosity as the film becomes thinner even when no such change occurs. How is the effect of changing surface tension allowed for in calculating the viscosity?

Prof. B. V. Deryaguin and **Prof. V. V. Karasev** (*Inst. Phys. Chem., U.S.S.R. Acad. Sci., Moscow, U.S.S.R.*) (*communicated*): In reply to Mason, first the direct effect of surface tension " gradient " $\partial\sigma/\partial h$ is extremely small or even zero in all our earlier experiments and in those under discussion. The reasons are: (i) this effect is absent for one-component films and in general very weak for organic liquids because in such systems the adsorption or surface activity at the liquid air interface is very small. (ii) The effect mentioned by Mason actually depends on the derivative

$$\frac{\partial\sigma}{\partial l} = \frac{\partial\sigma}{\partial h}\frac{dh}{dl}$$

where l is tangential distance along the film surface. Due to the factor dh/dl which is of the order of magnitude of 10^{-3}-10^{-4} this derivative is vanishingly small.

In reply to Scriven, our " model " of boundary viscosity, which depends on the wall distance independently of the distance from the film/air interface, is based on the very short-range action of the latter interface in comparison with that of liquid/solid interface. The independence of our results on the heights of asperities gave no opportunity for the estimation of asperity heights and profiles of our substrates.

In reply to Smith and Cameron, our dimethylsiloxane fluids contained no constituents that would react with our stainless steel. Besides such reactivity contradicts the decreasing of viscosity of dimethylsiloxane in the immediate proximity of the substrate.

Dr. S. G. Ash (*Thornton Res. Centre, Chester*) said: In the analysis of the density measurements and heats of wetting presented in the paper of Ash and Findenegg, the surface excess of cyclohexane was assumed to be zero. It is possible to analyze the same results without making this assumption, and so calculate the surface excess mass of cyclohexane. The surface excess mass of a pure liquid in contact with Graphon is calculated from the experimentally determined quantities using the equation:

$$\frac{S\Gamma_1}{\rho_1^l} = \frac{m_1^l}{m_1^g \rho_1^l} - \frac{v_1^p}{m_1^g} + \frac{1}{\rho_g}, \tag{1}$$

where Γ_1 is the surface excess mass of liquid; S the specific surface area of the Graphon; ρ_1^l the bulk liquid density; m_1^l, m_1^g the mass of liquid and Graphon in the pyknometer, v_1^p the volume of the pyknometer; ρ_g the absolute density of Graphon. Similarly, for a second liquid,

$$\frac{S\Gamma_2}{\rho_2^l} = \frac{m_2^l}{m_2^g \rho_2^l} - \frac{v_2^p}{m_2^g} + \frac{1}{\rho_g}. \tag{2}$$

All quantities in these expressions for Γ are measured except the density ρ_g of Graphon. This can be eliminated from eqn (1) and (2) to give a relationship between Γ_1 and Γ_2.

On the basis of the solid-like surface zone model, we equate the total heat ΔH_w of wetting of Graphon by a pure liquid to the sum of the contributions arising from the liquid-Graphon interactions H_w, and the structuring of the surface zone, $n^s H_f$ (n^s = number of moles of component in surface zone; H_f = heat of fusion of component). We identify liquid 1 with an n-alkane and liquid 2 with cyclohexane, and assume that H_w is the same in both systems. Then,

$$(\Delta H_w)_1 = H_w + n_1^s (H_f)_1; \tag{3}$$

$$(\Delta H_w)_2 = H_w + n_2^s (H_f)_2. \tag{4}$$

The quantity H_w can be eliminated from eqn (3) and (4) to give an expression relating n_1^s and n_2^s. The surface excess mass is related to the capacity of the solid-like surface zone by

$$\Gamma = n^s M (1 - \rho^l / \rho^s), \tag{5}$$

where M is the molecular weight and ρ^s the solid density of the hydrocarbon. The values of n_1^s, n_2^s (and Γ_1, Γ_2) can be calculated from the set of simultaneous eqn (1)-(5). This analysis is equivalent to ascribing the discrepancy between the structural contribution to the heat of wetting and the structural contribution calculated from the density measurements to the presence of a solid-like surface zone of cyclohexane (see fig. 2, 3 and 4 of the paper by Ash and Findenegg).

The calculated thickness of the solid-like surface zone of cyclohexane is $h_s = 0.5 (\pm 0.5)$ Å, ($T = 25$-$40°C$), and $\Gamma_{\text{cyclohexane}} = 0.4 \times 10^{-9} (\pm 0.4 \times 10^{-9})$ g cm^{-2}.

The absolute values of the surface excess mass of other liquids can be calculated from the excess mass quoted in the paper (relative to $\Gamma_{\text{cyclohexane}} = 0$) by the expression

$$\Gamma_{\text{abs}} = \Gamma_{\text{rel}} + 0.5 \times 10^{-9} \, \rho^{\text{l}}.$$

The structural contribution to the total heat of wetting of Graphon by cyclohexane is

$$(n^s H_f)_{\text{cyclohexane}} = 1.4(\pm 1.4) \text{ erg cm}^{-2},$$

which is only $1.3(\pm 1.3)$ % of the total heat.

Prof. J. Th. G. Overbeek (*University of Utrecht*) said: With regard to the paper by Ash and Findenegg, if a water-vapour interface has a contact angle on Graphon it is possible that near points where the Graphon particles touch or come very close together a vapour bubble is formed bounded by menisci with their convex sides towards the vapour. At a radius of curvature of about 1 μm such a meniscus could withstand a pressure difference of about 1 atm, thus bringing the liquid phase at 1 atm and the vapour phase at a very low pressure in equilibrium with one another. This would explain the negative excess volume of water (and of benzene).

Dr. G. H. Findenegg (*University of Vienna, Austria*) (*partly communicated*): We agree with Overbeek that the negative surface concentration of water at the Graphon/liquid interface could be caused by small vapour bubbles. From our results the vapour volume would be 2.8×10^{-3} cm^3 g^{-1} Graphon and hence there could be a very large number of small bubbles (10^7 or more per g Graphon) which would explain the good reproducibility of our results. We do not think that the surface concentration of benzene can be explained by the same effect because Graphon is completely wetted by benzene.

Dr. J. W. White (*Phys. Chem. Lab., Oxford*) said: With regard to the paper by Ash and Findenegg on the basis of recent neutron inelastic scattering measurements [1,2] I can comment on the *range* and *magnitude* of intermolecular forces likely between graphite and adsorbed alkanes and between the alkanes themselves. First, as to the range, for pyrolytic graphite the forces between the basal planes and their range has been measured by Dolling and Brockhouse.[3] The dispersion curve for lattice vibrations perpendicular to the sheets was measured and is a strong test of the range of the interatomic potential.[4] An almost exact sine curve was obtained which fits the theoretical model for nearest neighbour forces only. The contribution of next nearest neighbour planes to the total (attractive) force constant was only *ca.* 1.5 %. Our recent measurements of phonons along the [110] crystallographic direction in polyethylene and in the basal planes of hexadecane crystals indicates again that the lateral intermolecular forces are almost exclusively between nearest neighbours.

The magnitude of the force also comes from the neutron experiments and may be characterized by the elastic (Young's) moduli derived from the limiting slopes of the dispersion curves near the centre of the Brillouin zone (i.e., for long wavelength vibrations). To give a sense of scale the modulus of diamond ((110) planes), $E = 250 \times 10^{10}$ dyn cm^{-2} (25×10^{10} N m^{-2}) is mentioned. For the direction perpendicular to the graphite basal planes $E \approx C_{33} = 39 \times 10^{10}$ dyn cm^{-2} (3.9×10^{10} N m^{-2}). From the polyethylene data we find that $E = 6 \times 10^{10}$ dyn cm^{-2} (6×10^9 N m^{-2}) for paraffin chains separated by 4.1 Å (0.41 nm). The lowest modulus we

[1] J. F. Twisleton and J. W. White, *Mol. Phys.*, to be submitted.
[2] J. F. Twisleton, *Thesis*, (Oxford University, 1970).
[3] G. Dolling and B. N. Brockhouse, *Phys. Rev.*, 1962, **128**, 1120.
[4] A. J. E. Foreman and W. M. Lomer, *Proc. Phys. Soc. B*, 1957, **70**, 1143.

have found so far for these systems is in the hexadecane basal plane where $E = 1.5 \times 10^{10}$ dyn cm^{-2} (0.15 Nm^{-2}). By assuming an additivity these numbers serve as a guide for the orders of magnitude to be expected for the adsorption force and even may be used to estimate the wetting enthalpies. It emerges that these nearest neighbour estimates exceed the measured enthalpies by about an order of magnitude. This suggests the need for an upwards reassessment of the strength of pairwise forces rather than to postulate long-range interactions. The neutron data also strongly suggest that a change to " head stacked " adsorption (with the hydrocarbon perpendicular to the graphite) would only occur at very high coverage because of the advantageously high " mean energy " of attachment to graphite rather than to other alkanes.

Finally, there are possible neutron scattering experiments which could determine the intermolecular constants on Graphon surfaces. Neutron diffraction experiments would give the average dimensions of a group of ordered molecules from the sharpness of diffraction, and inelastic scattering from phonons in the adsorbed layers would give not only a measure of this (again from widths) but a measure of the modified intermolecular forces and therefore the mean intermolecular separations transverse to the chain axis.

Prof. R. J. Good (*Bristol University*) said: The results of Ash and Findenegg point to an important conclusion about the use of alkanes to determine the value of γ_s from contact angle data. If spreading pressure Π_e is negligible, then [1]

$$\gamma_s = \gamma_l(1 + \cos\theta)^2/4\Phi_{sl}^2: \tag{1}$$

For a non-polar liquid on a non-polar solid,[2]

$$\Phi_{sl} = (I_lI_s)^{\frac{1}{2}}/\tfrac{1}{2}(I_l + I_s), \tag{2}$$

where I is ionization energy. γ_s, when determined in this manner using liquid alkanes, has been described as the dispersion component of the surface tension of the solid.[3] It is better [4] identified as one half the free energy of cohesion across the specified crystallographic plane. In the present case, it could be written, γ_{0001}, i.e., referring to the basal plane of graphite. When the solid is covered by an adsorbed film, γ_s is one half the free energy of cohesion of the film-covered solid across the plane which is specified as

solid | monolayer (head-tail) | | monolayer (tail-head) | solid,

i.e., across the surface indicated by the double vertical lines. It is a property of the solid only; and Φ_{ls} should be very nearly constant for a series of alkanes in a single solid. It has recently been found [5] that for numerous low-energy surfaces' γ_s determined in this manner shows a trend with chain length of the alkane. Fowkes [3] attributes the trend in γ_s to the variation in density of the hydrocarbon with chain length. I prefer the conclusion that there is a trend in Φ with chain length, which is not represented by eqn (2).

Now, eqn (2) is based on the premise that differences in the structure of the liquid at the liquid/solid interface and at the liquid/vapour interface can be neglected. While this hypothesis can be valid for a liquid such as cyclohexane, the data reported by Ash and Findenegg show that it is not so for alkanes on graphite. Consequently there should be a trend of Φ with chain length for alkanes against graphite.

[1] R. J. Good and L. A. Girifalco, *J. Phys. Chem.*, 1960, **64**, 561.
[2] R. J. Good, *Adv. Chem. Series*, 1964, **43**, 74.
[3] F. M. Fowkes, *Ind. Eng. Chem.*, 1964, **56**, 40.
[4] R. J. Good, in *Wetting*, (Soc. Chem. Ind., London, 1967).
[5] A. W. Neumann and R. J. Good, to be published.

Accordingly, I suggest that Findenegg and his coworkers perform volumetric studies on other solids, for which contact angle measurements are available. Teflon would be an interesting substrate.

Dr. G. Peschel (*Universität Würzburg*) said: The fact that Ash and Findenegg used cyclohexane as a reference liquid showing no surface zone anomalies might be incompatible with our results at first sight for we found a disjoining pressure in thin layers of cyclohexane between fused silica plates. In one case the plates were fully covered with hydroxyl groups, in the other case the plates were free from hydroxyl groups, achieved by baking out the plates at about 850°C. In the first case, the disjoining pressure ranging between the melting point and about 15°C was larger than in the second case where the surfaces were less polar. It may be that Graphon surfaces only produce a negligible disjoining pressure in adjacent cyclohexane layers so that a pronounced discrepancy between his and our work is non-existent.

Mr. A. J. Groszek (*B.P. Co. Ltd., Sunbury*) said: With regard to the paper by Ash and Findenegg, I have established recently that long-chain n-paraffins are adsorbed very strongly from solutions in various volatile solvents on the basal plane surface of graphite. For all the graphites investigated the long-chain n-paraffins saturate the basal planes at low solution concentration (ranging from *ca.* 0.001 mol % for n-C_{32} and 10 for n-C_{16}) with the formation of distinct plateaux regions.[1] From the amount of adsorption of n-C_{32} it was inferred that the n-paraffin molecules lie flat on the basal plane surface. The data indicate that the n-C_{32} molecules are closely packed so that each H atom in a methylene group in contact with the surface occupies a carbon hexagon or $5.6 Å^2$. Similar results have been obtained for n-hexadecane at higher concentrations.[2, 3] The heat of displacement of n-heptane by n-hexadecane suggested strongly that at a mol fraction of about 0.1 n-hexadecane forms a monolayer composed of molecules lying flat on the basal planes as indicated by Aveyard.[3] The heats subsequently increase, unlike the amount of adsorption, to the value agreeing with the high heat of immersion of Graphon in n-heptane discussed by Ash *et al.*

The strong preferential adsorption of n-paraffins is very characteristic for graphite and has not been observed for solids having polar surfaces, e.g., the high heats of immersion are not observed for silica gel and carbon black.[4] In view of the fact that the strongly adsorbed n-paraffins are orientated parallel to the surface, it is most unlikely that such molecules would leave the graphite surface to produce closely-packed monolayers orientated normal to the surface. A much more energetically favourable situation would be for n-paraffin molecules to form several layers of molecules orientated parallel to the surface. The packing of n-paraffins in such multilayers would contribute to the heat of adsorption in the same way as in the film composed of vertically orientated molecules, but there would be no necessity for re-orientation of the first strongly adsorbed monolayers.

Another point concerns more specifically the difference in the heat of wetting of Graphon in n-heptane and n-hexadecane. We have determined this difference with the use of the flow-calorimeter for Graphon and other graphites by displacing n-heptane gradually with n-hexadecane and summing up all the heats of preferential adsorption so obtained.[2]

[1] A. J. Groszek, *Proc. Roy. Soc. A*, 1970, **314**, 473.
[2] G. I. Andrews and A. J. Groszek, *3rd Conf. Industrial Carbons and Graphite*, (Imperial College, London, 1970).
[3] R. Aveyard, *Trans. Faraday Soc.*, 1967, **63**, 2778.
[4] L. C. Robert, *Compt. rend.*, 1963, **256**, 655.

The difference obtained for a sample of Graphon in these experiments was 93 erg/cm^2, but for a more perfect graphite composed of thin flakes having a thickness of about 100 Å, the difference was much higher at 130 erg/cm^2. The important point arising from these experiments is that the heat of wetting of graphite in liquid paraffins depends on the type of graphite used and that the ordering of n-hexadecane is increased when the basal plane in graphite becomes more extensive.

Dr. J. K. Padday (*Kodak Ltd., Harrow*) said: Values γ_{SL} of the systems n-decane in contact with stearic, lauric or perfluorodecanoic acid monolayers of table 1 of the paper by Bailey and Price appear to be calculated from the measured values of angles of contact and γ_{SV} but with the important distinction that γ_{SV} refers to room air presumably in the absence of n-decane. Bangham and Razouk [1] have pointed out that the Young equation for this type of system should be written in the form

$$\gamma_{LV} \cos \theta = \gamma_S - \gamma_{SL} - \Pi,$$

where Π represents the lowering of the specific surface free energy of the solid by adsorption of vapour of the spreading liquid. The results of Bailey and Price indicate that in their system Π is very small or zero. Whilst this is not surprising with n-decane spreading on a monolayer of perfluorodecanoic acid the contact angle of which is large, it is surprising with the two fatty acid surfaces. The good agreement between advancing and receding contact angles of these two systems suggests that n-decane does not penetrate or otherwise alter the structure of monolayers of lauric or stearic acid on mica and that the systems behave ideally both in a thermodynamic and a physical sense.

Dr. A. I. Bailey (*Stuttgart*) said: In reply to Padday, the interfacial energy terms γ_{SV} in our work always refer to the modified surface, i.e., to a mica surface covered with adsorbed vapour films and/or adsorbed films of fatty acids. In the work of Bangham and Razouk, γ_S refers to the surface energy of the bare solid. The films were deposited by refraction from solution, i.e., after formation of the oriented monomolecular film, the solvent has no further tendency to adsorb and Π is negligible. When the adsorption takes place during the formation of the interfaces such as, e.g., takes place when clean mica is cleaved in a solution of polar material in non-polar solvent, then, as might be expected Young's equation is also not valid.

In reply to Cameron, we have not so far investigated films formed from solvents of different chain lengths. The effects described by Cameron are fascinating, and we will investigate them.

Dr. H. E. Ries (*Chicago*) said: I should like to know if, in any of the experiments performed by Bailey, Price and Kay, monolayers have been deposited on the mica by the Langmuir–Blodgett technique. I am afraid that other methods do not necessarily give close-packed films and that angles of contact can be somewhat misleading in this connection. By control of the surface pressures during Langmuir–Blodgett transfer, the nature of the transferred film is well established. [2]

Dr. A. I. Bailey (*Stuttgart*) said: In reply to Ries, we made several attempts to use Langmuir–Blodgett films. Our experience was that it was almost impossible to keep the specimens absolutely dust free during the time required for the deposition. It is also relatively easy to keep a few ml of solution dust free than the entire contents

[1] D. M. Bangham and R. I. Razouk, *Trans. Faraday Soc.*, 1937, **33**, 1459.
[2] H. E. Ries, Jr., and D. C. Walker, *J. Colloid Sci.*, 1961, **16**, 361.

of a trough.   In this experiment dust particles between the sheets cause distortions of the mica which affect the energy conditions.   Previous work by Courtney-Pratt in which the thickness of solvent deposited acid films was measured interferometrically showed that the thickness was consistent with the molecules having extended chains oriented perpendicular to the surface.   Since only a third of the molecules are acid molecules, if they alone occupied the surface, gaps between the molecules or tilting would result in a smaller measured thickness, so it is reasonable to suppose that the remainder of the film consists of adlineated solvent molecules, the whole film being close-packed or very nearly so.   The contact angles with water are low but consistent with our hypothesis of partial overturning of the polar molecules.   In any case, the contact angles and the energy measurements apply to the same film and so are consistent with each other.

Dr. B. A. Pethica (*Unilever Res., Port Sunlight*) said: In the paper by Bailey, Kay and Price it is proposed that for certain systems the Young equation be modified by addition of an extra term to include entropy effects.   To use their attractive terminology, " well-behaved " systems obey the Young equation.   There is no thermodynamic justification for putting extra terms in Young's equation, and the *apparent* failure of the equation is to be sought in the bad behaviour of the fatty acid film + water + mica system in which the equation is said to fail.   In this system the results were variable, which could be explained by migration of the fatty acid to the water vapour interface.   The consequent reduction in the interfacial tension would also explain the apparent failure of Young's equation.

Dr. A. I. Bailey (*Stuttgart*) said: In reply to Pethica, let us consider what would happen if the fatty acid migrated away to the water/vapour interface.   Each molecule which leaves would expose a region of clean mica/clean water interface which has a high interfacial energy.   During measurements of γ_{SL} we are not really concerned with what happens at the water/vapour interface.   This interface is several cm away from the specimen and the new solid/liquid interface is always formed in the bulk of the medium.   Capillary rise measurements made on water spread with lauric acid, for example, give the same values as clean water.   Now the absolute values of the measured interfacial energies are low and not consistent with this picture.   It seems also that the use of a reduced value of the interfacial energy of water, to restore the balance of Young's equation, only works if one assumes that the monolayer is both present on the solid surface and available to reduce the interfacial energy of the water. The overturning of some of the adsorbed molecules accounts for both the low values of γ_{SL} and the existence of the hysteresis of contact angle effects.

Dr. A. Cameron (*Mech. Eng. Dept., Imperial College*) said: I am always interested in the possibility of finding different methods of studying the effects of matching chain lengths of additives and carriers, such as we discussed in the paper by Askwith, Cameron and Crouch.[1]   Have Bailey *et al.* any further experimental evidence when the surfactant and carrier are precisely matched?.   It would be interesting if they could carry out some tests.   Their method is ideally suited for such studies.

Dr. K. W. Miller (*Dept. of Pharmacology, Oxford*) said: The results of Bailey's model building are most interesting and instructive.   The effect of the non-polar part of the monolayer on the adjacent water structure seems to be proportional to the area of hydrocarbon exposed.   A more sensitive measure of these structural

<hr>

[1] *Proc. Roy. Soc. A*, 1966, **291**, 500.

changes is the entropy, but it would be most difficult to obtain this by model building. The entropy of dissolving non-polar gases in water is known, however, with a fair degree of accuracy, and this entropy is closely related to the surface area of the gases dissolved.[1] This observation lends further credibility to Bailey's findings.

Her model building also illustrates the way that " hydrophobic bonding " originates, not so much from the hydrophobic surface itself as from the hydrophilic nature of the water molecule ; indeed, hydrophilic bonding would be a more accurate term. Bailey takes this into account by orientating her surface water molecules with their hydrogen bonds towards the bulk water. This constraint is equivalent to introducing more order into the system, thus decreasing its entropy. In statistical mechanical terms, this entropy decrease may be assigned to the loss at the surface of some of the many configurations that water molecules are free to take up in the bulk solvent. Present ideas about the nature of these configurations are of necessity crude, but the opposing requirements of packing closely around the hydrophobic surface whilst maintaining the maximum possible degree of hydrogen bonding will evidently give rise to the decrease in entropy that is usually, rather loosely, ascribed to an increase in structure.

Dr. A. I. Bailey (*Stuttgart*) said: A general comment on the model building described in our paper might answer some of the questions which arose in private discussion. The models were made in order to help to visualize what might happen when molecules, which form part of the normal water structure, encounter a hydrophobic entity. The only assumptions that we made with regard to the water molecules at the hydrocarbon interface were (i) that they should be hydrogen-bonded and participate in cluster formation, and (ii) that they should occupy all the space there unless prevented from doing so by the shape of clusters in the neighbourhood. These simple assumptions led to the re-orientation of molecules at the interface and to constraints on the shape of the clusters which contain these molecules. Hence, owing to the exclusion of some generally available states, the system has a lower entropy than for pure water. The increased order is not in the nature of a true phase change such as would result from a change in the intermolecular attraction, but rather it is due to a disturbance of the normal water structure. Perhaps the term " icebergs " used by some workers has been rather misleading. Similar effects presumably must take place also at the surface of water. The model building helps to give an idea of how, with the minimum of assumptions, the entropy changes observed by other workers, might arise. These ideas must represent an oversimplification, however, since they provide no explanation as to why some molecules are more hydrophobic than others.

[1] K. W. Miller and J. H. Hildebrand, *J. Amer. Chem. Soc.*, 1968, **90**, 3001.

Measurement of Forces between Colloidal Particles

By L. M. Barclay and R. H. Ottewill

School of Chemistry, University of Bristol, Bristol, England

Received 30th April, 1970

Apparatus has been constructed for the measurement of the pressure created by disperse systems as a function of the volume concentration of the disperse phase. Experiments with sodium montmorillonite as the colloidal system have enabled the force to be obtained as a function of the distance between the plates down to distances of the order of 10 Å. The results have been compared with those expected theoretically on the basis of the DLVO theory. The forces obtained at distances of less than 50 Å are much greater than those predicted by the theory and the additional force appears to arise from solvation effects in the thin liquid film between the particles. This has been confirmed by carrying out measurements in the presence of a non-ionic surface-active agent, a known stabilizing agent, where the repulsive forces due to solvation are enhanced.

Information on the stability of disperse systems can be obtained by both kinetic [1-3] and equilibrium methods.[4-7] In kinetic studies the electrolyte is added to the system and the rate at which the stability is lost is examined, i.e., the rate of flocculation is determined.[8] Although this approach has provided considerable information, the interpretation of the kinetic data is limited. There are therefore considerable advantages to studying a disperse system in its own environment under essentially equilibrium conditions. One method of carrying out such a study is to measure the pressure developed in the system as a function of the distance of separation of the particles. Some swelling pressure studies [4-9] have been used hitherto to obtain information of this sort. The type of apparatus previously used has been considerably refined and the present communication describes the apparatus used and some studies carried out using sodium montmorillonite dispersions in the presence and absence of a non-ionic surface active agent. Studies carried out using monodisperse polystyrene latex particles will be described in a later publication.

EXPERIMENTAL

MATERIALS

The distilled water used was doubly-distilled from an all-Pyrex apparatus. Sodium chloride was B.D.H. a.r. material which was roasted before use.

n-Dodecyl hexaoxyethylene glycol monether ($C_{12}E_6$) was prepared by the Williamson ether synthesis.[10] The critical micelle concentration determined by surface tension measurements was 7.25×10^{-5} M at 25°C (compare 7.8×10^{-5} M [11]).

The montmorillonite was a sample of montmorillonite no. 26 (Bentonite) from Clay Spar, Wyoming, as prepared for the American Petroleum Institute Clay Mineral Standards Project no. 49. About 30 g of the dry material was dispersed in 1 l. of distilled water containing 1 ml of 30 % v/v hydrogen peroxide per 100 g of clay. After standing for 24 h, the sample was centrifuged to remove impurities such as silica particles. Conversion to sodium montmorillonite was achieved by redispersing the clay particles in molar sodium chloride solution and allowing the dispersion to stand for several days. The sodium montmorillonite was separated by centrifugation at 38,000 g for 1 h. For compression studies the centrifuged material was dispersed in 10^{-1}, or 10^{-4} M, sodium chloride solution and then dialyzed against that solution for a long period.

ADSORPTION STUDIES

A dispersion of *ca.* 1.5 % w/w sodium montmorillonite which had been extensively dialyzed against twice-distilled water was used as a stock dispersion. Various weights ranging from 0.4 to 1.5 g of this stock dispersion were added to a series of $C_{12}E_6$ solutions having concentrations in the range 10^{-4}-10^{-3} M and a volume of 25 ml; these solutions were contained in stoppered tubes. The mixtures were equilibrated at $25 \pm 0.05°C$ for 3 days with frequent shaking. The surface tension of the mixture was used to determine the equilibrium concentration of $C_{12}E_6$. A comparison of the surface tensions of the supernatants of a series of mixtures which had been centrifuged at 38,000 g with those of the original mixtures showed them to be identical. Hence the centrifugation procedure was dispensed with in later experiments.

APPARATUS FOR COMPRESSION STUDIES

The cell used for pressure measurements which was constructed of stainless steel is shown diagrammatically in fig. 1. All sharp edges were rounded off to prevent damage

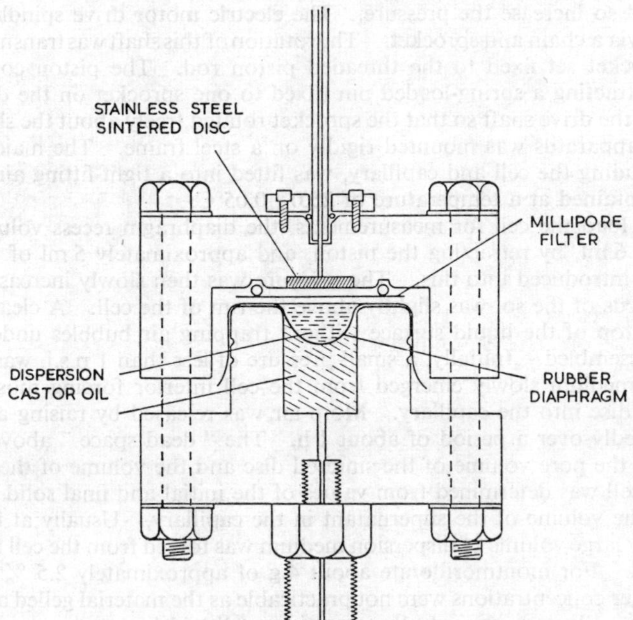

FIG. 1.—Cross-sectional diagram of compression cell.

occurring to the rubber diaphragm and the Millipore filters. Similarly, the end of the glass capillary, which was inserted into the cell, was bevelled in order to avoid chipping the tip when assembling and dismantling the cell. A PTFE O-ring placed between the capillary tip and the end of the steel orifice also minimized damage to the capillary. The capillaries were constructed from calibrated lengths of 3 mm diam. (inside) Viridia glass tubing. The stainless steel disc (porosity no. 3) fitted exactly into a recess in the top of the cell.

Samples of rubber latex for the cell diaphragm were selected from a large sheet, 0.007-0.010 in. thick (Holdfast Rubber Dam) by stretching pieces in front of a light source. Only sections free from bubbles within the rubber were used.   A calibration curve was constructed 'for each diaphragm' of the force applied by the stretched rubber to the sol.

Millipore MF filters (47 mm diam., 0.1 μ pore diam.) were used as membranes; the Viskase dialysis tubing used by some authors [9] was found to be unsuitable as a membrane and led to a very slow attainment of equilibrium.   Before use, approximately 4 l. of distilled water at 60°C were passed through each filter in order to remove impurities such as non-ionic surface-active agents and polyhydric alcohols.   The removal of surface-active material was monitored by measuring the surface tension of the filtrate.   After thorough washing the filters were dried at 70°C between filter papers held between perforated metal plates.

In order to generate pressure in the system, a Budenberg hydraulic gauge tester was used with caster oil as the hydraulic fluid.   The pressure applied was read directly on direct mounting Budenberg bronze tube Bourdon gauges.   The gauges were used to cover the ranges 0-20, 0-160 and 0-1600 p.s.i.   The 0-20 p.s.i. gauge was calibrated in position on the complete apparatus by connecting a mercury manometer to the hydraulic system.   The other two gauges were calibrated using a dead weight tester.

In order to maintain the selected pressure constant, a pressure compensating mechanism was built into the apparatus so that compensation occurred as the dispersion medium was forced out of the sol compartment.   For this purpose the pressure gauges were used as regulators.   A knife-edge metal contact was attached to the glass face of each gauge and a platinum contact to each of the gauge needle-pointers.   For each gauge the knife-edge contact and a suitable point on the chassis of the gauge were connected to an electric motor (Parvalux, 2 rev/min) which, by means of a chain drive mechanism, could rotate the piston drive wheel and so increase the pressure.   The electric motor drive spindle was connected to a drive shaft via a chain and sprocket.   The rotation of this shaft was transmitted via another chain and sprocket set fixed to the threaded piston rod.   The piston could be operated manually by retracting a spring-loaded pin (fixed to one sprocket on the drive shaft) from the key-way on the drive shaft so that the sprocket rotated freely about the shaft (see plate 1).

The whole apparatus was mounted rigidly on a steel frame.   The major section of the apparatus, including the cell and capillary, was fitted into a tight-fitting air-thermostat box which was maintained at a temperature of 25.0 ± 0.05°C.

In order to load the cell for measurements, the diaphragm recess volume was initially adjusted to *ca.* 6 ml, by retracting the piston, and approximately 5 ml of the degassed sol to be used was introduced into this.   The pressure was then slowly increased in the system until the meniscus of the sol was slightly above the rim of the cell.   A clean millipore filter was placed on top of the liquid surface without trapping air bubbles underneath it.   The cell was then assembled.   Initially, a small pressure of less than 1 p.s.i. was applied so that the dispersion medium slowly emerged from the cell interior forcing most of the air out of the sintered disc into the capillary.   More air was released by raising and lowering the pressure repeatedly over a period of about 1 h.   The " dead space " above the membrane which included the pore volume of the sintered disc and the volume of the exit aperture in the top of the cell was determined from values of the initial and final solid contents in conjunction with the volume of the supernatant in the capillary.   Usually at the beginning of each run a fairly large volume of dispersion medium was forced from the cell from a comparatively dilute sol.   For montmorillonite about 4 g of approximately 2.5 % w/w dispersions were used ; higher concentrations were not practicable as the material gelled at concentrations above 3 % w/w.   Since a relatively large volume of liquid had to be removed on the first compression the first equilibrium pressure reading often took up to 24 h.   The meniscus of the liquid in the capillary was observed with a cathetometer and equilibrium at a given pressure occurred when there was no further change in the position of the meniscus.

RESULTS

COMPRESSION STUDIES WITH MONTMORILLONITE

The curves of equilibrium pressure against distance of plate separation for montmorillonite in 10^{-4} and 10^{-1} M sodium chloride solutions are shown in fig. 2.

Plate 1 : Photograph of apparatus showing cell and compression mechanism.

[*To face page* 140.

Experimentally, the first compression on each sample gave a larger repulsion at a given distance up to a pressure in the region of 20 atm.   After the first compression, however, subsequent compression and decompression experiments gave coincident results as can be seen from fig. 2.   The distance between the plate surfaces, H_0, was calculated assuming a specific surface area of 800 m^2/g and using the expression $H_0 = 2V/mA$,

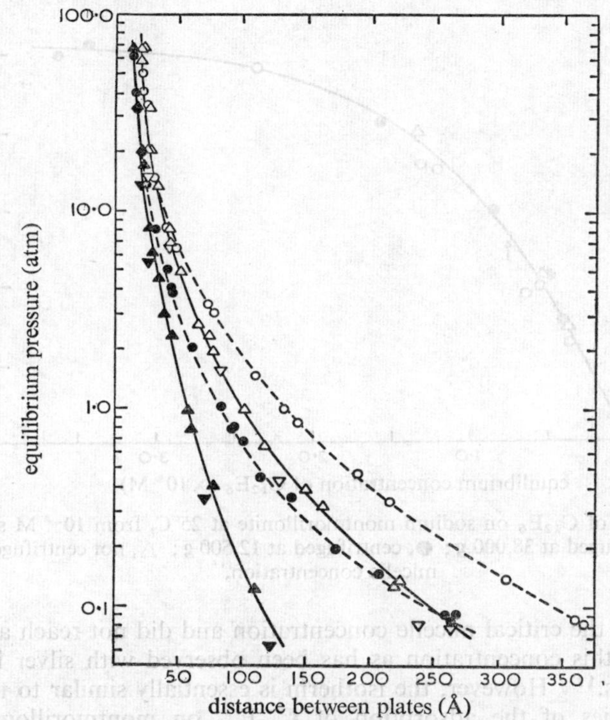

FIG. 2.—Equilibrium pressure against distance between plate surfaces for sodium montmorillonite dispersions at 25°C.   In 10^{-4} M sodium chloride solution: ○, first compression; ▽, decompression; △, subsequent recompression.   In 10^{-1} M sodium chloride solution: ●, first compression; ▼, decompression; ▲, subsequent recompression.

where V = volume of liquid in a sol containing m g of clay and A = specific surface area.   A check on the interplate separation distance was obtained by carrying our a low-angle X-ray diffraction examination * of two vacuum concentrated samples.   The particles in these samples had not been deliberately orientated and hence the distances obtained were probably the interparticle distances in oriented domains.   The X-ray results are compared with those obtained from the surface area in table 1.

TABLE 1.—INTERPARTICLE SPACING DISTANCES

% w/w of montmorillonite in sample	H_0 calc. from first order X-ray pattern	H_0 calc. from surface area
37.4	40.8 Å	42.0 Å
34.2	45.5 Å	48.2 Å

The X-ray results, based on a plate thickness of 8.5 Å, are in good agreement with those calculated using a surface area of 800 m^2/g.

* We thank Dr. S Clunie for this determination.

ADSORPTION OF $C_{12}E_6$ ON MONTMORILLONITE

The adsorption isotherm obtained for the adsorption of $C_{12}E_6$ onto montmorillonite from water at 25°C is shown in fig. 3.   The adsorption of $C_{12}E_6$ continued

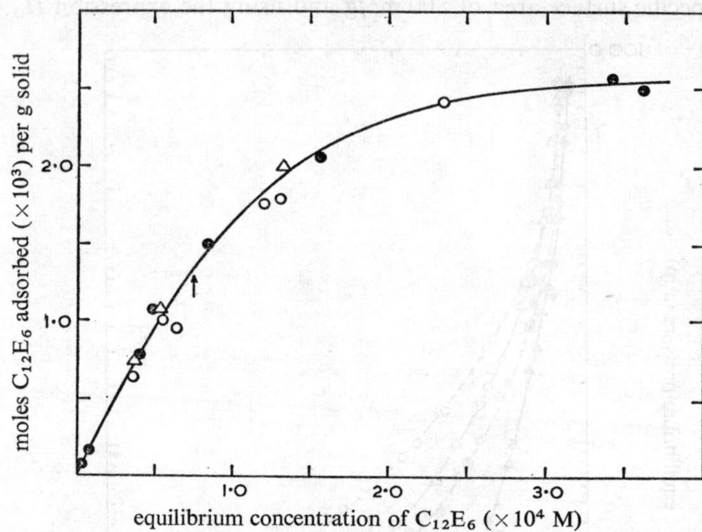

FIG. 3.—Adsorption of $C_{12}E_6$ on sodium montmorillonite at 25°C, from 10^{-4} M sodium chloride solution.   O, centrifuged at 38,000 g; ●, centrifuged at 12,500 g; △, not centrifuged.   ↑, critical micelle concentration.

to increase above the critical micelle concentration and did not reach a steady value at or just below this concentration as has been observed with silver iodide [12] and polystyrene latices.[11]   However, the isotherm is essentially similar to that found by Schott [13] in studies of the adsorption of $C_{12}E_{14}$ on montmorillonites.   At the critical micelle concentration (7×10^{-5} M) the adsorption of $C_{12}E_6$ was 1.16×10^{-3} mol/g, which on the basis of a surface area of 800 m²/g corresponds to an area per adsorbed molecule of 144 Å.²   This would correspond to a horizontal extended orientation of the $C_{12}E_6$ molecule.

COMPRESSION CURVES FOR MONTMORILLONITE IN THE PRESENCE OF $C_{12}E_6$

The equilibrium pressure against distance curves obtained for montmorillonite in the presence of 7×10^{-5} M $C_{12}E_6$ and 10^{-4} M sodium chloride are given in fig. 4. A pronounced difference occurred between the first and second compression curves, but subsequent compression and decompression points all fell on the same curve. A feature of the curves is that although the electrolyte concentration was maintained at 10^{-4} M at pressures below 20 atm the equilibrium distance was reduced in the presence of $C_{12}E_6$.   Above 20 atm the system was more expanded in the presence of $C_{12}E_6$.

DISCUSSION

The two basic problems encountered in the present work were the accurate measurement of the equilibrium pressure and the estimation of the interparticle spacing.   The use of a servo-mechanism allowed the applied pressure to be maintained accurately until the liquid expelled from the cell had reached an equilibrium height in the capillary.   However, the pressures observed in the compression of a sodium montmorillonite dispersion for the first time were always higher than those observed on subsequent compression and decompression cycles.   This effect was attributed to a grain pressure in which edge-face contacts between the clay plates occurred leading to a card-house structure of the type described by van Olphen.[14]   At the high pressures presumably the plates re-aligned to a parallel arrangement although the possibility of some domains could not be excluded.   The hysteresis was largest at low equilibrium pressures for both the electrolyte concentrations examined.   The displacement at

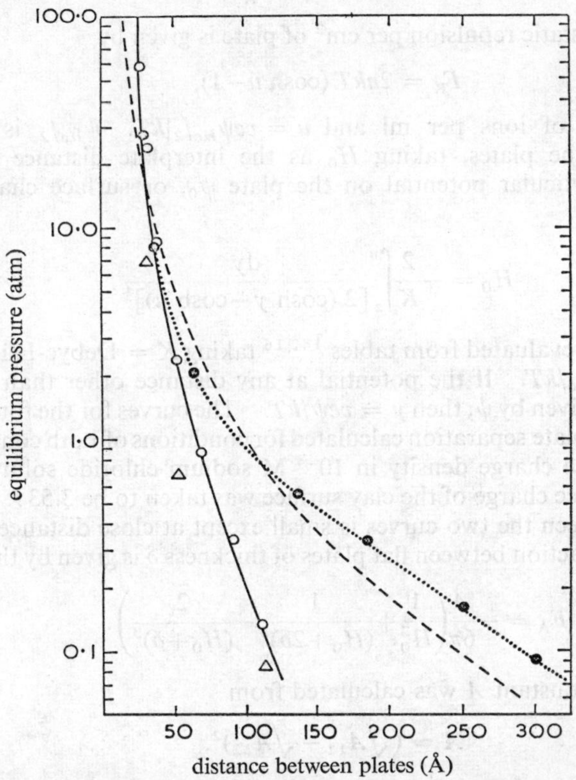

FIG. 4.—Equilibrium pressure against distance between plate surfaces for sodium montmorillonite in 10^{-4} M sodium chloride solution in the presence of an equilibrium concentration of 7×10^{-5} M $C_{12}E_6$. ... ● ..., first compression ; △, non-equilibrium decompression ; ○, subsequent recompression. - - -, curve in 10^{-4} M sodium chloride solution.

0.1 atm pressure was 130 Å in 10^{-1} M sodium chloride solution and 90 Å in 10^{-4} M sodium chloride solution ; the larger electrical repulsion in 10^{-4} M sodium chloride solution clearly aided the dispersion.   In all the experiments the second and subsequent compression curves gave the same results and there was good agreement between the results obtained with different samples.   We assume, therefore, that the grain pressure was largely eliminated after the first compression.

Since separation of the montmorillonite layers into basic sheets occurs, the surfaces can be considered as molecularly smooth.   On the basis of a sheet thickness of 8 Å, the low angle X-ray diffraction results agreed closely with those calculated from the surface area.   The agreement was sufficiently satisfactory (table 1) to conclude that although some error is involved in the determination of the interparticle spacing, this is probably small. Thus, the experimental evidence suggests that the system studied involved flat plate–flat plate interactions through the liquid medium. Hence, it is of considerable interest to compare the results obtained with those expected on the basis of the theory of colloid stability put forward by Deryaguin and Landau [15] and Verwey and Overbeek.[16] The theory involves a consideration of the electrostatic repulsion F_R and the van der Waals attraction F_A acting between the plates so that the total force can be written

$$F = F_R + F_A.$$

The force of electrostatic repulsion per cm² of plate is given by

$$F_R = 2nkT\,(\cosh u - 1),$$

where n = number of ions per ml and $u = ve\psi_{H_0}/2|kT$. $\psi_{H_0}/2$ is the potential mid-way between the plates, taking H_0 as the interplate distance which can be evaluated for a particular potential on the plate ψ_0, or surface charge, using the integral

$$H_0 = -\frac{2}{K}\int_z^u \frac{dy}{[2\,(\cosh y - \cosh u)]^{\frac{1}{2}}}.$$

This integral can be evaluated from tables [14, 16] taking K = Debye-Hückel reciprocal length and $z = ve\psi_\delta/kT$. If the potential at any distance other than at the surface or the mid-point is given by ψ, then $y = ve\psi/kT$. The curves for the force of repulsion against distance of plate separation calculated for conditions of both constant potential and constant surface charge density in 10^{-4} M sodium chloride solution are shown in fig. 5. The surface charge of the clay surface was taken to be 3.53×10^4 e.s.u./cm². The difference between the two curves is small except at close distances.

The force of attraction between flat plates of thickness δ is given by the expression [16]

$$F_A = -\frac{A}{6\pi}\left(\frac{1}{H_0^3} + \frac{1}{(H_0+2\delta)^3} - \frac{2}{(H_0+\delta)^3}\right).$$

The net Hamaker constant A was calculated from

$$A = (\sqrt{A_{11}} - \sqrt{A_{22}})^2,$$

where A_{11} = Hamaker constant of the particle and A_{22} that of the medium. A_{11} was taken as 2.0×10^{-12} erg, the value for silica,[17] and the value for water (A_{22}) was taken as 5.6×10^{-13} erg.[18] The thickness of the montmorillonite plates was taken as 8.0 Å. Calculations showed that the attractive energy became very small for such thin plates at interparticle distances greater than 30 Å and that the effect of retardation, calculated using the expression developed by Hunter,[19] was negligible.

The curve of total force against interparticle distance is also shown in fig. 5, which shows that the force at a distance greater than 40 Å arises solely from electrostatic repulsion. At a distance of *ca.* 15 Å there is a maximum in the curve and hence at shorter distances than this the van der Waals force of attraction should predominate and the plates would coagulate into a primary minimum. This, however, was not observed experimentally. Up to the highest pressures exerted (approximately 100 atm) it was always possible to decompress and retrace the original compression curve, thus indicating the reversibility of the system. The pressure continued to increase with decreasing distance and in 10^{-4} M sodium chloride the distance between the plates at an applied pressure of 100 atm was only 17.5 Å. Thus, no evidence is found for the primary minimum effects expected and it seems therefore that an additional force of repulsion has to be considered for such systems. Deryaguin and Greene-Kelly [20] have suggested that structural boundary layers of water 30-200 Å thick exist between the silicate layers of swollen montmorillonite and contribute to the stabilisation of the particles at close distances. van Olphen [21] and Briant [22] have also indicated that hydration forces cannot be neglected in clay systems and evidence for hydration forces in stable black soap films has been given by Goodman et al. [23] The resemblance between the present results and those obtained by Goodman and coworkers is striking and would suggest that the effects observed at high pressures,

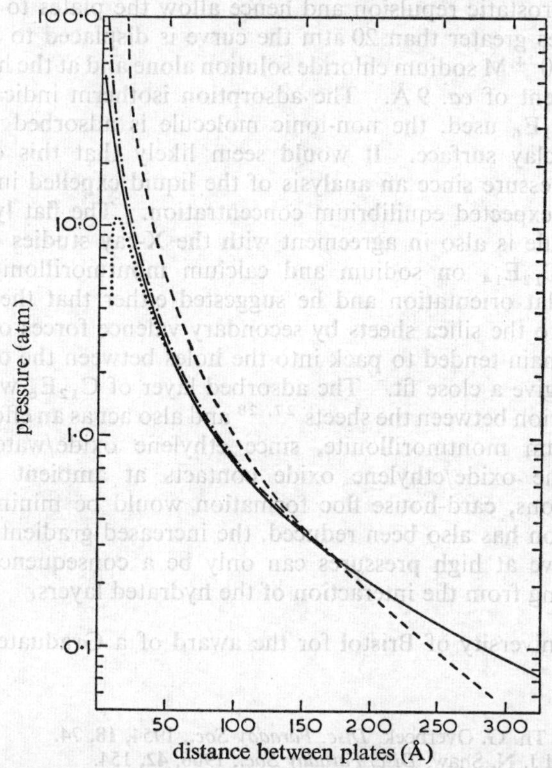

FIG. 5.—Pressure against distance between plates in 10^{-4} M sodium chloride solution. —, theoretical curve for electrical repulsion at constant potential (250 mV); –·–·– theoretical curve for electrical repulsion at constant charge (3.53×10^4 e.s.u./cm²); ...., total force between plates assuming a plate thickness of 8 Å. - - - -, experimental curve in 10^{-4} M sodium chloride solution.

in view of their reversibility, are due to solvation forces and not to grain pressure. An alternative possibility is that in the model of the electrostatic repulsion the electrical situation at close distances has been oversimplified.

At distances greater than 50 Å it is reasonable to postulate that the repulsion arises solely from electrostatic repulsion; this is in agreement with previous workers.[4, 9, 24, 25, 26] In the present work the predicted values of the pressure are larger than those found experimentally in 10^{-4} M sodium chloride solutions at distances greater than 150 Å, but are less than those measured at distances of less than 150 Å. In 10^{-1} M sodium chloride solutions the measured pressures were larger over the whole range examined than those predicted theoretically for the same distance of separation; this is not unexpected in view of the difficulties of the theory in the more concentrated electrolyte solutions.

RESULTS IN THE PRESENCE OF $C_{12}E_6$

The results shown in fig. 4 indicate that at an applied pressure of less than 20 atm the interparticle spacing is considerably less in the presence of $C_{12}E_6$. Although electrokinetic experiments were not possible with the montmorillonite plates, with silver iodide sols [12] and with polystyrene latex dispersions [11] a considerable drop in the electrokinetic potential occurs on the adsorption of $C_{12}E_6$. This would appreciably reduce the electrostatic repulsion and hence allow the plates to approach more closely. At pressures greater than 20 atm the curve is displaced to larger distances than those found in 10^{-4} M sodium chloride solution alone and at the highest pressures there is a displacement of *ca.* 9 Å. The adsorption isotherm indicates that at the concentration of $C_{12}E_6$ used, the non-ionic molecule is adsorbed in a horizontal orientation on the clay surface. It would seem likely that this orientation was maintained under pressure since an analysis of the liquid expelled into the capillary for $C_{12}E_6$ gave the expected equilibrium concentration. The flat lying orientation of the $C_{12}E_6$ molecule is also in agreement with the X-ray studies of Schott [13] on the adsorption of $C_{12}E_{14}$ on sodium and calcium montmorillonite. His results clearly indicated a flat orientation and he suggested either that the ethylene oxide groups were bound to the silica sheets by secondary valence forces or that the polyoxyethylene glycol chain tended to pack into the holes between the oxygen atoms of the silica surface to give a close fit. The adsorbed layer of $C_{12}E_6$ would reduce the van der Waals attraction between the sheets [27, 28] and also act as an effective dispersing agent for the sodium montmorillonite, since ethylene oxide/water contacts are preferred to ethylene oxide/ethylene oxide contacts at ambient temperatures.[11] Under these conditions, card-house floc formation would be minimized and since the electrical repulsion has also been reduced, the increased gradient of the pressure against distance curve at high pressures can only be a consequence of the strong repulsive forces arising from the interaction of the hydrated layers.

We thank the University of Bristol for the award of a Graduate Scholarship to L. M. B.

[1] H. Reerink and J. Th. G. Overbeek, *Disc. Faraday Soc.*, 1954, **18**, 74.

[2] R. H. Ottewill and J. N. Shaw, *Disc. Faraday Soc.*, 1966, **42**, 154.

[3] A. Watillon and A. M. Joseph-Petit, *Disc. Faraday Soc.*, 1966, **42**, 143.

[4] B. P. Warkentin, G. H. Bolt and R. D. Miller, *Soil Sci. Soc. Amer. Proc.*, 1957, **21**, 495.

[5] D. Tabor, *J. Colloid Interface Sci.*, 1969, **31**, 364.

[6] A. D. Roberts and D. Tabor, *Nature*, 1968, **219**, 1122.

[7] K. J. Mysels and M. N. Jones, *Disc. Faraday Soc.*, 1966, **42**, 42.

[8] R. H. Ottewill and J. A. Sirs, *Bull. Photometric Spectr. Group*, 1957, **10**, 262.

[9] G. H. Bolt, *Ph.D. thesis*, (Cornell University, 1964).

[10] J. M. Corkill, J. F. Goodman and R. H. Ottewill, *Trans. Faraday Soc.*, 1961, **57**, 1627.

[11] R. H. Ottewill and T. Walker, *Kolloid Z. Z. Polymere*, 1968, **227**, 108.

[12] K. G. Mathai and R. H. Ottewill, *Trans. Faraday Soc.*, 1966, **62**, 750, 759.

[13] H. Schott, *Kolloid-Z.*, 1964, **199**, 158.

[14] H. van Olphen, *An Introduction to Clay Colloid Chemistry* (Interscience, John Wiley, New York, 1963).

[15] B. V. Deryaguin and L. Landau, *Acta physicochim.*, 1941, **14**, 633.

[16] E. J. W. Verwey and J. Th. G. Overbeek, *Theory of the Stability of Lyophobic Colloids* (Elsevier, Amsterdam, 1948).

[17] W. Black, J. G. V. de Jongh, J. Th. G. Overbeek and M. G. Sparnaay, *Trans. Faraday Soc.*, 1960, **56**, 1597.

[18] H. R. Kruyt, *Colloid. Science*, (Elsevier, Amsterdam), vol. 1, 1952.
[19] R. J. Hunter, *Austral. J. Chem.*, 1963, **16**, 774.
[20] B. V. Deryaguin and R. Greene-Kelly, *Trans. Faraday Soc.*, 1964, **60**, 449.
[21] H. van Olphen, *T.A.P.P.I.*, 1968, **51**, 145A.
[22] J. Briant, *Compt. rend. IIIe Colloque l'A.R.T.F.P.*, 1968, p. 31.
[23] J. S. Clunie, J. F. Goodman and P. C. Symons, *Nature*, 1967, **216**, 1203.
[24] G. H. Bolt and R. D. Miller, *Soil Sci. Soc. Amer. Proc.*, 1955, **19**, 285.
[25] B. P. Warkentin and R. K. Schofield, *J. Soil. Sci.*, 1962, **13**, 98.
[26] R. Yong, L. O. Taylor and B. P. Warkentin, *Clays Clay Min.*, 1963, **13**, 268.
[27] M. J. Vold, *J. Colloid. Sci.*, 1961, **16**, 1.
[28] R. H. Ottewill, *Nonionic Surfactants* ed. M. J. Schick (Marcell Dekker), 1967, **1**, 627.

Boundary Layers Between Silver Iodide and Aqueous Solutions at Low Temperatures

By B. Vincent * and J. Lyklema

Laboratory for Physical and Colloid Chemistry, Agricultural University, de Dreijen 6, Wageningen, The Netherlands

Received 14th July, 1970

Potentiometric titration studies have been made on AgI suspensions over the temperature range 0-20°C, in the presence of $LiNO_3$, KNO_3 and $RbNO_3$. Points-of-zero-charge (p.z.c.), double-layer capacities and surface entropy data, derived from the titration results, indicate an increase in water " structure ", as the temperature tends to 0°C, around the p.z.c. region, and particularly on the positive side. For strongly negatively charged surfaces the charge ordering effect on the dipoles tends to break down this structuring. The results of flocculation studies on AgI sols have been combined with the titration data to yield information about changes in double layer parameters with temperature. The results fit the suggestions made regarding changes in water structuring at the interface.

During the last decades considerable progress has been made in the field of interfacial electrochemistry, especially with mercury electrodes. One of the important results has been that structural features of the liquid adjacent to the electrode, e.g., the direction of orientation of water dipoles, are reflected in measurable electrochemical quantities, such as the differential double-layer capacitance and the point of zero charge (p.z.c.). Hence, inference on the structural properties of the interphase can be derived from these and other electrochemical measurements.

Studies with non-metallic charge carriers are more scanty and, as a rule, less accurate, but they have the advantage that, for certain systems, stable dispersions can be made. In those cases the double-layer measurements may be amplified by stability studies. The AgI-system has received special attention because stable AgI sols can be made as well as reproducibly operating AgI electrodes. Moreover, its interfacial structural properties are of interest because AgI is a powerful cloud seeder.

Some preliminary information on the boundary layer structure of this system has now been obtained by combining double-layer and stability studies as a function of temperature.[1] In view of the expectation that structure formation in the interface would be promoted by a decrease in temperature, these measurements have now been extended to cover the temperature range down to 0°C.

EXPERIMENTAL

MATERIALS

AgI suspensions were prepared by addition of 0.1 M $AgNO_3$ to an equal volume of 0.1 M KI in the usual way.[2] Surface areas were determined for each suspension used by

* present address: Dept. of Physical Chemistry, University of Bristol, Cantock's Close, Bristol BS8 1TS ; or, I.C.I. Paints Division Ltd., Wexham Road, Slough, Bucks.

comparison of surface charge values in C g^{-1} from the experimental titration curve in 10^{-1} M KNO$_3$, with surface charge values in μC cm^{-2} from standard curves [3] for this system. Most of the experiments reported here were carried out on a suspension of surface area 1.34 m^2 g^{-1}. Sols were prepared by the similar addition of 2.0×10^{-2} M AgNO$_3$ to 2.2×10^{-2} M KI. Both suspensions and sols were aged at 80°C for 3 days.

All salts used were of A.R. quality : AgNO$_3$, KI, LiNO$_3$, Union Chimique Belge ; KNO$_3$, Baker ; RbNO$_3$, Merck. The RbNO$_3$ was recrystallized twice from water. All the other salts were used as supplied. Water was doubly-distilled and passed down a AgI column before use.

POTENTIOMETRIC TITRATIONS

The potentiometric titration technique by which (surface charge, pAg) or (surface charge, pI) curves may be obtained for aqueous AgI suspensions has been extensively discussed.[4, 5] The apparatus used here was essentially similar, except that the titration cell was constructed so that it could be completely immersed in a cryostat bath (Colora, Ultra Cryostat, KT 40 S). In this way the temperature of the cell was controlled to ± 0.1°C over a range 0-20°C. Potential measurements (± 0.2 mV) were made using a potentiometer (Knick Praisiziono pH Metre, type pH 34). The average reading of 4 or 5 Ag/AgI electrodes was taken. Calibration was carried out using standard solutions at pI 4.0, 5.0 and pAg 4.0, 5.0. The cell resistance was also checked periodically during a run (Phillips Bridge Gh 4249) to ensure that there was no blocking of the salt bridge capillary by AgI particles.

Titration runs were performed in the following manner. A sample of concentrated suspension of known composition was weighed in a graduated flask. This gave the weight of AgI without resort to drying. The suspension was washed into the titration vessel with 10^{-3} M salt solution. A set of titrations was then carried out in the following order :

(1) 20°C+$\rightarrow$−addition of KI ; (2) 10°C−$\rightarrow$+addition of AgNO$_3$; (3) 0°C+$\rightarrow$−addition of KI ; (4) 20°C−$\rightarrow$+addition of AgNO$_3$. At this point further inert electrolyte was added and the sequence repeated.

STABILITY MEASUREMENTS

These were performed using the " kinetic " method of Reerink and Overbeek.[6] The change in optical density (at $\lambda = 556$ nm) of the flocculating system was followed using a Vitatron spectrophotometer (Vitatron N.V., Dieren, The Netherlands). This instrument incorporates a magnetic stirrer device. The output was fed to a Kipp Micrograph pen-recorder (Kipp N.V., Delft, The Netherlands). The spectrophotometer cell-housing was thermostatted by pumping cooling liquid from the cryostat around its jacket. Temperature control was ± 0.5°C but at ambient temperature differentials of more than about 5°C it was necessary to enclose the apparatus within a dry-box to prevent condensation on the optical cell.

The flocculation runs were carried out in cylindrical optical cells (0.9 cm diam.). 0.5 ml sol (AgI concentration 10 mM) were added with the aid of a syringe, to 2 ml of salt solution in the cell which also contained a small glass-covered magnetic stirrer. Both the sols and salt solutions were adjusted to the necessary pI, and stored at the required temperature, prior to use.

RESULTS AND DISCUSSION

SOLUBILITY PRODUCT L AND POINT OF ZERO CHARGE (p.z.c.) OF AgI

pL values for AgI over the temperature range studied and at salt concentrations 10^{-3}-10^0 M were calculated from the standard electrode potentials. These are shown in fig. 1 together with the values obtained by other authors.[1, 7] There is an apparent continuous increase in the solubility of AgI with increase in temperature and with increase in inert electrolyte concentration.

The p.z.c. (pAg°) values (see table 1) was taken for each system as the mutual intersection point of the experimental "charge" (i.e., charge with respect to an arbitrarily chosen reference point) against pAg curves.[1] This point was generally well defined to within 0.2 pAg units, except that the 10^0 M curves tended to cross the $\sigma_0 = 0$ line at slightly higher pAg value.

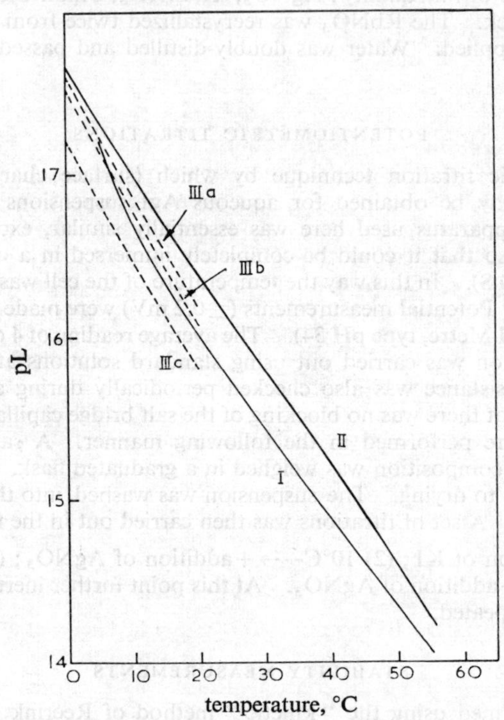

Fig. 1.—Solubility product (expressed as pL) of AgI as a function of temperature. I, Lyklema,[1] 10^{-1} M (with 10^{-3} M K biphthalate present); II, Honig and Hengst,[7] 10^{-4} M; III, this work, (a) 10^{-3} M, (b) 10^{-1} M, (c) 10^0 M.

Table 1.—pAg values for AgI

authors	ref.	method	salt	0°C	10°C	20°C
1, this work		titration	$LiNO_3$	7.10	6.40	5.80
			KNO_3	6.95	6.20	5.65
			$RbNO_3$	6.75	6.10	5.50
2, Honig and Hengst	7	suspension effect	KNO_3	5.51	5.51	5.52
3, Fairhurst	8	electrophoresis	KNO_3	—	5.54	5.40

The three sets of data show apparent discrepancies. This work indicates a significant rise in pAg° with decreasing temperature. Those of Honig and Fairhurst are virtually independent of temperature over this range. Titration results, however, yield true p.z.c. values, whereas the suspension effect and electrophoresis give iso-electric points (i.e.p.). At higher temperatures the p.z.c. values of Lyklema[1] and the i.e.p. of Honig again diverge (e.g.,at 60°C; p.z.c. = 5.45; i.e.p. = 4.38).* The

---

* Lyklema's measurements[1] were made in the presence of K-biphthalate and biphthalate ions tend to absorb specifically at the p.z.c. which would in turn tend to move the p.z.c. towards a higher pAg° value (e.g., at 20°C, p.z.c. = 5.72). This specific adsorption would presumably however decrease with increasing temperature, whereas the divergence between i.e.p. and p.z.c. increases.

difference between the p.z.c. and the i.e.p. appears to be minimal at about 20°C. As the solubility product itself shows no break around 20°C structural changes in the solid AgI can hardly be invoked to explain the observed behaviour.

A tentative explanation for the increase in pAg° on going from 20 to 0°C, as observed by us, is as follows. At 20°C the p.z.c. is very asymmetrical: $pAg° = 5.6$, $pI° = 10.6$, $pL (= pAg° + pI°) = 16.2$. Much less iodide is needed than silver to make the surface uncharged. At the same time the adsorption of Ag^+ at the positive side of the p.z.c. is very strong, whereas positive AgI sols are rather unstable. One way of explaining these facts is to assume that at 20°C Ag^+ absorbs largely in an associated form, i.e., as $AgNO_3$. In the first place this would support the result that only relatively small amounts of I^- are required to make the surface negative; in the second place it explains the high adsorption at the positive side without concomitant rise in potential. At 0°C the p.z.c. is less asymmetrical: $pAg° = 6.9$, $pI° = 10.5$, $pL = 17.4$. At the same time the adsorbability of Ag^+ at the positive side is reduced (see titrations). The implication is that at 0°C these $AgNO_3$ ion pairs are more dissociated, than at 20°C. Perhaps, therefore, due to an increase in water " structuring " at the interface the NO_3^- ions are less easily adsorbed in the Stern layer. This explanation is corroborated by the surface excess entropy calculations (see later).

POTENTIOMETRIC TITRATION RESULTS

The titration curves for $LiNO_3$, KNO_3 and $RbNO_3$ are given in fig. 2, 3 and 4 respectively. At low inert electrolyte concentrations (10^{-3} M,) $(d\sigma_0/dT)_{pAg}$ is negative. This is expected, since here the diffuse layer term dominates the total double-layer capacity. There is a tendency, however, noticeable at low negative surface charges, and particularly so on the positive side, for $(d\sigma_0/dT)_{pAg}$ to reverse sign in 10^{-1} and 1 M salt. By comparison with the curve at higher temperatures,[1] $(d\sigma/dT)_{pAg}$ becomes zero around 20°C, i.e., the same region where the anomalous increase in pAg° begins. The differential double layer capacity values at the p.z.c. are presented in table 2.

TABLE 2.—DIFFERENTIAL DOUBLE LAYER CAPACITIES AT p.z.c. ($\mu F/cm^2$)

i.e.c.	20°C			10°C			0°C		
	$LiNO_3$	KNO_3	$RbNO_3$	$LiNO_3$	KNO_3	$RbNO_3$	$LiNO_3$	KNO_3	$RbNO_3$
10^0 M	18.0	24.8	30.0	19.2	23.5	29.0	17.7	20.5	22.5
10^{-1} M	15.0	17.5	23.0	16.6	17.0	18.0	14.0	16.0	16.0
10^{-2} M	9.5	10.2	11.6	10.7	11.5	12.5	9.5	10.0	10.0
10^{-3} M	5.4	6.0	6.2	6.0	6.0	6.0	6.6	6.6	6.6

The apparent specificity with regard to the nature of the cation, particularly at higher salt concentrations, where the capacity is dominated by the Stern layer term, increasing in the order $Li^+ < K^+ < Rb^+$ indicates some absorption of cations at the p.z.c. as well as anions. This is also reflected in the slight dependency of the p.z.c. on the nature of the cation.

The decrease in capacity with lowering of temperature, particularly from 10 to 0°C, again points to desorption of NO_3^--ions. (Since the decrease is relatively smaller for $LiNO_3$ this might indicate some desorption of the strongly hydrated Li^+ ions also). Desorption of NO_3^- ions on lowering the temperature from 20 to 0°C was also indicated, on the positive side of the zero point of charge, from a preliminary components-of-charge analysis of the type suggested by Lyklema.[9]

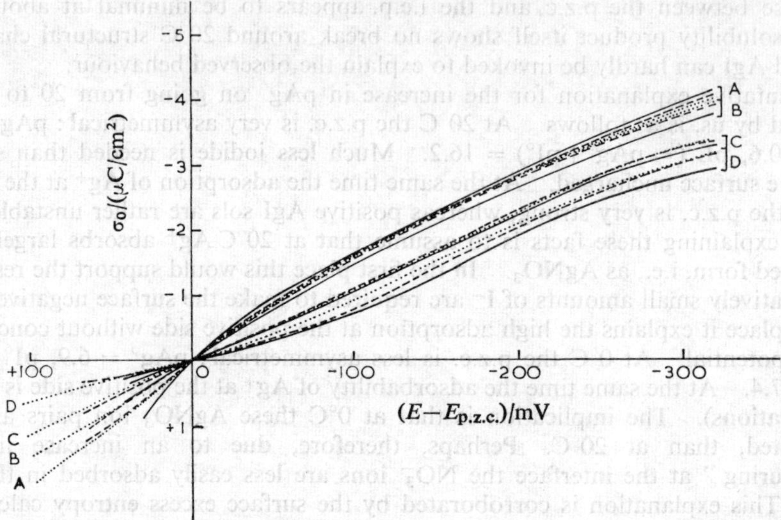

FIG. 2.—Surface charge-potential curves for AgI in the presence of various concentrations of LiNO₃:
presence of various concentrations of LiNO₃: A, 10⁰ M; B, 10⁻¹ M; C, 10⁻² M; D, 10⁻³ M.
—, 20°C; --- 10°C; .... 0°C.

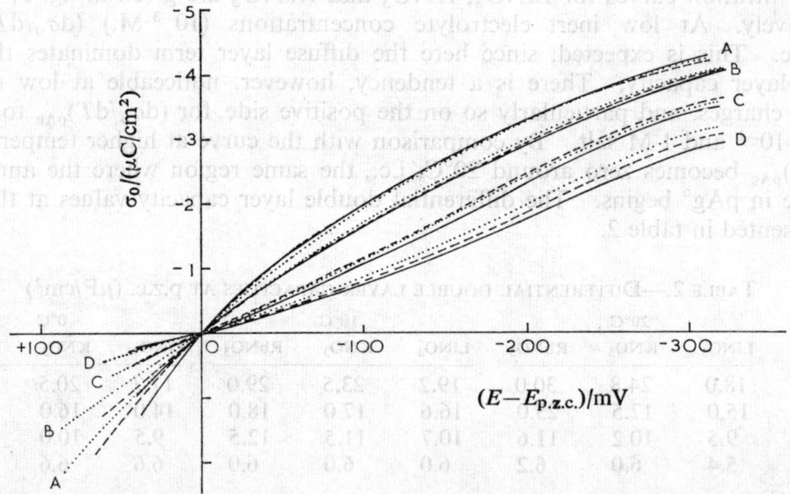

FIG. 3.—Surface charge-potential curves for AgI in the presence of various concentrations of KNO₃:
A, 10⁰ M; B, 10⁻¹ M; C, 10⁻² M; D, 10⁻³ M.　—, 20°C; ---, 10°C; ..., 0°C.

The differential capacity is much less temperature dependent at high negative
surface charge when any water " structuring " would be opposed by the orientating
effect on the water dipoles (e.g., in 10^{-1} M RbNO₃, $\psi_0 = -250$ mV: $C = 12.8$
(20°C), 13.0 (10°C), 13.2 (0°C) μF/cm²).

INTERFACIAL ENTROPIES

The derivative of the excess entropy of the AgI/solution interface with respect to
the surface potential ψ_0 can be calculated from the temperature dependence of the

surface charge using either one of two equations derived previously by Bijsterbosch and Lyklema.[10]

$$(\partial\eta^\sigma/\partial\psi_0)_{T,a_s} = -s^\alpha(C_{I^-}/F)+(\partial\sigma_0/\partial T)_{pAg,a_s}-(\partial\sigma_0/\partial\mu_s)_{pAg,T}(-s_s^\circ+R\ln a_s)-$$
$$(C/F)(-s_{Ag^+}^\circ+R\ln a_{Ag^+}), \qquad (1)$$

$$(\partial\eta^\sigma/\partial\psi_0)_{T,a_s} = s^\alpha(C_{Ag^+}/F)+(\partial\sigma_0/\partial T)_{pI,a_s}+(\partial\sigma_0/\partial\mu_s)_{pI,T}(-s_s^\circ+R\ln a_s)+$$
$$(C/F)(-s_{I^-}^\circ+R\ln a_{I^-}) \qquad (2)$$

In these equations the interfacial excess entropy η^σ is defined by

$$\eta^\sigma = S^\sigma-s^\alpha T_{AgI}-s^\beta\Gamma_w, \qquad (3)$$

where

$$S^\sigma = (S-S^\alpha-S^\beta)/A, \qquad (4)$$

s denotes molar entropies, the superscipts α and β apply to the solid (AgI) and liquid (W) phase respectively and subscript s refers to the salt. C_{I^-} and C_{Ag^+} are the contributions of the iodide and silver ions, respectively, to the total differential capacitance C. The way in which one splits C up into its component parts is immaterial.[10]

Eqn (1) and (2) are equivalent, the only difference being that in (1) the Ag^+ ion is taken to be the potential-determining species, whereas in (2) it is the I^- ion. The equivalence of the two equations was corroborated by the computations: although individual terms differed sometimes by a hundredfold the final values for $(\partial\eta^\sigma/\partial\psi_0)_{T,a_s}$ agreed within a few percent. The general behaviour of $(\partial\eta^\sigma/\partial\psi_0)_{T,a_s}$ is to a large extent determined by the $(\partial\sigma_0/\partial T)_{pAg,a_s}$ term. Further details on the method of computation are given in ref. (10).

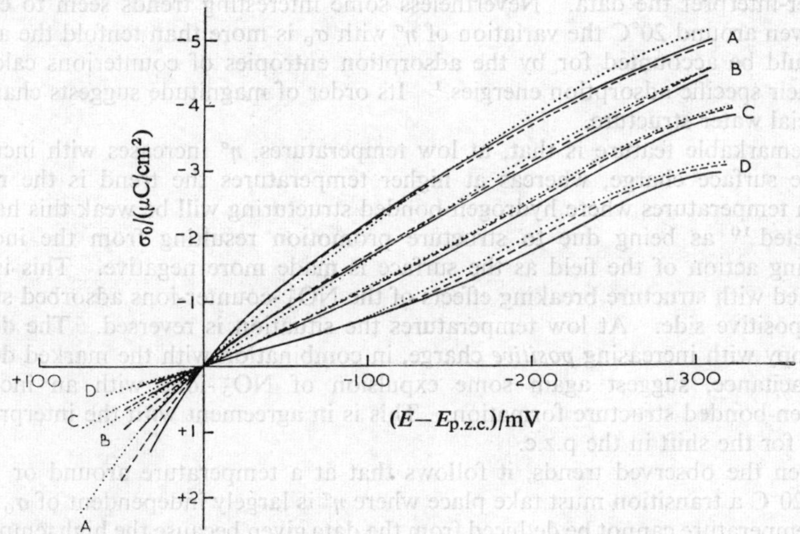

FIG. 4.—Surface charge-potential curves for AgI in the presence of various concentrations of $RbNO_3$: A, 10^0 M; B, 10^{-1} M; C, 10^{-2} M; D, 10^{-3} M. ——, 20°C; ---, 10°C; ..., 0°C.

The values obtained for $(\partial\eta^\sigma/\partial\psi_0)_{T,a_s}$ were plotted as a function of ψ_0, graphically integrated, and finally plotted as a function of σ_0 using the p.z.c. as the reference point. The results are given in fig. 5 together with some results obtained previously at higher temperatures in the presence of 10^{-3} M K biphthalate.[10] The data shown are for $LiNO_3$, since this would be expected to reveal structural changes in the interfacial water more strongly than KNO_3 or $RbNO_3$, because the masking effect of specific adsorption on the negative side is least for the Li^+ ion. (The curves for KNO_3 and $RbNO_3$ are in fact similar but the trends are less marked.)

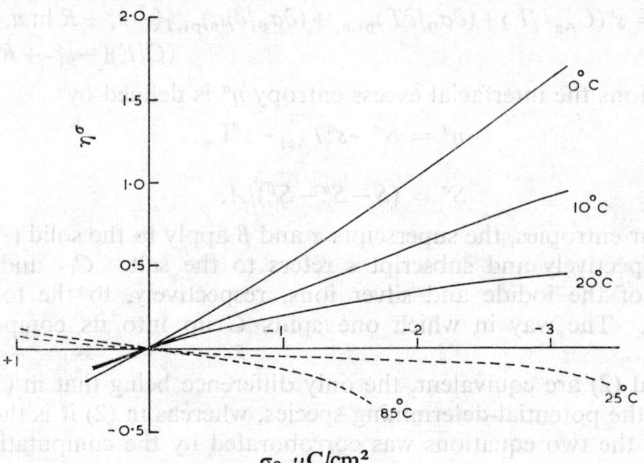

FIG. 5.—Interfacial excess entropy with respect to the point of zero charge, as a function of surface charge for decimolar solutions of $LiNO_3$: —, current work; ---, previous work [10] in which 10^{-3} K-biphthalate was also present.

Interfacial entropies are composite and complex quantities. One should therefore not over-interpret the data. Nevertheless some interesting trends seem to emerge. First, even around 20°C the variation of η^σ with σ_0 is more than tenfold the amount that could be accounted for by the adsorption entropies of counterions calculated from their specific adsorption energies.[1] Its order of magnitude suggests changes in interfacial water structure.

A remarkable feature is that, at low temperatures, η^σ increases with increasing negative surface charge, whereas at higher temperatures the trend is the reverse. At high temperatures where hydrogen-bonded structuring will be weak this has been interpreted [10] as being due to structure promotion resulting from the increased polarizing action of the field as the surface is made more negative. This is to be combined with structure breaking effects of the NO_3^--counter-ions adsorbed strongly at the positive side. At low temperatures the situation is reversed. The decrease in entropy with increasing *positive* charge, in combination with the marked decrease in capacitance, suggest again some expulsion of NO_3^--ions with an increasing hydrogen-bonded structure formation. This is in agreement with the interpretation offered for the shift in the p.z.c.

Given the observed trends, it follows that at a temperature around or slightly above 20°C a transition must take place where η^σ is largely independent of σ_0. (The exact temperature cannot be deduced from the data given because the high temperature curves are obtained in the presence of biphthalate). At this temperature, with increasing negative charge, the two opposing trends, i.e., the increase in dipole ordering and decrease in water structuring, would tend to cancel each other.

STABILITY

The stability of a hydrophobic sol, measured as the flocculation concentration (c_f in mmol/l.), reflects the charge and potential distribution in the electrical double layer. As a consequence, stability data can serve as an additional source of information. However, its applicability with respect to the detection of structure formation is limited because stable sols can only be made at sufficient high surface charge, i.e., in the region where any effect of structure formation is relatively weak.

There are two ways in which structure formation would be expected to affect c_f: (i) electrostatically, because if there is structure formation, the extent of counterion adsorption, and hence the potential of the outer Helmholtz phase ψ_d is changed; (ii) kinetically, because the rate of counterion transfer from the diffuse part of the double layer to the Stern layer during the encounter of the particles as required by DLVO-theory is reduced.

In fig. 6 and 7 flocculation concentrations under various conditions, and the corresponding double layer parameters, are shown. The flocculation concentration apparently decreases with decreasing temperature at pI 4.6, but the opposite is true at pI 6.6. The surface charge shows a slight maximum at around 10°C for all three cations at both pI. The trends in c_f and σ_0 are not parallel. This is not expected, *a priori*, since c_f reflects the charge *distribution* whereas σ_0 is the total charge. In order to analyze these data further Stern potentials, specific adsorption, inner layer capacities and dielectric constants have been calculated using an analysis to be discussed elsewhere,[11] but which may be summarized briefly as follows. From the flocculation data, and an ionic components of charge analysis [9] on well-established potentiometric titration data, for KNO_3 at 20°C, a value of $A = 2.40 \times 10^{-20}$ J was established for the Hamaker constant for AgI in aqueous systems. This value was then used to calculate the Stern potentials, and hence the other double-layer parameters, from the flocculation concentrations and titration data.

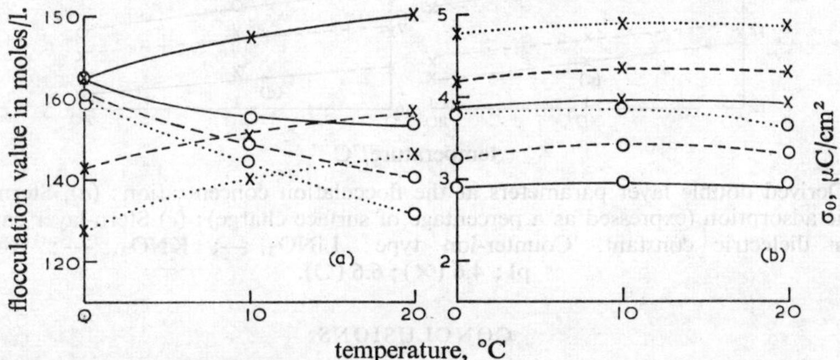

FIG. 6.—Experimental parameters under various conditions: (*a*), flocculation concentrations; (*b*), surface charge. counter ion type: $LiNO_3$, —; KNO_3, ---; $RnNO_3$, ... pI: 4.6($\times$); 6.6($\bigcirc$).

The specific adsorption curves (expressed as a percentage of the surface charge) indicate little or no dependence on temperature at pI 4.6, but at pI 6.6 there appears to be a definite fall-off in specific adsorption of cations on lowering the temperature from 20 to 0°C. Thus at highly negative surface charges (as at pI 4.6), the specific adsorption is largely governed by electrostatic forces and is therefore largely temperature independent over this range, but at lesser negative surface charges (as at pI 6.6) there is at least some effect from water structuring, leading to some desorption of counter ions, on lowering the temperature.

The Stern-layer capacity falls slightly with increasing temperature and is in the order $Rb^+ > K^+ > Li^+$, as expected, but is also in the order pI 6.6 > pI 4.6. Since specific adsorption increases with decreasing pI, and there are probably no significant changes in inner layer thickness, this could reflect changes in the inner layer dielectric constant. (At such strong specific adsorptions (~ 50 %) the occupancy of the Stern layer is such that the polarizabilities of the adsorbed ions themselves probably contribute significantly to the dielectric constant of the inner layer.) The mean value at 20°C, 7.5 ± 1.5, corresponds well with the value of about 6 quoted for the inner layer at the Hg + water interface.[12] The mean value rises to about 8.0 ± 1.5 at 0°C, which again supports the idea of increased hydrogen-bonded structuring of the interfacial water, although only slight at these negative charges.

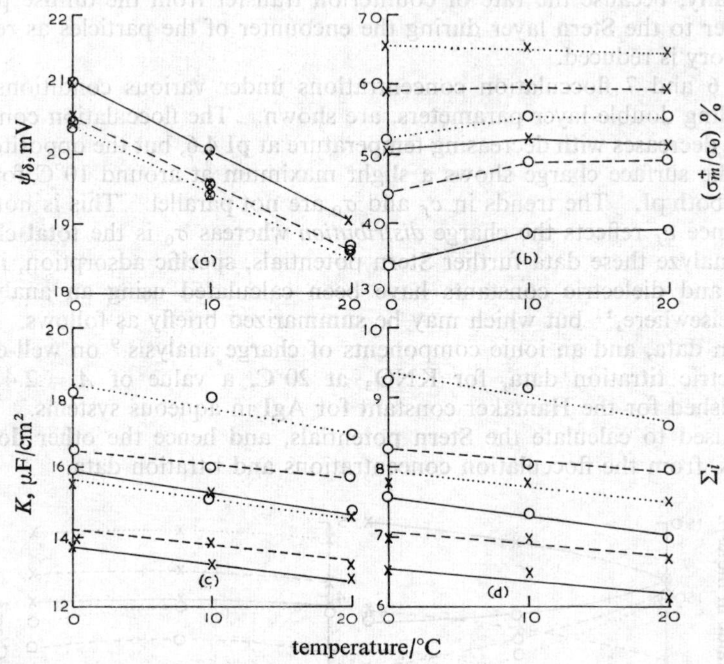

FIG. 7.—Derived double layer parameters at the flocculation concentration: (a), Stern potential; (b), specific adsorption (expressed as a percentage of surface charge); (c) Stern-layer capacity; (d), Stern-layer dielectric constant. Counter-ion type: $LiNO_3$, —; KNO_3, ---; $RbNO_3$, ... pI: 4.6 ($\times$); 6.6 ($\bigcirc$).

CONCLUSIONS

It is generally observed that adsorption increases with decreasing temperature. For the adsorption of I^--ions on AgI this trend was confirmed.[1] The more detailed study of the adsorption of potential determining ions, reported in this paper, shows that this trend is not followed at temperatures between 0 and 20°C. At the negative side of the p.z.c. the adsorbed amount (and hence the surface charge σ_0) is almost independent of temperature, whereas on the positive side σ_0 tends to decrease with decreasing temperature. The effect is that of increased structure formation in the aqueous layer adjacent to the AgI surface at less negative or more positive charges, together with consequent desorption of NO_3^--ions. It is fully corroborated by surface excess entropy calculations.

Flocculation experiments are less informative because only data at high negative surface charges are available. In this region the structure formation effect is less pronounced. However, the stability trends observed at lower negative surface charge agree well with the theory suggested above. There remains the discrepancy between p.z.c.-values from titration and i.e.p.-values from the suspension effect and electrophoresis. The experimental data seem to be well established. Unfortunately we cannot offer a satisfactory explanation for this.

We thank the Royal Society for the provision of an Overseas Fellowship to one of us (B.V.), and also thank Miss Olga van Hiele for technical assistance.

[1] J. Lyklema, *Disc. Faraday Soc.*, 1966, **42**, 81.
[2] see e.g., B. H. Bijsterbosch and J. Lyklema, *J. Colloid Sci.*, 1965, **20**, 665.
[3] *Int. Crit. Tables*, to be published.
[4] J. Lyklema and J. Th. Overbeek, *J. Colloid Sci.*, 1961, **16**, 595.
[5] G. L. Mackor, *Rec. Trav. Chim.*, 1951, **70**, 763.
[6] H. Reerink and J. Th. Overbeek, *Disc. Faraday Soc.*, 1954, **18**, 74.
[7] E. P. Honig and J. H. Th. Hengst, *J. Colloid Interface Sci.*, 1969, **30**, 109.
[8] A. L. Smith, private communication of unpublished work by D. Fairhurst.
[9] J. Lyklema, *Trans. Faraday Soc.*, 1963, **59**, 418.
[10] B. H. Bijsterbosch and J. Lyklema, *J. Colloid Interface Sci.*, 1968, **28**, 506.
[11] B. Vincent, B. H. Bijsterbosch and J. Lyklema, to be published.
[12] J. O'M. Bockris, M. A. V. Devanathan and K. Müller, *Proc. Roy. Soc. A*, 1963, **274**, 55.

Boundary Layer near the Surface of a Solid Body and Low-Frequency Dielectric Dispersion

By S. S. Dukhin

Institute of Colloid and Water Chemistry, Academy of Sciences of the Ukrainian S.S.R.
Kiev, U.S.S.R.

Received 19th March, 1970

The paper deals with the possibility of studying the stagnant part of the boundary layer on the basis of effects due to polarization of the electrochemical double layer (DL). A theory is developed for polarization of the diffuse part of a thin DL of a spherical particle in a direct and alternating field, allowing for the effect of ion flow through the stagnant part of the DL. Formulae are obtained on the basis of this theory for electrophoresis complicated by polarization and for the large low-frequency dielectric dispersion $\bar{\varepsilon}(\omega)$. Calculation of the stagnant layer thickness and the ionic mobilities of this layer is shown to be possible on the basis of electrophoretic measurement on three fractions of spherical particles with identical values of ψ_d and ζ. Using only one fraction, we can obtain similar information from the low-frequency section of $\bar{\varepsilon}(\omega)$.

The advisability of investigating electrokinetic and related electro-surface phenomena in disperse systems in connection with the boundary layer problem has been acknowledged [1] and substantiated by systematic studies [2-4] which have already yielded valuable results.

We shall consider certain new means for studying the boundary layer by electric methods. Since terminology in this new field of research has not yet been established, we shall adhere to Deryaguin's model-system ideas and terminology. We shall call a boundary layer one in which anomaly of structure and the structural-sensitive properties of the liquid are observed. The fact that within a layer of some thickness the liquid loses fluidity and does not participate in hydrodynamic processes is only one of many manifestations [5] of this anomaly of liquid structure. This layer is called stagnant. Such effects as thermo-osmosis [6] and capillary osmosis [7] indicate that the thickness of the boundary layer may exceed that of the stagnant layer.

It is natural to compare the thickness of the stagnant layer with the distance from the surface to the slip plane, a conception based on electrokinetic phenomena. The stagnant layer thickness may then be estimated by the difference in the values of the Stern ψ_d and electrokinetic potentials. Simultaneous determination of the ζ and ψ_d potentials have been made for an oil-water interface,[9] with an absorption monolayer of ionogenic surfactant and for spherical micelles.[10] In both cases the slip plane coincides with the surface enveloping the hydrated charges of the long-chain ions. This does not, however, affect the possibility of their being a stagnant polymolecular layer of polar liquid on a solid surface.

Some difference between the ζ and ψ_d potentials may be due to the viscoelectric effect [11] and the roughness of the surface.[12] Later research,[10, 13] however, indicates that the viscoelectric constant was overestimated by more than one order in ref. (11). Hence, with moderate values of ψ_d and of the ionic force, the viscoelectric effect cannot cause a perceptible difference between the ψ_d and ζ potentials and hinder the estimation of the stagnant layer thickness.

Since the time of Freundlich, who introduced the concept of the stagnant layer, it was believed that within this layer the ions, as well as the liquid, were immobile, so that only the mobile part of the diffuse layer contributes to the surface conductance. This assumption is part of Bikerman's [14] surface conductance theory and the Overbeek-Booth double-layer polarization theory,[15] and has persisted up to recently.[16] There are no grounds for this assumption, and it should be discarded. It has long been noted that normal values of ionic mobilities are retained in hydrodynamically immobile gels.[17] The physical meaning of viscosity differs [18] depending on whether we consider the macroscopic flow of the system or the movement of small molecular particles through the same medium. The structural frame of the polymer excludes the possibility of macroscopic movement of liquid, but does not hinder the thermal motion of the liquid molecules and ions,[18] and consequently cannot prevent ion migration in an electric field. Similarly, structure formation of a polar liquid under the effect of a surface gives rise to a hydrodynamically stagnant layer. But the intensive thermal motion of the liquid and the ion is maintained, as evidenced from nuclear resonance data.

Accordingly, all diffuse layer ions are to be considered mobile, and we shall call the diffuse-layer charge the mobile charge in distinction to the electrokinetic charge, which is only part of the mobile charge. Simultaneous determination of the mobile and electrokinetic charges yields information about the ion content in the stagnant layer, from which the difference between the ψ_d and ζ potentials and the stagnant layer thickness may be calculated with the aid of the Gouy-Chapman theory.

Information about the mobile charge can be obtained from surface conductance measurements, but the question then arises as to the possible deviation from the normal of ionic mobilities in the stagnant layer. The difference between the mobile and electrokinetic charges may also be due to surface roughness. This difference should not, however, be sensitive to temperature variations, as is the case in structure formation of the liquid, which determines the stagnant layer thickness.

Fridrikhsberg [2] measured three magnitudes characterizing the electric properties of $BaSO_4$ microcrystal surfaces : the surface conductance, streaming potential and ionic adsorption. Comparing the results, he concluded that the tangential electromigration ionic currents permeate the stagnant layer, as well as the mobile part of the double layer, that the magnitude of ionic mobility in it is close to the bulk magnitudes of the mobilities, and its thickness is approximately 15 Å.

The precision of Fridrikhsberg's method of composite electro-surface measurements is decreased because capillary-porous systems were used. It is difficult to substantiate rigorously the possibility of calculating the specific surface conductance κ^σ and the ζ potential from macroscopic measurements on capillary-porous systems. Moreover, when the contribution of the surface conductance to total conductance of the system becomes perceptible, a necessary condition for measuring κ^σ, there is considerable double-layer polarization, a substantial effect of which cannot be allowed for practically.

However, the effect of double-layer polarization, which complicates investigation of the electrokinetic properties of capillary-porous system, affords new opportunities for studying the stagnant layer in dilute dispersions. Development along these lines was, apparently, retarded because the cumbersome mathematical apparatus in the Overbeek-Booth theory of double-layer polarization of spherical particles made it impossible to take into account the difference between the ψ_d and ζ potentials, manifested in the effect of the ionic stream through the stagnant layer on polarization and electrophoretic movement.   The mathematics of double-layer polarization of particles of regular shape has recently been considerably simplified [20, 21] in the important special case of a thin double layer.   The new mathematical technique permits extension of the theory, and in particular, to base theory on various models of the colloid micelle, taking into account the difference between the ψ_d and ζ potentials.

Under the effect of an external electric field, tangential flows of diffuse layer ions arise, which redistribute them over the surface; the double layer is deformed and polarized, and thus departs from the original spherical symmetrical structure.

Steady tangential flows of double-layer ions are maintained through ion exchange with the contiguous volume of electrolyte.   The steady exchange of cations and anions is possible only when concentration differences arising beyond the double layer as well.   A change in electrolyte concentration along the outer double-layer boundary causes a change in its thickness, the latter becoming less when the electrolyte concentration is higher, and increases with a decrease in electrolyte concentration.

An essential simplification in thin double-layer theory[21] is that although the double layer as a whole is in a non-equilibrium state, equilibrium is maintained locally between the given double-layer area and the contiguous volume of electrolyte.   For each double layer area, i.e., at fixed θ, Boltzmann's formula retains its validity, connecting the ionic concentration inside the double layer $C^{\pm}(X,\theta)$ with the concentration at its outer boundary $c^{\pm}(\theta)$ and with the change in potential across the double layer $\phi(X,\theta)$:

$$C^{\pm}(X,\theta) = c^{\pm}(\theta) \exp\left\{\mp z^{\pm} e\phi(X,\theta)/kT\right\}, \tag{1}$$

where θ is the angle with the external field E, X is the distance to the surface, $z^{\pm}$ is the electrovalence.

The deviation of $\phi(X,\theta)$ from the initial spherically symmetrical $\phi^{\pm}(X)$, caused by the concentration change $c^{\pm}(\theta)$, can be expressed in general form through $c^{\pm}(\theta)$ on the basis of the local equilibrium between the given double-layer area and the contiguous volume of electrolyte.   Functions are derived in ref. (20)-(21) describing the spatial potential and charge distribution in the polarized double layer and the contiguous electrolyte volume, involved in the process of exchange with the double layer.   This yielded a formula [22] for electrophoresis velocity complicated by double-layer polarization U_e.   We present the formula for dimensionless electrophoretic mobility in a symmetrical electrolyte, $z^{+} = z^{-} = z = 1$:

$$\tilde{u}_e = \frac{6\pi\eta_1 e}{\varepsilon_1 kT}\frac{U_e}{E} = \tfrac{3}{2}\tilde{\zeta} -$$

$$\frac{3\tilde{\zeta}[(2+6m)\sinh^2(\tilde{\zeta}/4)+q]+6\ln\cosh(\tilde{\zeta}/4)[(2+6m)\sinh(\tilde{\zeta}/2)-3m\tilde{\zeta}+2g]}{\kappa a+(8+24m)\sinh^2(\tilde{\zeta}/4)-24m\ln\cosh(\tilde{\zeta}/4)+4q}, \tag{2}$$

where

$$q = p(\cosh(\tilde{\psi}_d/2)-\cosh(\tilde{\zeta}/2)), \quad g = p(\sinh(\tilde{\psi}_d/2)-\sinh(\tilde{\zeta}/2)),$$

$$\tilde{\psi}_d = (e\psi_d/kT), \quad \tilde{\zeta} = (e\zeta/kT).$$

η_1 and ε_1 are the viscosity and dielectric constant of the liquid, p is the ratio of the diffusion coefficients in the stagnant layer and in the bulk, m is the dimensionless parameter given in ref. (15), (20), a is the radius of the sphere, κ^{-1} is the double layer thickness.

The second term, always negative, characterizes the decrease in velocity due to polarization.   Factors q and g characterize the effect of ionic currents through the stagnant layer on polarization and electrophoresis. If the electrophoretic movement is measured on three fractions of spherical particles with surfaces of identical nature (i.e., ζ and ψ_d is the same for all three fractions and κa is known) we obtain three equations with three unknown, from which we determine ζ, ψ_d and p.

Since with double-layer polarization the particle acquires an induced dipole moment, the dielectric constant of the suspension should differ from that of the medium not only because of the trivial effect due to the difference between the dielectric constants of the medium and the particle. If the frequency of the alternating field is not great, so that local equilibrium can be established between the given double layer area and the contiguous electrolyte volume, the polarization mechanism in the alternating field is the same as in the static, which enabled the author together with Shilov [23] to extend the thin double-layer polarization theory to the case of an alternating field of moderate frequency. Polarization of the particle proved to be associated with the ion concentration decrease along the outer double-layer boundary, which varies periodically in time with a certain lag behind the imposed field. This lag in phase is mathematically expressed in the fact that polarizability, i.e., the ratio of the dipole moment to the external field α^*, is a complex variable, if the time dependence of the alternating field is described by the exponent of the complex argument.

Owing to the presence of conductance, the continuous phase (electrolyte) is also polarized with a perceptible lag in phase, so that its dielectric constant is also complex, $\varepsilon_1^* = \varepsilon_1 - i4\pi K_1/\omega$ being proportional to the conductance of the continuous phase K_1.

The dielectric increment caused by introduction into the continuous phase of colloidal particles with volume-ratio n may be represented by means of the Maxwell-Wagner theory [24]:

$$\Delta\bar{\varepsilon}^* = 3n\varepsilon_1^*\alpha^*. \tag{3}$$

On multiplying Im ε_1^* by Im α^*, a component of Re $\Delta\bar{\varepsilon}^*$ of the suspension arises, which increases considerably with decrease in ω, since the latter is accompanied by an unlimited increase in Im ε_1^*. The complicated formula presented in ref. (25) for the static dielectric increment Re $\Delta\bar{\varepsilon}_l^*$, i.e., Re $\Delta\bar{\varepsilon}^*$ with $\omega\to0$, is simplified for highly charged particles, when exp $(\tilde{\psi}_d/2)\gg1$:

$$\frac{\text{Re }\Delta\bar{\varepsilon}_l^*}{n} = \frac{9\varepsilon_1}{4}\left[\frac{\exp(\tilde{\psi}_d/2)+3m\exp(\tilde{\zeta}/2)}{1+(2/\kappa a)\{\exp(\tilde{\psi}_d/2)+3m\exp(\tilde{\zeta}/2)\}}\right]^2. \tag{4}$$

A huge rise in Re $\Delta\bar{\varepsilon}^{-*}$ was observed in the low-frequency band by Schwann et al.,[26] working with monodisperse suspensions of spherical polystyrene particles in an aqueous solution of KCl. For the special case of $a = 0.094\ \mu$m, $n = 0.3$ the experimentally found relationships Re $\bar{\varepsilon}^*(\omega)$ (curve 1) and Im $\bar{\varepsilon}^*(\omega)$ (curve 2) are presented in fig. 1. Assuming $\psi_d = \zeta$, $p = 1$, $\kappa a = 60$ (in accordance with the electrolyte conductance given in ref. (26), and using the measured value of Re $\bar{\varepsilon}^*(\omega)$ at $\omega\to0$, equal to 2370, we arrived at the conclusion that $\tilde{\psi}_d = 3.5$ for the particles. Accepting this value of $\tilde{\psi}_d$ we then calculated Im $\bar{\varepsilon}^*(\omega)$ (curve 3). A certain discrepancy between the experimental and theoretical curves of Im $\bar{\varepsilon}^*(\omega)$ may be due to the fact that at the volume-ratio of the particles used in the experiment their diffusion atmospheres greatly overlapped, a complication which cannot be readily taken into consideration in the theory, and hence was neglected.

For the interpretation of their experimental data, Schwann and collaborators used Schwarz's hypothesis [24] about the peculiar behaviour of counterions which they called " bound ". These ions are said to be capable of redistribution under the effect

of the field along the particle surface without moving away from it, i.e., the possibility
of ion exchange with the disperse medium is excluded.   Such behaviour of the double-
layer ions has not yet been explained on double-layer theory; moreover, in later
papers the investigators note the inadmissibility of such an idealization.[28]   The
double-layer model accepted by Schwarz assumes that the outer part of the double
layer is formed by bound ions only and there is, accordingly, no diffusion layer.
Within the frame work of this model, Schwarz satisfactorily interprets the experimental results.

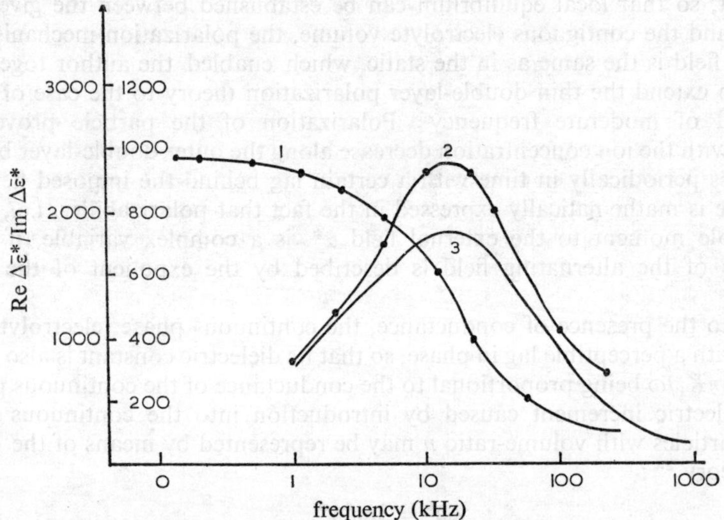

FIG. 1.—Dispersion of dielectric constant of dispersion polystyrene spherical particles; experimental
data [26] and theory of polarization of diffused part of double layer [25]; 1 and 2, Re $\Delta\bar{\varepsilon}^*$ and Im $\Delta\bar{\varepsilon}^*$
from data of ref. (26); 3, Im $\Delta\bar{\varepsilon}^*$ theoretical curve.[25]

Even if there is no diffuse atmosphere in an equilibrium double layer, as assumed
by Schwarz, on polarization of the bound counterion layer, a diffuse atmosphere
appears, locally compensating the polarization charge of the bound ions.   Since
Schwarz did not have at his disposal a theory of the polarization of the diffuse part
of the double layer, his theory did not take into account the potential jump in this
diffuse atmosphere and its effect on the tangential transfer of the bound ions.   The
correction of this error in a paper by Shilov and the author [25] dealing with Schwarz's
model indicates that in Schwarz's theory Re $\Delta\bar{\varepsilon}^*(\omega)$ is over-estimated by more than
one order, i.e., there can be no agreement between Schwarz's theory and Schwann's
experiment.

Ions bound according to Schwarz's theory affect $\Delta\bar{\varepsilon}^*$ but cannot contribute to ζ.
Hence, experimental proof of the equality of ζ and ψ_d potentials by the electrophoresis
data and dielectric measurements is at the same time proof of the non-existence of
Schwarz's bound ions.

Chelidze [29] measured U_e and $\bar{\varepsilon}^*(\omega)$ for almost monodispersed diluted latex
suspensions of nairite ($a = 0.13 \ \mu$m, $\kappa a = 15$) and chloroprene ($a = 0.45 \ \mu$m,
$\kappa a = 60$) in aqueous solutions obtaining values of 3400 and 1400 for Re $\Delta\varepsilon_i^*/n$.
According to these experimental data, ζ_{SM} calculated according to Smoluchowski's

formula, equals 2.3; $\tilde{\psi}_d$ and $\tilde{\zeta}$, calculated from the solution of the system of equations
(2) and (4) equal 3.5 and 3.5.   For chloroprene $\zeta_{SM} = 2.5$, $\tilde{\psi}_d = 3.2$, $\tilde{\zeta} = 3$.

The difference between ψ_d and ζ_{sm} and the agreement within limits of experimental error of ψ_d and ζ calculated from eqn (2) and (4) corroborate the correctness of the theory of double-layer polarization in a direct and alternating field, its effect on electrophoresis and $\Delta\bar{\varepsilon}^*$. Along with the difference of more than one order between the experimental values [26] of Re $\Delta\bar{\varepsilon}_l^*$ and those calculated from the refined formula for $\Delta\bar{\varepsilon}^*$ derived for the Schwarz model,[25] Chelidze's experiments [29] indicate that Schwarz's model does not agree with the facts, at least with respect to the investigated systems, and that $\Delta\bar{\varepsilon}^*(\omega)$ measurements constitute a promising method for measuring the ψ_d potential. In an experimental verification of the DLVO theory [13, 30] or the electroviscous effect on monodisperse suspensions it is advisable to measure the large low-frequency dielectric dispersion rather than electrophoresis.

In addition, agreement of ζ and ψ_d indicates that in the investigated systems boundary layers are either absent or so thin as compared to κ^{-1} that they cannot be detected with the present experimental precision. To secure detection and measurement of the boundary layer thickness by the recommended method, latices should be used for which the presence of boundary layers is more probable (those studied in ref. (5), for instance) and κ^{-1} should be decreased by raising the electrolyte concentration. If not only Re $\Delta\bar{\varepsilon}_l^*$ but the entire low-frequency section of the dispersion curve is used, it is possible to determine the change in ionic mobilities in the boundary layer along with ζ and ψ_d.

[1] J. Th. G. Overbeek, *Pure Appl. Chem.*, 1965, **10**, 359.

[2] D. A. Fridrikhsberg, and V. Y. Barkovsky, *Kolloid Zhur.*, 1964, **26**, 722.

[3] N. Bondarenko, S. Nerpin, *Bulletin RILEM*, 1964, **29**, 13; N. Bondarenko, S. Nerpin, *Int. Congr. Pure Appl. Chem.* (Moscow, 1965), thesis A and B.

[4] S. V. Nerpin and A. F. Chudnovsky, *Soil Phys.*, (Nauka, Moskow, 1967), chap. 13.

[5] G. A. Johnson, S. M. A. Lecchini, E. Y. Smith, J. Clifford and B. A. Pethica, *Disc. Faraday Soc.*, 1966, **42**, 120; B. V. Deryaguin, *Disc. Faraday Soc.*, 1966, **42**, 109.

[6] B. V. Deryaguin and G. P. Sidorenkov, *Doklady A.N. S.S.S.R.*, 1941, **32**, 622.

[7] B. V. Deryaguin, G. P. Sidorenkov, E. A. Zubashchenko and E. V. Kiseleva, *Kolloid Zhur.*, 1947, **9**, 335.

[8] ref. (4), chap. I, §9.

[9] D. A. Haydon, *Proc. Roy. Soc. A*, 1960, **258**, 319.

[10] D. Stighter, *J. Phys. Chem.*, 1964, **68**, 3600; *J. Colloid Interface Sci.*, 1967, **23**, 379.

[11] J. Lyklema and J. Th. G. Overbeek, *J. Colloid Sci.*, 1961, **16**, 501.

[12] J. J. Bikerman, *J. Chem. Phys.*, 1941, **9**, 880.

[13] A. Wattilon and A. M. Joseph-Petit, *Disc. Faraday Soc.*, 1966, **42**, 143.

[14] J. J. Bikerman, *Z. phys. Chem. A*, 1932, **163**, 378.

[15] J. Th. G. Overbeek, *Kolloid Beihefte*, 1943, **54**, 287; F. Booth, *Proc. Roy. Soc. A*, 1950, **203**, 514.

[16] P. H. Wiersema, A. L. Loeb and J. Th. G. Overbeek, *J. Colloid Interface Sci.*, 1966, **22**, 78.

[17] T. Graham, *Ann.*, 1862, **121**, 1; R. Taft and L. Malm, *J. Phys. Chem.*, 1939, **42**, 499.

[18] H. Moraweta, *Macromolecules in Solution* (Wiley Interscience, N.Y., chap. 2 § A5.

[19] B. V. Deryaguin and S. S. Dukhin, *Kolloid Zhur.*, 1969, **31**, 350.

[20] S. S. Dukhin, *Sbornik issledovaniya v oblasti poverkhnostynykh sil*, (Nauka, Moscow, 1967), p. 335.

[21] S. S. Dukhin and V. N. Shilov, *Kolloid Zhur.*, 1969, **31**, 706.

[22] S. S. Dukhin, ref. (20), p. 364; S. S. Dukhin and N. M. Semenichin, *Kolloid Zhur.*, 1970, **32**, 360.

[23] V. N. Shilov and S. S. Dukhin, *Kolloid Zhur.*, 1970, **32**, 117.

[24] K. W. Wagner, *Arch. Electrotechn.*, 1914, **2**, 371.

[25] V. N. Shilov and S. S. Dukhin, *Kolloid Zhur.*, 1970, **32**, no. 2, S. S. Dukhin, *Dielectric Properties of Disperse System in Surface and Colloid Science*, ed. Egon Matijevic, (Wiley, N.Y., 1970), vol. 3, 293.

[26] P. H. Schwan, G. Schwarz, J. Maszuk and H. Pauly, *J. Phys. Chem.*, 1962, **68**, 2626.

[27] G. Schwarz, *J. Phys. Chem.*, 1962, **68**, 2636.

[28] I. P. MacTague and J. H. Gibbs, *J. Chem. Phys.*, 1966, **44**, 4295.

[29] T. L. Chelidze and V. N. Shilov, *Kolloid Zhur.*, in press.

[30] R. H. Ottewill and J. N. Shaw, *Disc. Faraday Soc.*, 1966, **24**, 154.

GENERAL DISCUSSION

Dr. L. M. Barclay (*B.P. Ltd., Penarth*) and **Dr. R. H. Ottewill** (*University of Bristol*) (*communicated*): In our paper we have reported measurements of the force of interaction between montmorillonite plates as a function of electrolyte concentration and the distance of separation between the plate surfaces. The other model which is of interest theoretically is that for interaction between spherical particles. For this case the curve of potential energy V against distance H_0 of surface separation has the general form shown in fig. 1a. Three features of the curve are of particular interest, the secondary minimum S, the point of inflexion I, and the primary maximum M. Differentiation of the appropriate expressions for potential energy [1] of interaction between spherical particles leads to a curve of force $(-\partial V/\partial H_0)$ against H_0. This has the form shown in fig. 1b and it is clear that the maximum force occurs at

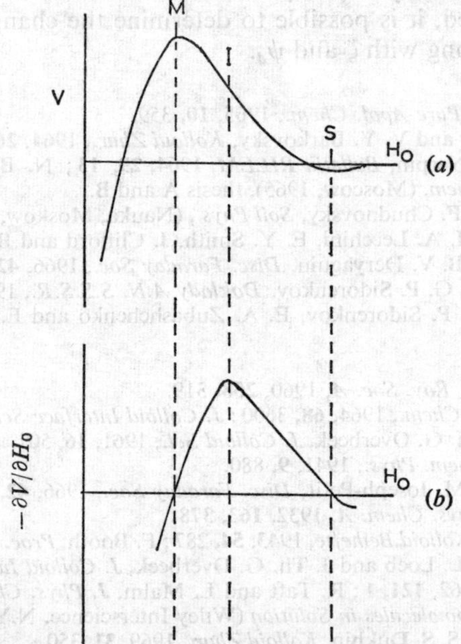

Fig. 1.—Potential energy (*a*) and force (*b*) againt distance for interaction between two spherical particles (schematic).

the distance at which the point of inflexion occurs on the potential energy curve; at the primary maximum and the secondary minimum the force is zero. Thus, the interaction between spherical particles should lead experimentally to a force against distance curve showing these features, the distances and magnitudes of the force being dependent on particle size and ionic strength. An extensive series of measurements have been made using the apparatus described in our paper with monodisperse

[1] E. J. W. Verwey and J. Th. G. Overbeek, *Theory of Stability of Lyophobic Colloids*, (Elsevier, Amsterdam, 1948).

latex preparation as the disperse system. The results obtained with a latex having a model diameter of 1845 Å are shown in fig. 2. It is clear from this curve that very little pressure is exerted by the particles until the distance of surface separation is about 320 Å. The pressure then rises with decreasing distance and appears to reach a maximal at a separation distance of ca. 230 Å. The distance axis on fig. 2 has been calculated on the basis of hexagonal close-packing between spherical particles. Although this assumption is supported by electron-microscope examinations of highly compressed, then dried, latex dispersions it is not yet certain whether a completely ordered hexagonal close-packed system is maintained at various volume concentrations, or whether ordered domains are formed at the lower pressures; the packings now being examined by optical methods. The effect of a change in diameter of the particles from 1760 to 10 990 Å is shown in fig. 3. The smaller particle size shows a gradual increase of pressure until a volume concentration of ca. % is reached, then the pressure rises rapidly. The larger particle size, however, shows a pressure rise at ca. 60 % volume concentration which is almost perpendicular to the volume-concentration axis. The potential-energy curves for the smaller particle size latex at this ionic strength (10^{-3}) do not show a secondary minimum of any significance, whereas those for the larger particles do. It therefore seems highly probable that the volume concentration at which the pressure starts to rise rapidly with distance corresponds with the packing together of the particles into a secondary minimum structure. From the theoretical calculations, the force of repulsion rises very rapidly for particles at small distance, whereas the force for particles at larger distances shows a resemblance to those expected. Quantitative conclusions can be reached for dispersions of spherically shaped particles once the distance of surface separation has been unequivocally established.

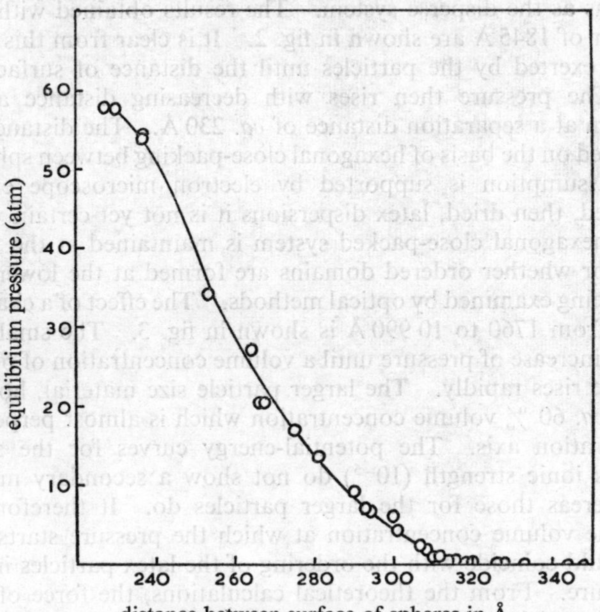

Fig. 2.—Pressure against distance between particle surfaces.   Polystyrene latex at pH 10 and 25°C; diameter of particle = 1845 Å.

Prof. J. Th. G. Overbeek (University of Utrecht) said: One expects the best fit between theory and experiment for larger distances between the plates. According to fig. 5 (and fig. 4) of the paper by Ottewill et al. the calculated repulsion is larger than the one found in the experiments. The fit might be improved by introducing the Stern-correction which would lower the calculated repulsion. Did he try this and does it tend to reasonable values for the capacity of the Stern layer?

Dr. R. H. Ottewill (University of Bristol) said: In reply to Prof. Overbeek, we have made some calculations of the force of interaction including an allowance for the Stern layer and thus does take over the final long distances. Unfortunately this procedure does involve a somewhat arbitrary choice of the Stern adsorption potential. The capacity obtained for the inner layer is about 12 μF.

Prof. G. M. Bell (Chelsea College) and **Dr. S. Levine** (Manchester University) said: These comment on the work of Barclay and Ottewill may also have some relevance to other papers. Denoting the spherical, van der Waals' and structural terms in the force per unit area between plane parallel surfaces by V_R, V_A and A respectively we agree with Barclay and Ottewill that A is not likely to be significant at separations of more than 50 Å and think that this limit could be considerably lower. Existing theoretical values of V_A/V_R are far from exact and differences between these values and measured pressures are not reliable estimates of V_A/V_R. The remainder of our remarks will deal with V_R.

The DVO calculation of V_R excludes the ionic environment (or discharge) effect which depends on the difference between the ionic solution in the

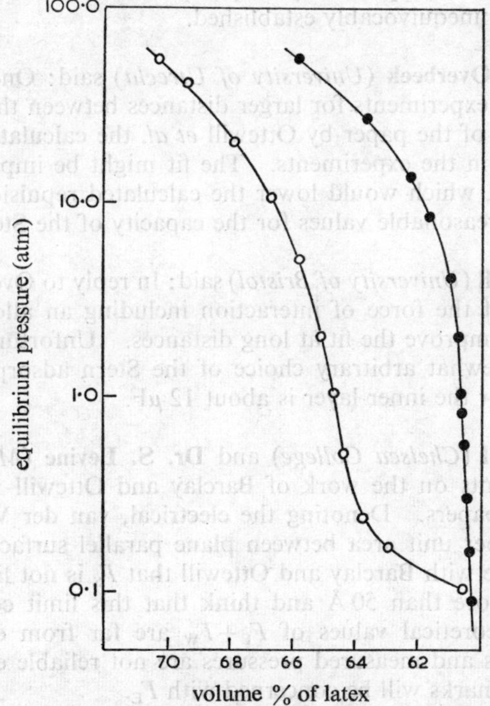

Fig. 3.—Pressure against volume concentration of polystyrene latex at 25°C; ○, 1760 Å diam.; ●, 10 990 Å diam.

latex preparations as the disperse system. The results obtained with a latex having a modal diameter of 1845 Å are shown in fig. 2. It is clear from this curve that very little pressure is exerted by the particles until the distance of surface separation is about 320 Å. The pressure then rises with decreasing distance and appears to reach a maximum at a separation distance of *ca.* 230 Å. The distance axis on fig. 2 has been calculated on the basis of hexagonal close-packing between spherical particles. Although this assumption is supported by electron microscope examinations of highly compressed, then dried, latex dispersions it is not yet certain whether a completely ordered hexagonal close-packed system is maintained at the various volume concentrations, or whether ordered domains are formed at the lower pressures; the packing is now being examined by optical methods. The effect of a change in diameter of the particles from 1760 to 10 990 Å is shown in fig. 3. The smaller particle size shows a gradual increase of pressure until a volume concentration of 64 % is reached; then the pressure rises rapidly. The larger particle size material, however, shows a pressure rise at *ca.* 60 % volume concentration which is almost perpendicular to the volume concentration axis. The potential-energy curves for the smaller particle size latex at this ionic strength (10^{-5}) do not show a secondary minimum of any significance, whereas those for the larger particles do. It therefore seems highly probable that the volume concentration at which the pressure starts to rise rapidly with distance could coincide with the ordering of the latex particles into a secondary minimum structure. From the theoretical calculations, the force of repulsion rises very rapidly for particles with diameters of 10,000 Å, whereas the rise for particles with diameters of the order of 1,000 Å is gradual. The curves obtained bear a close resemblance to those expected theoretically. More quantitative conclusions can be reached for dispersions of spherically shaped particles once the distance of surface separation has been unequivocably established.

Prof. J. Th. G. Overbeek (*University of Utrecht*) said: One expects the best fit between theory and experiments for larger distances between the plates. According to fig. 5 (and fig. 4) of the paper by Ottewill *et al.* the calculated repulsion is larger than the one found in the experiments. The fit might be improved by introducing the Stern-correction, which would lower the calculated repulsion. Did he try this, and does it lead to reasonable values for the capacity of the Stern layer?

Dr. R. H. Ottewill (*University of Bristol*) said: In reply to Overbeek, we have made some calculations of the force of interaction including an allowance for the Stern layer and this does improve the fit at long distances. Unfortunately, this procedure does involve a somewhat arbitrary choice of the Stern adsorption potential. The capacity obtained for the inner layer is about 12 μF.

Prof. G. M. Bell (*Chelsea College*) and **Dr. S. Levine** (*Manchester University*) said: These comments on the work of Barclay and Ottewill may also have some relevance to other papers. Denoting the electrical, van der Waals' and structural terms in the force per unit area between plane parallel surfaces by F_E, F_W and F_S respectively we agree with Barclay and Ottewill that F_S is not likely to be significant at separations of more than 50 Å and think that this limit could be considerably lower. Existing theoretical values of $F_E + F_W$ are far from exact and differences between these values and measured pressures are not reliable estimates of F_S. The remainder of our remarks will be concerned with F_E.

The DLVO evaluation of F_E excludes the ionic environment (e.g., discreteness-of-charge) effects which depend on the difference between the actual situation in the vicinity of an ion and the mean field situation. These include the image effect due to polarization of the dielectric interface by the ion field and the self-atmosphere effect which for a finite ion includes the " cavity " effect due to the displacement of mean charge by the ionic volume. The physical meaning of these omissions can be appreciated from the fact that for an uncharged air/electrolyte interface the DLVO theory would give uniform ionic concentrations and hence zero excess surface tension. Unfortunately, the introduction of ionic environment and other corrections considerably complicates the DLVO theory and for two interacting charged interfaces adequate numerical values of F_E are not so far available although calculations have been made by Sanfeld [1] and others. Some idea as to possible effects may be obtained from the authors' work [2] on the force between uncharged plates in an electrolyte medium. This is based on a model like that of Onsager and Samaras,[3] but with two interfaces instead of one, and also uses local values of the Debye-Hückel constant κ in the screening terms rather than the bulk value. It is found that

$$F_W = P_1 + P_2 + P_3,$$

where P_1 is the ideal osmotic pressure term, while P_2 and P_3 are due directly to image-self-atmosphere effects. With a point-ion model, the form of these terms is unaltered for symmetrically charged plates but the numerical values alter owing to the changed ionic concentrations. For charged or uncharged plates at separations $2h$ large compared with κ^{-1} the terms $P_2 + P_3$ give an h^{-3} law repulsion. However, the equivalent Hamaker constant is only 3×10^{-14} erg so that this repulsion will be cancelled by the van der Waals attraction unless retardation is appreciable. As pointed out by Pethica,[4] P_1 is attractive for uncharged plates owing to the depletion of ionic concentration between the plates, but for charged plates at sufficiently high potential it is repulsive as in DLVO theory where $F_E = P_1$. The sum $P_2 + P_3$ is always repulsive and may account for the excess repulsion found by Barclay and Ottewill for 0.1 M electrolyte and at some separations for 0.0001 M electrolyte. Caution is necessary as to the overall effect since ionic depletion between the plates due to imaging will reduce the magnitude of the repulsive term P_1. In a complete theory, ionic volume and polarization effects would also have to be considered [1, 5, 6] as well as the effect of the thinness of the plates on image potentials.

The standard DLVO theory of F_E assumes a " reservoir " large enough for the chemical potential of each species to remain effectively constant as the surfaces are moved. This does not seem to be the case in Barclay and Ottewill's experiments where in the initial state nearly all the fluid is between the plates and is then " squeezed out ". Another point is that both the constant potential and the constant surface charge assumptions are artificial. However, the use of a realistic adsorption isotherm would give results lying between the two curves so that if the latter are close together, as in the calculations for 0.0001 M electrolyte, the error is likely to be small.

Dr. L. Barclay (*B.P. Penarth*) and **Dr. R. H. Ottewill** (*University of Bristol*) said: We appreciate the points made by Bell and Levine. We agree that the assumptions of constant charge and constant potential are somewhat artificial, particularly as it

[1] A. Sanfeld, *Thermodynamics of Charged and Polarized Layers* (Wiley-Interscience, London, 1968).
[2] G. M. Bell and S. Levine, *J. Chem. Phys.*, 1968, **49**, 4584.
[3] L. Onsager and N. N. T. Samaras, *J. Chem. Phys.*, 1934, **2**, 528.
[4] B. A. Pethica, *Expt. Cell. Res. Suppl.*, 1961, **8**, 123.
[5] G. M. Bell and S. Levine, *Chemical Physics of Ionic Solutions*, B. E. Conway and R. G. Barradas, ed. (John Wiley & Sons, Inc., 1966). pp. 409-461.
[6] S. Levine and G. M. Bell, *Disc. Faraday Soc.*, 1966, **42**, 69.

is conceivable that the extent of adsorption in the Stern layer could be a function of the distance of separation between the plates at distances as close as those studied.

Prof. H. van Olphen (*Arnhem, Netherlands*) said: With regard to the paper by, Barclay and Ottewill, the alternative approach to swelling measurements on expanding clay systems is to allow individual flakes to swell under no constraints, and to obtain the equilibrium unit layer distance from X-ray diffraction patterns.[1-3] As expected from double-layer theory, equilibrium distances decrease with increasing electrolyte concentration of the medium. At each equilibrium distance, repulsive and attractive forces must cancel each other. However, the calculated double-layer repulsion at each equilibrium distance observed appears to be considerably greater than the van der Waals attraction. If a specific adsorption potential is assumed for the inter-layer counter ions to reduce the repulsion to equal the van der Waals attraction, the adsorption potential would have to vary with electrolyte concentration, which is unlikely. An alternative trivial explanation of the observations is that swelling is limited by cross-linking of parallel particles by non-parallel ones. Estimates of edge-to-face linking forces of bentonite particles derived from rheological observations show that relatively few cross-links would be necessary to limit the swelling as observed.[4]

Dr. R. H. Ottewill (*University of Bristol*) said: The procedure suggested by van Olphen is an interesting alternative to the one that we have employed, but there may be some advantages in starting with a dilute system in which the particles are well dispersed. We consider that in the first compression edge-face links are broken down. However, once the system has been subjected to a pressure of *ca.* 80 atm, then most of the plates would be likely to take up a parallel arrangement. This contention appears to be supported by the fact that after the first compression the pressure can be decreased to *ca.* 1 atm and then recompressed to a high value several times with a high degree of reproducibility in the results. It is probable that the whole system will not be composed of parallel plates; rather that there will be domains with a parallel arrangement of plates. Negative adsorption measurements of the chloride ion gave a value of 690 m^2/g which would suggest that a high degree of dispersion into fundamental plates had occurred. The possible variation of Stern potential with electrolyte concentration could be a deficiency of the theory and it would be interesting to calculate the force of interaction with allowance for the discreteness of the charges. Moreover, it is conceivable that at such close distances of approach the extent of adsorption could be a function of the distance of plate separation.

Dr. B. A. Pethica (*Unilever Res., Port Sunlight*) said: Fundamental to the arguments of the paper by Barclay and Ottewill is the question as to whether the pressures measured by their apparatus reflect the characteristics of interacting plates every-where surrounded by the fluid phase, or whether there is in the cell a gel-like matrix of interacting particles equivalent to a solid structure capable of resisting mechanical deformation. The question can be settled in a variety of ways, the most direct being to relate the pressure-volume data directly to the vapour adsorption isotherm in the same colloid system. Similarly, the general utility of the electrostatic repulsion calculations could be illuminated by a direct Donnan calculation, using the same charge density assumptions as were made in the calculations given in the paper.

[1] K. Norrish, *Disc. Faraday Soc.*, 1954, **18**, 120.
[2] K. Norrish and J. A. Rausell-Colom, *Clays and Clay Minerals*, 1962, **10**, 123.
[3] D. E. Andrews, P. W. Schmidt and H. van Olphen, *Clays and Clay Minerals*, 1967, **15**, 321.
[4] H. van Olphen, *J. Colloid Sci.*, 1962, **17**, 660.

A further thermodynamic question relates to the effect of pressure on the c.m.c. of the polyoxyethylene surfactant, which was used at 7×10^{-5} M. The micelle point at atmospheric pressure is quoted as 7.25×10^{-5} M in one place, and 7×10^{-5} M in another. In either case, the concentration in the compression experiments is close to the c.m.c., and it would be necessary to know the effects of pressure on the c.m.c. and on the adsorption for a full analysis of the pressure effects in the suspensions.

Dr. Th. F. Tadros (*Plant Protection Ltd., Bracknell, Berks*) said: Does the adsorption of $C_{12}E_6$ on montmorillonite above the CMC correspond to vertically oriented chains or does Ottewill obtain adsorption of micelles? Has he measurements of equilibrium pressure in presence of $C_{12}E_6$ at high salt concentrations where the double layer repulsion term tends to zero, leaving V_A and ΔG_s?

Dr. L. M. Barclay (*B.P. Penarth*) and **Dr. R. H. Ottewill** (*University of Bristol*) said: The first part of Pethica's question has essentially been covered in the reply to van Olphen. Pethica is quite correct in stating that the same results should be obtained from measurements of water vapour adsorption on montmorillonite. The comparison is, however, not as easy as it might seem. First, as observed by many authors,[1-3] the water vapour adsorption isotherms on montmorillonites depend on the source of the montmorillonite. For example, in the work of Barshad, a type II isotherm was obtained with a sodium montmorillonite from Otay, California, and a type III isotherm with a sodium montmorillonite from Clay Spur, Wyoming. The results therefore are only meaningful if obtained on exactly the same material, preferably on the same batch. A second problem with comparing published work on water adsorption with our data is that the pressure measurements have been made over a range which corresponds to p/p_0 values in the range 0.9-1.0. The published isotherms are very short of data in this region and it is well known that it is a difficult range in which to measure adsorption. Conversion of published data into pressure against distance curves does give curves resembling our own, but the adsorption data are not precise or extensive enough to allow a quantitative comparison. We have not carried out any Donnan calculations.

Where the effects of pressure on c.m.c. have been measured, the c.m.c. increases with pressure up to 1,000 atm. I am not aware of any similar measurements with non-ionics and we cannot measure this directly in our apparatus. It was possible to analyze $C_{12}E_6$, however, in the solution expelled into the capillary. The concentration of this was that expected from the isotherm and hence we concluded that the effect of pressure on the adsorption equilibrium in the range examined, was negligible. Where the effects of pressure on c.m.c have been measured,[4, 5] the c.m.c. increases with pressure up to 1,000 atm.

Dr. L. Barclay (*B.P. Penarth*) and **Dr. R. H. Ottewill** (*University of Bristol*) said: In reply to Tadros, the adsorption isotherm for $C_{12}E_6$ on montmorillonite is an interesting one since it differs in form from those found on hydrophobic substrates such as silver iodide,[6] Graphon [7] and polystyrene.[8] With these materials the

[1] I. Barshad, *8th Nat. Conf. Clays and Clay Minerals*, 1959, **8**, 84.
[2] W. A. White, *3rd Nat. Conf. Clays and Clay Minerals*, 1955, **3**, 186.
[3] A. C. Zettlemoyer, G. J. Young and J. J. Chessick, *J. Phys. Chem.*, 1955, **59**, 962.
[4] R. F. Tuddenham and A. E. Alexander, *J. Phys. Chem.*, 1962, **66**, 1839.
[5] S. D. Hamann, *J. Phys. Chem.*, 1962, **66**, 1359.
[6] K. G. Mathai and R. H. Ottewill, *Trans. Faraday Soc.*, 1966, **62**, 750, 759.
[7] J. M. Corkill, J. F. Goodman and J. R. Tate, *Trans. Faraday Soc.*, 1966, **62**, 979.
[8] R. H. Ottewill and T. Walker, *Kolloid-Z. Z. Polymere*, 1968, **227**, 108.

isotherm reaches a saturation value close to the c.m.c. With montmorillonite, adsorption continues to increase above the c.m.c. despite the fact that constant activity of the solute is normally assumed to be reached at the c.m.c. In fact, a limiting adsorption value of 2.5×10^{-3} mol/g is only attained at about an equilibrium concentration of 3×10^{-4} M $C_{12}E_6$. On the basis of a surface area of 800 m^2/g this corresponds to an area per molecule of 53 Å^2. It is tempting to conclude that this corresponds to a vertically oriented layer since this value compares favourably with the figure of 55 Å^2 found for adsorption of $C_{12}E_6$ of Graphon and one of 40 Å^2 obtained on polystyrene where it is clear that vertically orientated monolayers are formed. However, in the work of Schott [1] on the adsorption of $C_{12}E_{14}$, who used X-ray methods to determine the distance between the sheets, two steps were found, one at a distance of 4.4 Å and the other at 8.4 Å. The first distance corresponded to a single layer of surface-active agent molecules lying flat on the surface and the second to a bi-layer of molecules. This evidence appeared to indicate attachment of the ethylene oxide groups to the hydrophilic silica layers and this would argue against reorientation to form a vertically orientated layer. It is probably possible for the hydrocarbon chains to associate, however, and thus form dimers on the surface and this would give an area per molecule not drastically different from that found.

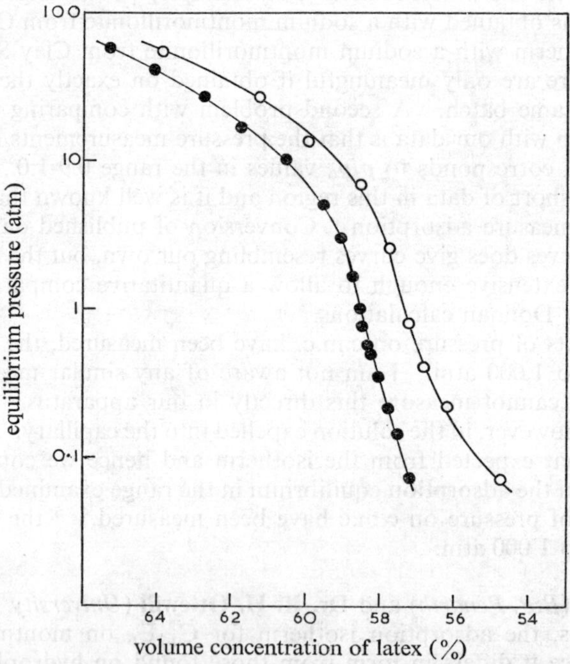

Fig. 1.—Equilibrium pressure against volume concentration for a latex, with an adsorbed layer of dodecylhexaoxyethylene glycol monoether ($C_{12}E_6$): ○, pH 10 without added sodium chloride; ●, pH 9.6 with 0.5 M sodium chloride.

We have not measured the interaction between montmorillonite sheets with adsorbed $C_{12}E_6$ at high salt concentrations. However, we have carried out an experiment of this type using polystyrene latices [2] and the results are given in fig. 1.

[1] H. Schott, *Kolloid-Z.*, 1964, **199**, 158.
[2] L. Barclay, *Ph.D. Thesis*, (University of Bristol, 1970).

Two curves are presented, one for particles having a diameter of 1,845 Å and an adsorbed layer of $C_{12}E_6$ at pH 10.0 without added salt, and the other in the presence of 0.5 M sodium chloride. The addition of salt enables the particles to come closer together as would be expected from the decrease of electrical repulsion, but there is still a very strong repulsion and there is a considerable increase in the gradient at low pressures. The fact that the system remains a stable dispersion can be attributed to " steric stabilization " and this is confirmed by kinetic studies of stability. However, until we have precise measurements of the distances between the spherical particles, we cannot tell whether overlap of the adsorbed layers actually occurs.

Dr. G. Peschel (*Universität Würzburg*) said: With regard to the paper by Ottewill *et al.*, the disjoining pressure between colloidal particles in disperse systems might be sensibly dependent on the diameter of the particles as some evidence shows; e.g., Jura and Harkins, in applying their well-known method for determining the surface area of powders, condensed water from saturated water vapour in titanium dioxide particles which had a mean diameter in the μ region. The obtained layer thickness is of only 5 molecular diameters, which cannot be regarded as a pronounced long-range orientation effect. Zorin [1] on the other hand carried out a similar experiment but using a macroscopic flat glass surface. He found layer thicknesses of about 100 molecular diameters. Therefore, additional information concerning the repelling forces between the particles might be gained by a variation of the particle diameters in Ottewill's apparatus. Are such experiments possible?

Dr. A. L. Smith (*Chem. Dept., Liverpool Polytechnic*) said: In the paper of Vincent and Lyklema, it is stated that titration experiments yield true p.z.c. values which might therefore (by implication) be expected to differ from the isoelectric points (i.e.p.) determined electrokinetically. However, the (σ_0, E) plots from titrations will only have a common intersection point (used to determine the p.z.c.) when there is no specific adsorption, in practice at low electrolyte concentrations. Under these circumstances, the p.z.c. and i.e.p. coincide. There thus seems to be a real experimental discrepancy between authors revealed by table 1, not explicable by specific adsorption. It does not seem necessary to invoke the adsorption of silver nitrate ion pairs, of doubtful existence in bulk solution, to explain the higher differential capacities observed on the positive side of the p.z.c. and still less to explain the asymmetry of the p.z.c. While some specific adsorption of nitrate ions is to be expected, and is indeed revealed by the small shifts of the p.z.c. to higher pAg values in 1 mol dm^{-3} KNO$_3$, the fluoride ion produces differential capacities on the positive side almost as high as those in the presence of nitrates even though no shifts in p.z.c. can be detected.[2] Specific adsorption of the strongly hydrated fluoride ion is moreover not expected and it is hardly likely that the fluoride and nitrate ions would have similar tendencies to form ion-pairs with the silver ion.

The higher capacities on the positive side could be at least partly due to a continuation on the positive side of the rise in capacity K of the Stern layer, evident from all published capacity measurements, as the AgI surface becomes less negatively charged. This variation of K, which cannot on the negative side be an apparent effect due to specific adsorption (of cations) since this would only make the variation larger, could have its origin in water dipole re-orientation,[3] variation in the thickness

[1] Z. A. M. Zorin and N. V. Churayev, *Kolloid Zhur.*, 1968, **30**, 371.
[2] J. Lyklema, *Trans. Faraday Soc.*, 1963, **59**, 418.
[3] S. Levine, G. M. Bell and A. L. Smith, *J. Phys. Chem.*, 1969, **73**, 3534.

and/or effective dielectric constant of the inner layer or possibly solid-state effects.[1]

The two effects cited as giving higher total differential capacities on the positive side, viz., specific adsorption of anions or higher values of K, will have opposite effects on the Stern plane potential ψ_d, the first decreasing ψ_d and the second increasing it. This should be reflected in the electrokinetic ζ potential. Experimental electrophoretic mobilities of positively charged AgI are not noticeably smaller than those on the negative side at the same value of $|\psi_0|$ and, while not excluding some specific adsorption of nitrate ions, are certainly not consistent with adsorption of Ag^+ " largely in the associated form as $AgNO_3$ ".

Dr. B. Vincent (*University of Bristol*) and **Prof. J. Lyklema** (*Wageningen*) said: The first point Smith makes, viz., that there is a real experimental discrepancy between p.z.c. and i.e.p. measurements is virtually a restatement of our remark at the end of the paper. However, we did not imply that this difference was due to specific adsorption of one of the added ions. In fact, table 1 shows that there *are* some specific ionic effects but these are small as compared to the difference between p.z.c. and i.e.p.

In connection with the interpretation of the high capacitances on the positive side of the p.z.c., at 20°C the capacitance in KF is definitely lower than that in KNO_3,[2] although even in KF it is much higher than on mercury. High capacities at the positive side of the p.z.c. have also, though to a lesser extent, been observed by Iwasaki and De Bruyn for Ag_2S [3] but not for mercury. (The high capacitances found on oxidic materials, such as TiO_2 or Fe_2O_3 have a different cause and should not be considered in this connection.) Given the fact that there seems to exist a relationship between Ag^+ as the potential-determining ion and high capacitances, one is almost automatically led to an interpretation in terms of specific interactions involving Ag^+ adatoms. As moreover the high surface charge does not produce a high ζ-potential (it is comparable to that at the negative side, as confirmed by Smith in his remark) it is logical to postulate considerable counterion adsorption in the Stern-layer as well. If, on the other hand, an explanation is sought in terms of factors like dipole orientation and dielectric constant variation, i.e., factors that are not primarily related to the presence of Ag^+-adatoms, it is not clear whether high capacitances are not observed on mercury as well. It was for these reasons, and others given in the paper, that we drew attention primarily to typical chemical effects, occurring in the Stern layer, thereby not excluding other, but secondary possibilities.

Finally, the fact that we are concerned with a trend that is typical for AgI, or at least for silver salts, implies at the same time that we should not make recourse to bulk phenomena (like the extent of ion-pairing in solution) to interpret our experimental facts, because, if they were due to bulk phenomena, they should occur in *all* double layers, in disagreement with established facts.

Dr. Th. F. Tadros (*Plant Protection Ltd., Bracknell, Berks*) said: In the paper by Vincent and Lyklema, they have assumed that Ag^+ adsorbs at 20°C largely in an associated form, i.e., as $AgNO_3$, whereas at 0°C, $AgNO_3$ ion-pairs are more dissociated. The association constant is given by the Fuoss equation,

$$K_A = (4\pi Na^3/3000) \exp (e^2/DakT),$$

[1] E. P. Honig, *Trns. Faraday Soc.*, 1969, **65**, 2248.
[2] J. Lyklema and J. Th. G. Overbeek, *J. Colloid Sci.*, 1961, **16**, 595.
[3] I. Iwasaki and P. L. de Bruyn, *J. Phys. Chem.*, 1958, **62**, 594, fig. 8.

where a is the distance of closest approach between the paired ions and the other terms have their usual meaning. They mentioned in their paper that as a result of increased hydrogen-bonded structuring from 20 to 0°C, D increases from 7.5 ± 1.5 to 8.0 ± 1.5. However, this increase in D does not outweigh the decrease in T and the final result would be higher K_A at the lower temperature contrary to their explanation. Do I understand that they are speaking in terms of structure-stabilized ion-pairs?[1]

Dr. B. Vincent (*Bristol University*) and **Prof. J. Lyklema** (*Wageningen*) said: We agree with Tadros that a decrease in ion pairing with decreasing temperature is not what one would intuitively expect. However, the experimental facts do show a definite *decrease* in Ag^+-adsorption with *decreasing* temperature and neither is this trend *a priori* expected, at least not on the basis of classical energetic or electrical considerations. It was this trend, in conjunction with a few auxiliary considerations, that led us to think of increased structure formation with decreasing temperature. The anomalous behaviour of the association of Ag^+ and NO_3^- in the Stern layer, as tentatively postulated by us must be looked upon in a similar fashion: it is largely due to—presumably structural—causes that are not accounted for by the Fuoss equation. In this connection, the slight increase of the inner layer dielectric constant with decreasing T (fig. 7d) in our paper applies to *negatively* charged AgI-surfaces, whereas the structure formation takes place mainly at the *positive* side of the p.z.c.

Dr. S. Levine (*Manchester University*) and **Dr. A. L. Smith** (*Liverpool College of Technology*) (*communicated*): The tentative explanation given by Vincent and Lyklema for their increase with temperature in the pAg of the AgI suspension at the p.z.c. is not convincing. Recently Honig[2] has queried the neglect by colloid chemists of the diffuse layer in the solid AgI phase due to lattice (Frenkel) defects and a further study of this problem has been made by Levine *et al.*[3] Some of the theory in these papers seems relevant to the shift with temperature in the pAg at the p.z.c. We use the subscripts, c, b and s to denote the solid phase, the (surface) phase boundary and the aqueous solution phase respectively. Then the condition of uniform electrochemical potential μ_{Ag^+} of the Ag^+ ion yields the relations

$$\mu_{Ag^+} = \mu_s^\circ - 2.303kT\,\mathrm{p}Ag = \mu_b^\circ + e_0\psi_b + kT \ln v_b = \mu_c^\circ + e_0\psi_c + kT \ln n_c.$$

Here μ_c°, μ_b° and μ_s° are functions of pressure and temperature; ψ_c and ψ_b are the electrostatic potentials inside the AgI crystal and on the AgI surface (taking the potential zero in the interior of the solution), n_c is the volume density of Frenkel defects in the solid phase and v_b the surface density of corresponding defects on the AgI surface (responsible for the excess or deficit of Ag^+ ions on this surface): e_0, k, and T have their usual meanings. At the p.z.c., in the absence of specific adsorption of indifferent electrolyte, we may assume $\psi_b = \chi_b$ (the χ-potential) the potential drop at the p.z.c. due to water dipole orientation, mainly in the Stern inner region.[4] We note that $\psi_c \neq \psi_b$ at the p.z.c. because there would still be a surface charge equal in magnitude but opposite in sign to the diffuse layer charge inside the solid phase. The surface defects may have various origins, e.g., vacancies in steps, kinks on a step etc., and we assume such defects lead to an excess or deficit of both Ag^+ and I^- ions at the surface. Here n_c, v_b and χ_b will all be functions of temperature. Also

[1] e.g., R. M. Diamond, *J. Phys. Chem.*, 1963, **67**, 2513.
[2] E. P. Honig, *Trans. Faraday Soc.*, 1969, **65**, 224; *Nature*, 1970, **225**, 537.
[3] P. L. Levine, S. Levine and A. L. Smith, *J. Colloid Interface Sci.*, 1970, **34**, 549.
[4] S. Levine, G. M. Bell and A. L. Smith, *J. Phys. Chem.*, 1969, **73**, 3534.

$\mu_s^o - \mu_b^o$, which is partly due to the difference between the free energies of solvation of Ag^+ ion in the solution and on the surface, will depend on temperature. These considerations certainly suggest a dependence of pAg on temperature at the p.z.c.

Dr. B. Vincent (*Bristol University*) and **Prof. J. Lyklema** (*Wageningen*) said: We thank Levine for his suggestion to explain the observed shift in the p.z.c. on the basis of a variation of n_c and v_b with temperature. A quantitative study will be needed to evaluate to what extent the changes in the properties of the solid phase contribute to the observed shift. At any rate, these changes must be reversible with change in temperature because the p.z.c. measurements were reversible as a function of T. At the same time, variation of the solid state properties cannot be solely responsible for the observed trends. First, there are definite lyotropic effects in the p.z.c., in the adsorption of potential-determining ions and in the interfacial excess entropy (not given in fig. 5 but mentioned in the text of our paper). This proves that at least a (great) part of the effects has to do with the properties of the Stern layer. Moreover, the more or less pronounced transition around 20°C in, e.g., the surface excess entropy and in the double-layer capacitance at the positive side of the p.z.c., is not reflected in the solubility product, indicating that we are concerned not with a transition in the solid-phase properties but in the surface properties. In conclusion, there is now ample evidence that we are concerned with interfacial effects. There is no indication for solid phase effects as well but their existence cannot be excluded *a priori*. Quantitative evaluation would be desirable.

Nuclear Magnetic Resonance Studies of Water in Disperse Systems

By J. Clifford, J. Oakes and G. J. T. Tiddy

Unilever Research Laboratory, Port Sunlight, Cheshire, England

Received 5th May, 1970

The nuclear magnetic resonance relaxation times of water protons in aqueous suspensions of polystyrene lattices and in lamellar mesomorphic phases have been measured. The results indicate that the surfaces examined have no long range effect on water structure. The main effect is the binding of water molecules to charged surface groups though an additional effect occurs when pores are formed by flocculation of the polystyrene spheres or when the lamellar phase contains water layers less than 20 Å thick.

The effect of surfaces on the structure and properties of liquid water has been much discussed [1, 2] as a possible factor determining the behaviour of disperse systems. For example, it has been suggested that the effect of surfaces on water contributes to the stability of colloidal dispersions,[3] and that the state of water in biological systems is very different from that of ordinary bulk water.[4] However, much of the experimental evidence concerning water in disperse systems is indirect or incomplete, i.e., it has been obtained from work on inadequately characterized systems. Consequently there remains much doubt as to the nature and range of surface effects in aqueous dispersions.

The measurement of the nuclear magnetic resonance relaxation times of water protons is a direct method of determining the mobility of water molecules.[5] It has been extensively applied to the investigations of monolayer amounts of water adsorbed on surfaces [6] and to a wide variety of complex biological and other practical systems, but less often to well-characterized colloidal dispersions in which solid surfaces are in contact with bulk water. Most of the work which has been done on such systems (e.g., micelles,[7] silica dispersions,[8] clays,[9] mesomorphic phases of soaps [10]) has indicated that, unless the water is present in small or intermediate [11] pores, the perturbing effect of surfaces on water structure is confined to relatively few (1-3) molecular layers of water adjacent to the surfaces and that long-range effects are absent. However, investigations of dispersions of polyvinyl acetate spheres in water [3] indicated that (i) fastest flocculation rates were markedly slower than predicted by the Smoluchowski theory (ii) that the particles are surrounded by a 30 Å-thick layer of water with a viscosity 1,000 times greater than that of bulk water at the same temperature, and (iii) that as the concentration of particles is increased, a co-operative effect occurs, increasing the viscosity of water between particles, and that this becomes noticeable at distances of about 0.1 μm. It has been suggested that these effects were due to highly porous particle surfaces.[1] Because of this, and because the long-range effect seemed anomalous when considered in relation to results obtained in other systems, measurements have been made on well-characterized polymer sphere dispersions and are described here.

As it was found that water structuring effects were very small for water layers of the dimensions obtainable in these systems, the properties of water in lamellar mesomorphic phases were also investigated. In these systems water layers of from 50 to 8 Å thick can be obtained. The behaviour of water in such phases is of particular interest in that each water layer together with its two boundary monolayers of soap molecules resembles a thin soap film. Consequently investigations of the state of the water in these systems are relevant to studies of soap film properties.

EXPERIMENTAL

MATERIALS

POLYSTYRENE LATICES

Polystyrene latices were prepared by emulsion polymerization of styrene (with sodium decanoate as the emulsifying agent) by a method similar to that described by Ottewill.[12] Impurities were removed by filtration followed by dialysis. The dialysis is continued (for about 6 weeks) until the electrophoretic mobility of the particles at pH 9 remained constant. Full details of the preparation and characterization of the latices will be given in a later publication in which other properties of the dispersions will be described. For our work two preparations were used. Their characteristics are summarized in table 1.

TABLE 1

mean particle diam.	COO$^-$ groups per particle	area per COO$^-$ group Å
5 200 Å	740×10^3	119
890 Å	1.8×10^3	1 370

Histograms of particle size distributions are shown in fig. 1. The particle diameters were determined by electron microscopy and confirmed by light scattering. The concentrations of COO$^-$ groups were estimated from potentiometric titrations.

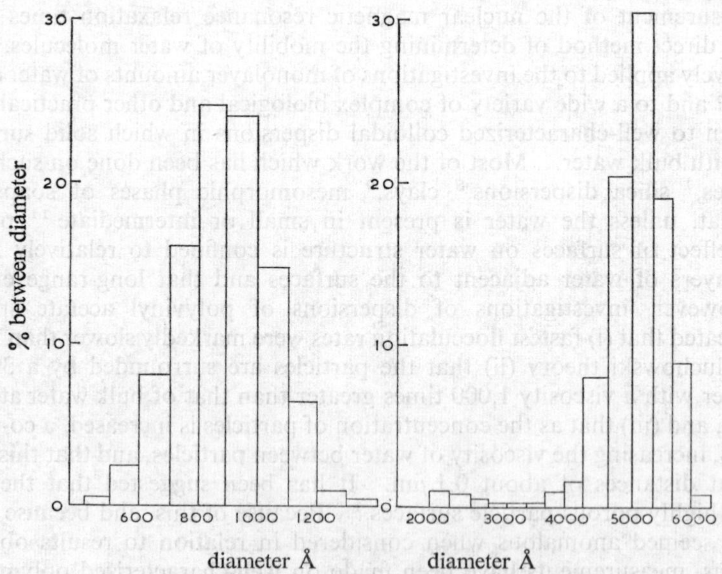

FIG. 1.—The size distribution histograms for the polystyrene particles.

All n.m.r. measurements were carried out at pH of between 7 and 8 where the COO⁻ groups are completely ionized. A small dependence of water proton spin-spin relaxation rate on pH was noted. This was the same as that observed for pure water, the rate being slightly greater at pH 6-8 than at higher and lower pH values (at pH values of less than 5 flocculation occurred). This effect, which is attributed to exchange modulated interactions with O^{17} nuclei does not appear to be affected by the presence of the colloid particles. For the measurements described, samples of a high solids content were required. These were obtained by concentrating dilute suspensions by means of centrifugation. Provided that concentration is to no more than about 20 % volume fraction of solid, this method is completely reversible. The latices could be redispersed by shaking—or more conveniently by means of ultra-sonic irradiation. The n.m.r. measurements on the water protons were time independent and were unaltered by cycles of concentration and dilution and by cycles of temperature variation. If concentrations of more than 20 % volume fraction of solid were prepared, irreversible flocculation occurred, and irreversible changes occurred in the n.m.r. properties of the suspensions. Such changes occurred at much lower volume fractions if other methods of concentration—evaporation of water in a stream of gas, or under vacuum, or freeze drying—were used. Consequently all the results described were obtained on systems concentrated by centrifugation.

LAMELLAR MESOMORPHIC PHASES

The system decanol+sodium caprylate+water was used to provide lamellar mesomorphic phases, as Eckwall [13] has shown that, in this system, the lamellar D phase is given by a wide range of compositions at room temperature. A triangular diagram, based on Eckwall's work, showing the D-phase region is given in fig. 2. The D-phase consists of bilayers of caprylate and decanol molecules separated by layers of water. Mixtures of the required compositions were made up, melted, mixed thoroughly and equilibrated at room temperature for at least one week before n.m.r. measurements were made. X-ray diffraction measurements were also made on the samples. These were in reasonable agreement with the measurements made by Eckwall (the small differences observed do not affect our interpretation of our n.m.r. results), and confirm the lamellar structure suggested by him for the D-phase.

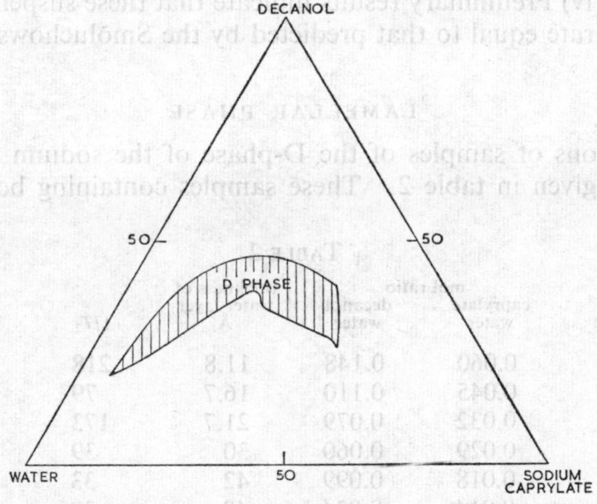

Fig. 2.—Lamellar D-phase compositions in the sodium caprylate+decanol+water system at 20°C.

MEASUREMENTS

Nuclear magnetic resonance relaxation times of water protons in these systems were measured with a Bruker Physik pulse spectrometer at a frequency of 60 MHz. The temperature of the samples was controlled to within 1°C with a gas flow thermostat. Spin-lattice relaxation times were measured with 90°-90° pulse programmes. Spin-spin relaxation times were measured by the Gill Meiboom [14] technique. Preliminary work shown that, for these samples, the results of spin-spin relaxation time measurements were dependent on pulse separations, decreasing as pulse separations are increased. This effect, which has been attributed to the diffusion of molecules in the magnetically inhomogeneous disperse system,[15] was eliminated by the use of very short pulse separations (80 μs). The precision of the measurements is better than $\pm 5\%$ for T_1, and for T_2 in the polymer latices, and $\pm 10\%$ for T_2 in the lamellar phase systems where separation of non exponential decay curves into two components is involved.

RESULTS

POLYMER LATICES

The results of the n.m.r. measurements on the polystyrene latices are shown in fig. 3, 4, 5 and 6. The relaxation rates $1/T_1$ and $1/T_2$ of the water protons are shown as functions of $V/(1-V)$, where V is the volume fraction of solid material. For all samples, exponential decay curves were obtained which could be characterized by a single relaxation time. (The polystyrene protons gave free induction decays with characteristic times $T_2 \sim 1 \times 10^{-5}$ s and $T_1 \sim 1.5$ s and did not interfere with the measurements on the water protons). Samples containing less than 20 % v/v of solid gave T_1 results which were identical with those of pure water within experimental error.

Some other observations were: (i) Samples with concentration of solid of more than 20 % were flocculated and could not be entirely redispersed by shaking or by ultrasonic irradiation. (ii) All samples freeze at temperatures between 0 and $\sim 2°C$, i.e., no observable n.m.r. signal remains below this temperature. (iii) At solid concentrations of below 20 % v/v the chemical shift of the water protons in the lattices is identical with that of bulk water within experimental error. At concentrations greater than this, a concentration dependent chemical shift is observed—about 1 p.p.m. to high field relative to water at the same temperature for a 75 % solid content system. (iv) Preliminary results indicate that these suspensions have a maximum flocculation rate equal to that predicted by the Smoluchowski theory.[16]

LAMELLAR PHASE

The compositions of samples of the D-phase of the sodium caprylate, decanol, water system are given in table 2. These samples containing both water and lipid

TABLE 2

sample	mol ratio caprylate water	decanol water	thickness of water layer Å	$1/T_2$	$1/T_{1W}$
1	0.060	0.148	11.8	218	1.48
2	0.045	0.110	16.7	79	1.04
3	0.032	0.079	21.7	173	0.57
4	0.029	0.060	30	39	0.52
5	0.018	0.099	42	33	0.49
6	0.014	0.034	48	30	0.45
7	0.045	0.135	43	172	0.58
8	0.049	0.147	15	135	1.7
9	0.074	0.143	16.5	206	1.95
10	0.102	0.142	9.7	426	2.63
11	0.129	0.141	8.3	891	3.20
12	0.160	0.142	8.0	1,211	2.0

protons and their relaxation behaviour was more complex than was observed for the polystyrene suspensions. The spin-spin relaxation curves were non-exponential and consisted of two components. One had an intensity proportional to the amount of non-aqueous component present and spin-spin relaxation times of from 2.5×10^{-5} to 5×10^{-5} s. The other, proportional to the fraction of water present in the sample, had relaxation rates $1/T_2$ which are given in table 2 and shown in fig. 7 as a function of mol caprylate/mol water. For some samples the slower decay curve was itself non-exponential. This is considered to be due to the presence of an additional lamellar phase, probably because of incomplete equilibration in these highly viscous systems. For these samples the spin-spin decay curve is resolved into its two components, and an average relaxation rate calculated from their proportions and relaxation rates is included in fig. 7 and table 2.

The spin-lattice relaxation curve was always exponential for all samples. Thus, the measured spin-lattice relaxation rate $1/T_1$, shown in fig. 8 as a function of mol caprylate/mol water is an average of the relaxation rates of the lipid protons and the water protons. This averaging is due to the combined effects of spin diffusion in the non-aqueous part of the sample and molecular diffusion in the water. As we are interested only in the water protons in the present study, the relaxation rate of the lipid protons was measured by using caprylate + decanol + deuterium oxide systems (the results are shown in fig. 8) and its effect on the average relaxation rate calculated and subtracted from the measured rate for each sample and the spin lattice relaxation rate of the water protons, $1/T_{1w}$ estimated and recorded in table 2 and given as a function of mol caprylate/mol water in fig. 9.

DISCUSSION

POLYSTYRENE LATICES

It has been shown [17] that if protons exist in two states exchanging so that on average n.m.r. relaxation rate is observed,[19] then provided that $T_A \gg T_B$, $S_A \ll S_B$, $P_A \gg P_B$,

$$\frac{1}{T} = \frac{1}{T_A} + \frac{P_B}{P_A}\left(\frac{1}{T_B + S_B}\right),$$

where T is the observed relaxation time, T_A and T_B are the relaxation times in the two states, P_A and P_B are the populations of the two states, and S_B is the average residence time of a proton in the state B.

If it is assumed that water affected by the particle surface exchanges with bulk water, then we may denote state A as bulk water and state B as water modified by the particles. Consequently, there should be a linear relationship between the observed relaxation rate $1/T$ and $V/(1-V)$, where V is the volume fraction of solid, provided that there is no interaction between the particles in their affect on water structure. Fig. 3-6 can be considered as describing two distinct concentration regions, 0-20 % solids content, where a non-flocculated dispersion is being examined, and more than 20 % solid content, where flocculation has occurred. Below 20 % solid content a linear relation between $1/T_2$ and $V/(1-V)$ is observed for both the 890 and 5,200 Å diam particles. There is no measurable effect of the particle surfaces on T_1 of the water protons, when solids content is less than 20 %.

The lack of any measurable decrease in T_1 and the small effect on T_2 for water protons in the non-flocculated systems indicates that there is little long-range influence of these surfaces on water structure.

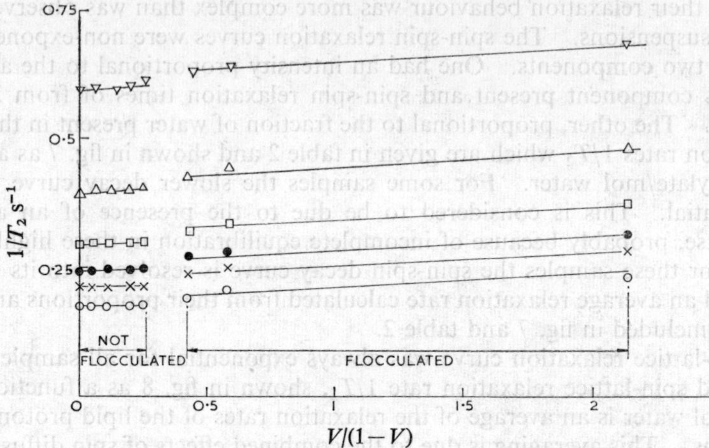

FIG. 3.—The dependence of the spin-spin relaxation rate of water protons $1/T_2$ on $V/(1-V)$, where V is the volume fraction of solid, for suspensions of 5,200 Å diameter polystyrene particles : ○, 58°C; ×, 47°C; ●, 37°C; □, 30°C; △, 20°C; ▽, 10°C.

On the basis of the Bloembergen, Purcell and Pound theory for dipolar n.m.r. relaxation in liquids, the observations could be explained either by a negligibly small effect of the surfaces on the motion of water molecules, or to the presence of tightly-bound water molecules with correlation times for molecular rotation of five or six orders of magnitude greater than those in normal water. The difference between T_1 and T_2 in these systems indicates that the second alternative is correct, but the magnitude of the T_2 effect shows that the amount of tightly-bound water must be small—much less than a monolayer of water molecules on the surface.

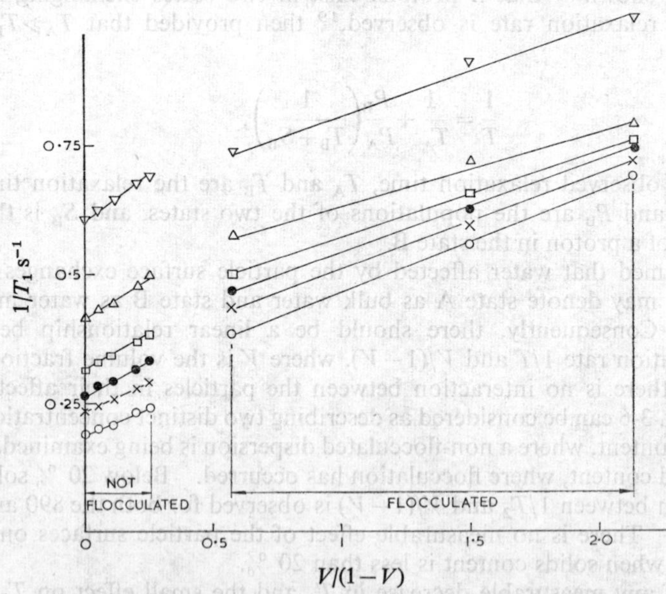

FIG. 4.—The dependence of the spin-lattice relaxation rate of water protons, $1/T_1$ on $V/(1-V)$ where V is the volume fraction of solid, for suspensions of 5,200 Å diameter polystyrene particles : ○, 58°C; ×, 47°C; ●, 37°C; □, 30°C; △, 20°C; ▽, 10°C.

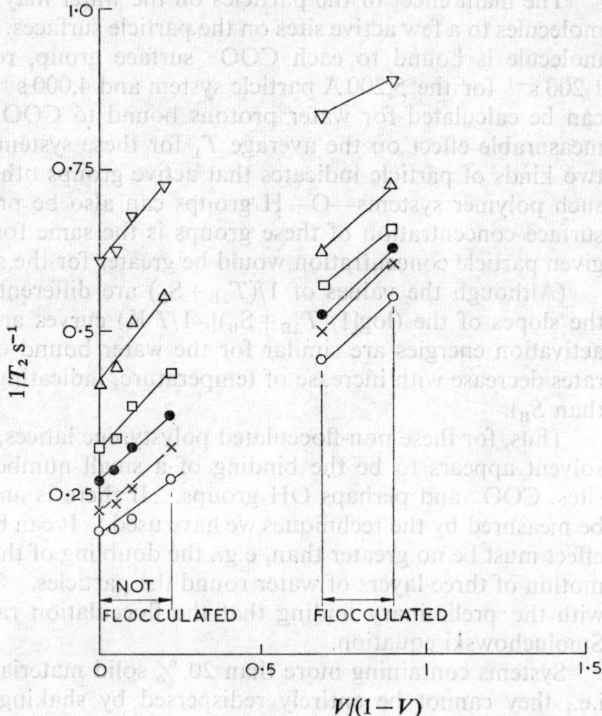

Fig. 5.—The dependence of the spin-spin relaxation rate of water protons, $1/T_2$ on $V/(1-V)$, where V is the volume fraction of solid, for suspensions of 890 Å diameter polystyrene particles: $\bigcirc$, 58°C; $\times$, 47°C; $\bullet$, 37°C; $\square$, 30°C; $\triangle$, 20°C; $\triangledown$, 10°C.

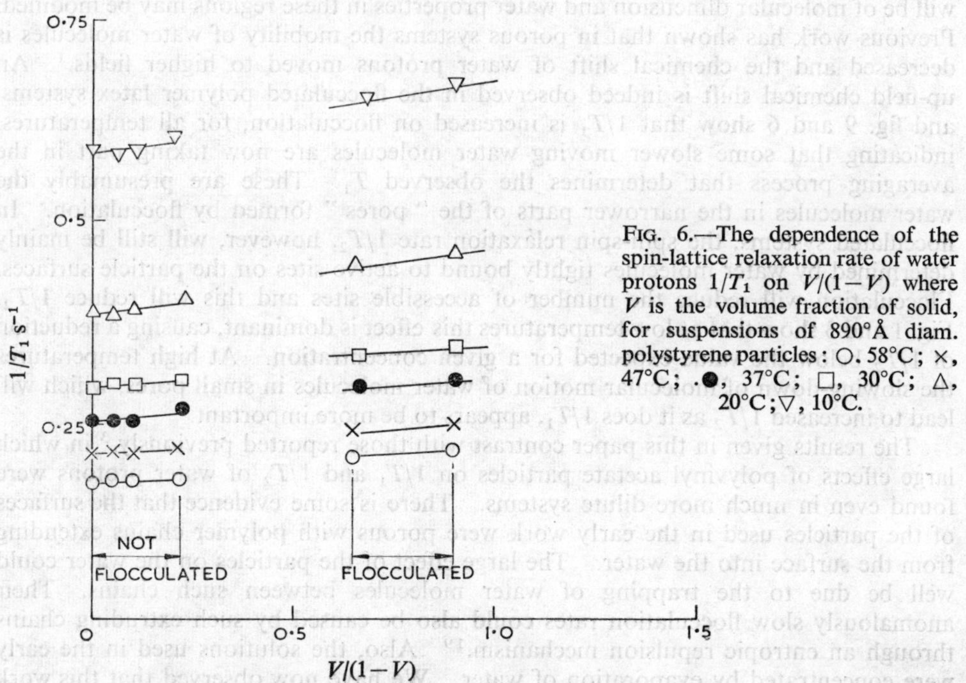

Fig. 6.—The dependence of the spin-lattice relaxation rate of water protons $1/T_1$ on $V/(1-V)$ where V is the volume fraction of solid, for suspensions of 890°Å diam. polystyrene particles: $\bigcirc$. 58°C; $\times$, 47°C; $\bullet$, 37°C; $\square$, 30°C; $\triangle$, 20°C; $\triangledown$, 10°C.

The main effect of the particles on the water may be due to the binding of water molecules to a few active sites on the particle surfaces.   If it is assumed that one water molecule is bound to each COO^- surface group, relaxation rates, $1/(T_{2B}+S_B)$ of $1,200$ s^{-1} for the $5,200$ Å particle system and $4,000$ s^{-1} for the 890 Å particle system, can be calculated for water protons bound to COO^- groups.   There would be no measurable effect on the average T_1 for these systems.   The difference between the two kinds of particle indicates that active groups other than COO^- are present.   In such polymer systems—O—H groups can also be present on the surface.[19]   If the surface concentration of these groups is the same for both particles the effect for a given particle concentration would be greater for the smaller particles, as is observed.

(Although the values of $1/(T_{2B}+S_B)$ are different for the different size particles the slopes of the $(\log[1/(T_{2B}+S_B)]$, $1/T$ K) curves are the same, indicating that the activation energies are similar for the water bound on both types of particle.   The rates decrease with increase of temperature, indicating that T_{2B} is substantially larger than S_B).

Thus, for these non-flocculated polystyrene latices, the main effect on the aqueous solvent appears to be the binding of a small number of water molecules at active sites, COO^- and perhaps OH groups.   If there is any other effect it is too small to be measured by the techniques we have used.   It can be calculated that any such extra effect must be no greater than, e.g., the doubling of the correlation time for molecular motion of three layers of water round the particles.   These conclusions are in accord with the preliminary finding that the flocculation rate of these particles obeys the Smoluchowski equation.

Systems containing more than 20 % solid material are at least partly flocculated, i.e., they cannot be entirely redispersed by shaking or ultrasonic irradiation.   In flocculated systems " pores " will be formed by contact between particles, and, in regions near where the contact takes place, the distance between the solid surfaces will be of molecular dimension and water properties in these regions may be modified. Previous work has shown that in porous systems the mobility of water molecules is decreased and the chemical shift of water protons moved to higher fields.[1]   An up-field chemical shift is indeed observed in the flocculated polymer latex systems, and fig. 9 and 6 show that $1/T_1$ is increased on flocculation, for all temperatures, indicating that some slower moving water molecules are now taking part in the averaging process that determines the observed T_1.   These are presumably the water molecules in the narrower parts of the " pores " formed by flocculation.   In flocculated systems, the spin-spin relaxation rate $1/T_2$, however, will still be mainly determined by water molecules tightly bound to active sites on the particle surfaces. Flocculation will reduce the number of accessible sites and this will reduce $1/T_2$. Fig. 1 and 3 show that at low temperatures this effect is dominant, causing a reduction of $1/T_2$ below the value expected for a given concentration.   At high temperatures the slowing-down of molecular motion of water molecules in small pores, which will lead to increased $1/T_2$ as it does $1/T_1$, appears to be more important.

The results given in this paper contrast with those reported previously [3] in which large effects of polyvinyl acetate particles on $1/T_1$ and $1/T_2$ of water protons were found even in much more dilute systems.   There is some evidence that the surfaces of the particles used in the early work were porous with polymer chains extending from the surface into the water.   The large effect of the particles on the water could well be due to the trapping of water molecules between such chains.   Their anomalously slow flocculation rates could also be caused by such extruding chains through an entropic repulsion mechanism.[19]   Also, the solutions used in the early were concentrated by evaporation of water.   We have now observed that this work causes flocculation at relatively low concentrations of solid. This may well account for the changes in slopes of the $(1/T, V/(1-V))$ curves which were observed, and which were attributed to long-range cooperative effects on structuring of the water between polyvinyl acetate latex particles.

LAMELLAR PHASES

The data given in table 2 and fig. 7 and 9 indicated that the relaxation rates of the water protons on the system are determined mainly by the ratio of COO^- groups to water molecules in the system. There is no such relationship between the [decanol]/[water] ratio and water proton relaxtaion rates. In these systems, also, eqn (1) should hold. If one assumes that the dominant effect on water mobility is binding

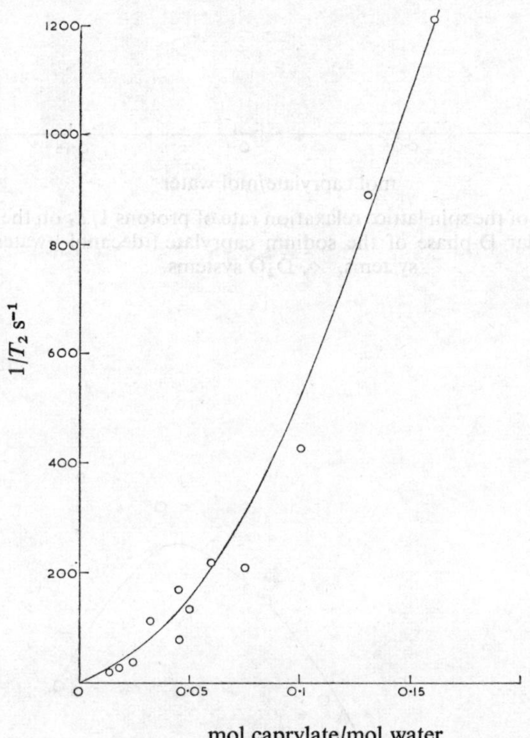

FIG. 7.—The dependence of the spin-spin relaxation rate of water protons, $1/T_2$ on the ratio mol caprylate/mol water, in the lamellar D-phase of the sodium caprylate+decanol+water system.

to COO^- groups then the observed relaxation rate $1/T$ should vary linearly with the ratio (mol caprylate)/(mol water). Fig. 9 and table 1, 2 show that this is so for water layer thicknesses of from 48 to 22 Å. Over this range the effect of the surfaces on the spin lattice relaxation rate of the water protons can be accounted for in the same way as in the polystyrene latices—by an exchange averaging of water tightly bound to COO^- groups, with water with the same molecular rotation and translation rates as normal bulk water. Exchange with water bound to —O—H groups and with —O—H protons will also occur, but it seems that this is not so important in these systems.

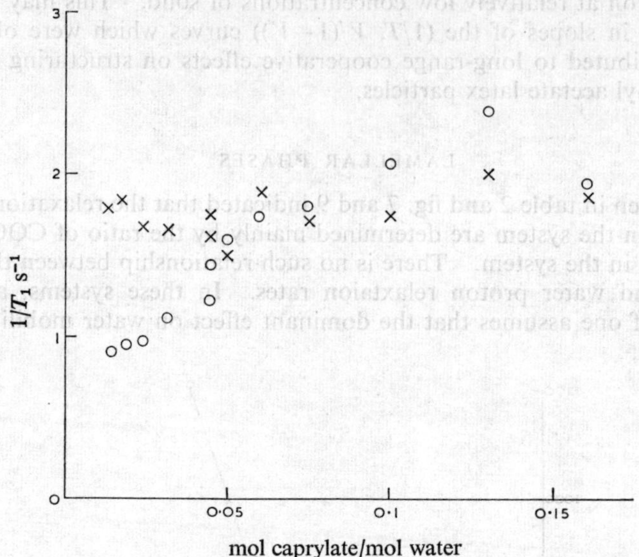

FIG. 8.—The dependence of the spin lattice relaxation rate of protons $1/T_1$ on the ratio mol caprylate/ mol water, in the lamellar D-phase of the sodium caprylate+decanol+water system. $\bigcirc$, H_2O systems, $\times$, D_2O systems.

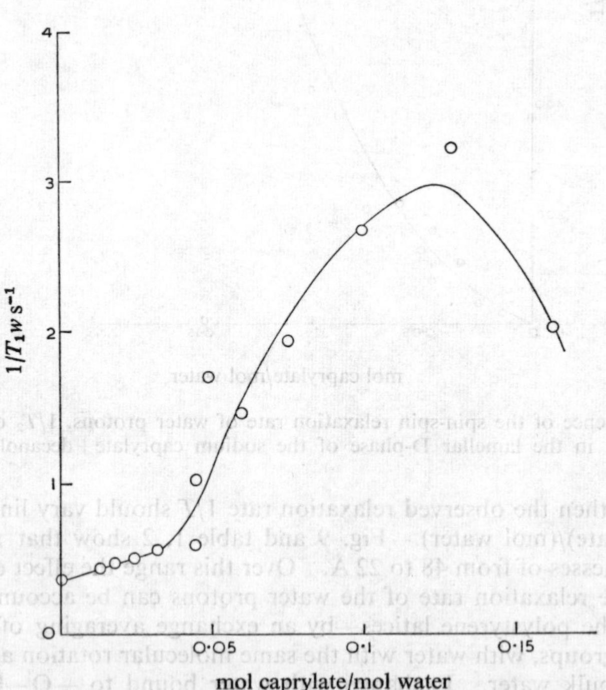

FIG. 9.—The dependence of the spin lattice relaxation rate of water protons $1/T_1W$ on the ratio mol caprylate/mol water, in the lamellar D-phase of the sodium caprylate+decanol+water system.

When water layer thicknesses of 17 Å or less are involved, however, there is a much greater effect on $1/T_1$ for water protons. Fig. 9 and table 2 show that there is first a marked increase in the relaxation rate, then a decrease as the thickness of the water layer is reduced, with a maximum $1/T_1$ at about 8.3 Å. This is exactly what would be expected on the basis of the Bloemburgen, Purcell and Pound theory of dipolar n.m.r. relaxation in liquids, if the water molecules moved more and more slowly as the water layer thickness is reduced below 20 Å. At the maximum value of $1/T_1$ at 8.3 Å the *average* correlation time for rotation of water molecules in the system will be 2×10^{-9} s, about three orders of magnitude longer than in ordinary water.

This effect is much greater than is observed in a solution with the same concentrations of Na^+ and COO^- groups in the form of sodium acetate, even for water layer thicknesses of more than 20 Å. From the gradients of $(1/T,$ mol $COO^-/$mol $H_2O)$ curves, molar relaxation rate enhancements $\Delta(1/T)$ can be calculated; where $\Delta(1/T) = (1/T)$ (molar solution)$-(1/T)$ (water). These are shown in table 3. (The

TABLE 3

system	$\Delta(1/T_1)$ at 20°C $l\,s^{-1}\,mol^{-1}$	$\Delta(1/T_2)$ at 20°C $l\,s^{-1}\,mol^{-1}$
5,200 Å diam particles (not flocculated)	—	17.4
890 Å particles (not flocculated)	—	71.4
lamellar D-phase of caprylate+ decanol+water systems; water layer thickness 20 Å	0.095	26.0
solution of sodium acetate	0.027	0.027

$\Delta(1/T_1)$ for the polymer latices are too small to be measured at the concentrations of COO^- groups attainable.)

$1/T$ value for solutions of sodium acetate does not vary much with concentration from 0-8 M.) The relatively small effect of CH_3COO^- groups in solution on water proton n.m.r. relaxation times is presumably due to the effect of the rotation of the acetate ion+water complex on the correlation times determining the dipolar relaxation rates of the water protons, an effect which would not occur in COO^- groups on surfaces.

The spin-spin relaxation rates of water protons in the lamellar D-phase system, shown in fig. 7 are also largely determined by the averaging between water molecules tightly bound to —COO^- groups and —OH groups and the remaining water molecules. The spin-spin relaxation rates for the bound water molecules are much higher than the spin-lattice relaxation rates, so that here the effect of water layer thickness on water structure is much less evident. Nevertheless, some increase in slope of the $(1/T_2,$ mol caprylate/mol water) curve is evident as the thickness of the water layer is decreased. The difference between the $1/T$ values given in table 3 for the polymer latices and those for lamellar phase are not surprising as systems with widely different concentrations of $-COO^-$ groups and —OH groups are being considered.

CONCLUSIONS

In the systems we have examined there appears to be no long-range effect of surfaces on water structure. For the polymer latices, measurable effects on water structure only appear when pores are formed, either as a result of the polymerization method or by flocculation. Normally, the main effect is the binding of water molecules to charged surface groups although the modification of one or two layers of water by uncharged surfaces cannot be excluded. In general, the effect of the modification of water structure by colloid particles on colloid stability for non-porous systems is likely to be small.

Similar conclusions can be drawn for the lamellar phase that we have investigated. For systems containing layers of water more than 20 Å thick, all the effects observed can be explained in terms of the binding of a few water molecules to —COO⁻ groups. Where the water thickness layer is less than 20 Å there is a marked reduction in water molecular mobility as normal water structure ceases to exist and is replaced by a more rigid structure determined by the interaction of water with charged surface groups, counter ions, and hydrogen-bonding surface groups. Thus, it is only for such relatively narrow water layers that special water structure effects need be allowed for when the properties of thin soap films are being considered.

The authors thank Dr. D. Nicholls who made the polystyrene latices, Dr. A. Lips and his colleagues who characterized them, and Dr. B. A. Pethica, for helpful discussions.

[1] J. Clifford and B. A. Pethica, *Hydrogen Bonded Solvent Systems*, ed. A. K. Covington, 1968, pp. 169-179.
[2] B. V. Deryaguin, *Disc. Faraday Soc.*, 1966, **42**, 109.
[3] G. A. Johnson, S. M. A. Lecchini, E. G. Smith, J. Clifford and B. A. Pethica, *Disc. Faraday Soc.*, 1966, **42**, 120.
[4] F. W. Cope, *Biophys., J*, 1969, **303**, 9.
[5] T. M. Connor, *Trans. Faraday Soc.*, 1963, **59**, 1574.
[6] D. E. Woessner, *J. Chem. Phys.*, 1963, **39**, 2783.
[7] J. Clifford, B. A. Pethica and W. A. Senior, *Ann. N.Y. Acad. Sci.*, 1965, **125**, 458.
[8] J. Clifford and S. M. A. Lecchini, *Soc. Chem. Ind. Monograph, no.* 25, *Wetting*, p. 174.
[9] T. H. Wu., *J. Geophys. Res.*, 1964, **69**, 1083.
[10] K. N. Lawson and T. J. Flautt, *J. Phys. Chem.*, 1968, **72**, 2066.
[11] M. M. Dubinin, *Quart. Rev.*, 1959, **9**, 101.
[12] R. H. Ottewill and J. N. Shaw, *Kolloid Z. Z. Polymere*, 1967, **218**, 34.
[13] L. Mindell, K. Fontell, H. Lehtinen and P. Ekwall, *Acta Polytechn. Scand.*, 1968, **74**, 1.
[14] D. Gill and S. Meiboom, *Rev. Sci. Instr.*, 1958, **29**, 688.
[15] J. R. Hansen and K. D. Lawson, *Nature*, 1970, **225**, 542.
[16] A. L. Smith, private communication.
[17] D. E. Woessner, *J. Chem. Phys.*, 1961, **35**, 41.
[18] E. L. Mackor, *J. Colloid Sci.*, 1951, **6**, 492.
[19] R. H. Ottewill, private communication.

Interlayer Water in Vermiculite: Thermodynamic Properties, Packing Density, Nuclear Pulse Resonance, and Infra-Red Absorption.

By J. Hougardy,* J. M. Serratosa,† W. Stone* and H. van Olphen‡

Laboratoire de Physica-Chimie Minérale, Instituto de Edafologia y Biologia Vegetal,
and Exploration and Production Research Division of Shell Development Company,
P.O. Box 481, Houston, Texas, U.S.A.

Received 16th *March,* 1970

Water vapour sorption isotherms on sodium and magnesium vermiculite of high charge density
were measured at 25 and 50°C.   Heats of immersion at various stages of hydration were measured
at 25°C.   According to X-ray observations during the sorption process, two discrete interlayer
hydrates are obtained with one and two monolayers of water between the unit layers of the crystallites.
The derived integral entropy of adsorption indicates a reduced freedom of motion of the interlayer
water molecules compared with that for water molecules in the liquid state.   Comparison of the
apparent density of the hydrated clay with the calculated crystallographic density indicates that interlayer water is slightly more densily packed than liquid water.   However, the bulk density of water in
a sodium vermiculite suspension is normal for the water in excess of the two hydration layers.

Nuclear pulse resonance results obtained on sodium vermiculite shows that water in the one-
layer hydrate is organized and some hypotheses are presented regarding this organization.   Water
molecules in the two-layer complex show the same degree of orientation only below $-65°C$.   Infra-
red absorption spectra for hydrated flakes of sodium vermiculite were obtained as a function of angle
between the (001) plane of the crystallites and the i.-r. beam.   The results indicate orientation of water
molecules in the one-layer hydrate at room temperature.

The system expanding clay + water offers unique possibilities for the experimental
study of the properties of water near a solid surface.   In the expanded state, the adsorbed water, intercalated between the unit layers of the silicate represents a considerable volume fraction of the system; the thickness of the intercalated water layers
can be measured by X-ray diffraction; oriented flakes can be used, for example, in
i.-r. studies, affording the determination of the disposition of the water molecules. In
suspension, water properties can be determined at distances beyond the thickness of
adsorbed water layers. Furthermore, the solid surface is flat and its structure is
reasonably well known, as well as its charge density which is determined by isomor-
phous ion substitutions within the unit layers.

A high charge density vermiculite was selected as a particularly suitable mineral
since it displays two well-separated discrete stages of hydration with one, respectively
two monolayers of water between the unit layers. Identical samples were used in the
participating laboratories. Previously published data on the system are presented in
summarized form only.

* present address: Laboratoire de Physico-Chimie Minérale, Heverlee-Louvain, Belgium.
† present address: Instituto de Edafologia y Biologia Vegetal, Madrid, Spain.
‡ present address: National Academy of Sciences, Washington D.C., U.S.A.

CHARACTERIZATION OF THE SYSTEM

The vermiculite sample was obtained near Llano, Texas.   Impurities were removed, and the counter ions occurring in the natural clay (primarily Mg^{2+}) were replaced by Na^+ ions by exhaustive treatment with NaCl.   The resulting Na-clay was washed with distilled water.   The Mg-clay was prepared from the Na-clay by treatment with $MgCl_2$ solutions.   The unit cell formula derived from chemical analysis is:

$$(Si_{5.28}Al_{2.72})(Al_{1.32}Mg_{4.58})O_{20}(OH)_4 - Na^+_{1.6}(resp. - Mg^{2+}_{0.8}).$$

Hence, the unit cell weight is 793.6 (776.2 for Mg).   The cation exchange capacity is 200 mequiv./100 g of dry clay (204 mequiv./100 g for Mg).   The unit cell dimensions are: $a = 5.21$ Å, $b = 9.18$ Å, $c = 9.82$ Å (9.3 Å for Mg).   The area available per ion on exterior surfaces is 60 Å^2 (120 Å^2 for Mg), and in the interlayer space 30 Å^2 (60 Å^2 for Mg).   The total unit layer surface area is 725 m^2/g (741 m^2/g for Mg). The surface density of charge is 26.7 μ C/cm^2.   The total exterior surface area from argon desorption isotherms, applying B.E.T. analysis is 3.5 m^2/g (3.1 m^2 for Mg), hence, the total area of the interlayer space is 361 m^2/g (369 m^2/g for Mg).

DENSITY OF WATER IN THE SYSTEM

Pyknometric density determinations on the sodium vermiculite + water system have been reported by Deeds *et al.*[1]   The apparent density of the clay was compared with the calculated crystallographic density.   The observed small difference of these values would indicate an estimated 2.5 % greater packing density of the water molecules in the interlayer space, assuming the normal density of water beyond two adsorbed water layers in bulk.   The latter assumption is supported by determination of the apparent density of the clay with two pre-adsorbed layers of water using n-decane as the displacement liquid.   This experiments yields a value which is identical with the apparent density of the clay in water.   Since it is unlikely that abnormal densities would occur in n-decane surrounding the hydrated clay particle, the water density would indeed be normal beyond about 5 Å from the surface according to these observations.   These results contradict those of Anderson and Low,[2] who concluded from experiments involving differential displacement of water in a bentonite paste by mercury (comparing injected mercury volume increments with displaced water volume increments) that water densities are up to 3 % low in the range between 0 and 60 Å from the clay surface.   If their results were correct, and would also apply to the vermiculite + water system, the apparent density of sodium vermiculite in water should have been about 20 % lower than that in the system prehydrated vermiculite + n-decane.   This difference would be far greater than the experimental error in the pyknometric data.

THERMODYNAMIC PROPERTIES OF THE SYSTEM

COMPOSITION OF THE HYDRATES

Adsorption-desorption isotherms for water vapour at 25 and 50°C were determined and (001) spacings were measured at successive stages of hydration by van Olphen.[3, 4] From the two-step isotherms, monolayer and two-layer coverages were derived from Langmuir plots.   The results are summarized in table 1.   The monolayer hydrate of sodium vermiculite contains 2 molecules of water for each sodium ion which is positioned midway between the unit layers.   In the two-layer hydrate each sodium ion is surrounded by almost 6 water molecules, 3 in a plane above and 3 in a plane below the midway sodium ion.   In the Mg-vermiculite monolayer hydrate, which is stable only below a relative pressure of 0.015, the water molecules are very loosely packed, and at higher presssures the formation of the two-layer hydrate is favoured over the filling of vacant positions in the monolayer hydrate.   In the two-layer

TABLE 1.—COMPOSITION OF HYDRATES

clay	(001) spacing	molec. ratio H_2O/ion	area per molec. H_2O
Na-vermiculite			
one-layer hydrate	11.8 A	2.0	13.5 A²†
two-layer hydrate	14.8 A	5.6-6.0	10.8-10.0 A²
Mg-vermiculite			
one-layer hydrate	11.6 A	3.5	17.0 A²
two-layer hydrate*	14.8 A	10·0	12.0 A²
	14.8 A	11.0	10.8 A²

* Two two-layer hydrates can be distinguished in the isotherm having identical X-ray spacings but different packing densities for water.

† allowing 3 A² for the area occupied by the sodium ion.

hydrate the water molecules are partially coordinated octahedrally around the Mg ions, and partially arranged in a hydrogen bonded network according to Bradley and Serratosa.[5]

HYDRATION ENERGY

Hydration energies were derived from the sorption isotherms.   For Na-vermiculite these are (at 25°C) 105×10^{-3} J/m² for the first layer, and 40×10^{-3} J/m² for the second layer.   These values represent the net energies of expansion consisting of the combined effects of adsorption, electrostatic layer interaction, and the van der Waals attraction between the unit layers.   The electrostatic attraction between the cations and the negatively charged unit layers amounts to $4\pi\sigma^2 x/\varepsilon$ in which σ is the charge density, $2x$ the separation of the unit layers, and ε the dielectric constant which will have a value between about 3 and 6.   Neglecting the van der Waals attraction which will be small with respect to the electrostatic attraction, the adsorption energy equals the sum of the net hydration energy and the electrostatic energy.   For the formation of the monolayer hydrate the adsorption energy amounts to 240 to 375×10^{-3} J/m² which corresponds with 10-15.5 kcal/mol of sodium ion.   For the formation of the two-layer complex from the dry clay the adsorption energy amounts to 21-32 kcal/mol of sodium ion.   These estimates show that the hydration of the clay calculated on a per ion basis is considerably smaller than the ion hydration energy in bulk solution.

HEATS OF IMMERSION

Heats of immersion were determined for the clays at various stages of hydration. For Na-vermiculite the average heats of hydration amount to 3.75 and 1.9 kcal/mol of water for the first and the second layer of water respectively.   For Mg-vermiculite these values are 8.0 and 3.5 kcal/mol of water respectively.

ENTROPY OF HYDRATION

Integral entropies of hydration were derived from isotherm and heat of immersion data.   Because of lack of detail in the isotherms in the small monolayer hydration region for the Mg clay, entropies could only be evaluated for the sodium clay. For both the one-layer and the two-layer adsorption processes the integral entropies are negative with respect to that of liquid water: $T(S-S_L)$ is between -2 and -3 kcal/mol of water. These values apply to changes occurring in the entire system and include changes in the adsorbed phase, the crystal phase, and the ions. Since the latter two are likely to result in entropy gains, the contribution to the entropy change by the water phase may be somewhat more negative than the net values. Hence, the water molecules in the adsorbed phase will have a higher degree of order than that existing in the bulk liquid, according to this interpretation.

NUCLEAR PULSE RESONANCE

Pulsed nuclear magnetic resonance experiments were carried out on powdered samples of sodium vermiculite. Particle diameters ranged from 7 to 0.5 μm. The Fe content of the sample was 1,710 p.p.m. as determined by flame absorption spectroscopy. The one-layer and two-layer hydrates were prepared by equilibration of the sample with water vapour at relative pressures of 0.18 and 0.66 respectively. Measurements were made in the temperature range from 50 to $-180°$C.

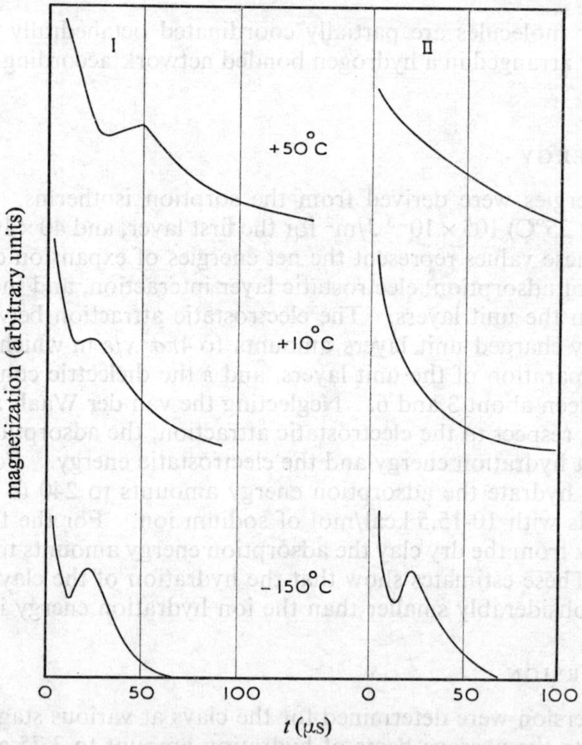

Fig. 1.—Free induction decay of powdered Llano Na-Vermiculite at various temperatures. Curve marked (I) for monolayer hydrate, marked (II) for two-layer hydrate.

The shape of the free induction decay signal following a 90° pulse shows some interesting features. For the monolayer hydrate, the signal shows a hump, within the free induction decay, irrespective of temperature (fig. 1). The hump shifts towards the origin of the signal as the temperature is lowered from ambient to $+5°$C and then stays at approximately the same position for lower temperatures. This effect (reflecting the presence of a characteristic doublet in the n.m.r. absorption curve [6,7]) is attributed to an averaged preferential orientation of the water molecules. This orientation effect which is observed at room temperature, is enhanced upon lowering of the temperature and levels off below $+5°C$. Due to the random orientation of the vermiculite flakes it was impossible to determine the orientation of the water intra-protonic vector with respect to the crystal faces. However, in an oriented specimen of hectorite clay, Woessner and Snowden [8] have shown from analogous experiments that the intraprotonic vector has a tendency to be parallel to the (001) planes. Experiments on oriented flakes of vermiculite are in progress.

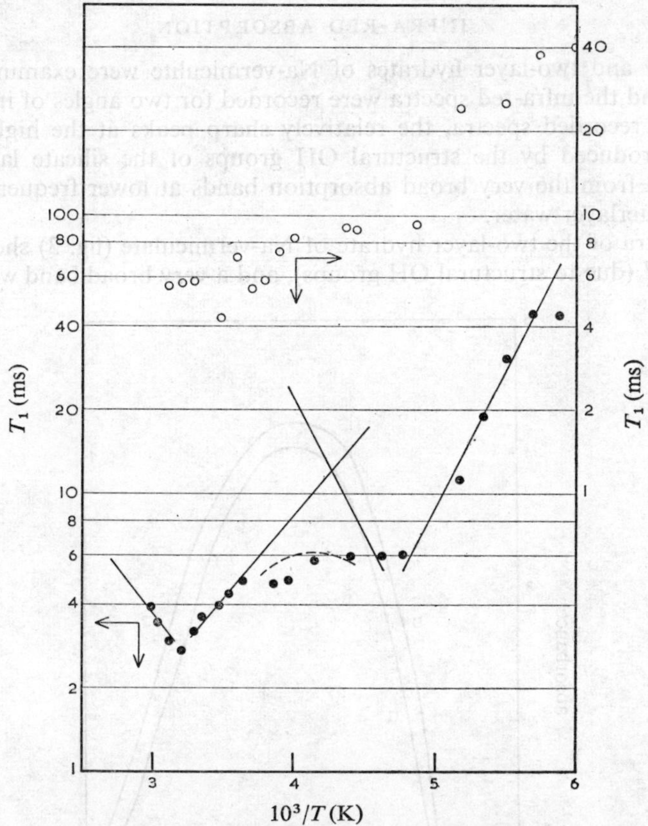

Fig. 2.—Spin lattice relaxation times of powdered Llano Na-Vermiculite at various temperatures. ●, two-layer hydrate, ——— lines drawn show positions of the two minima, - - - - - - calculated combination of the two relaxations, ○, one-layer hydrate.

For the two-layer hydrate of vermiculite, the decay of the signal is monotonic for temperatures between 20 and $-60°C$. Below this temperature a hump is again observed (fig. 1). It is concluded, therefore, that for the two-layer hydrate the orientation effect is obtained at a much lower temperature than for the monolayer hydrate. Apparently, the presence of a larger amount of water prevents to a large extent the preferential orientation.

In order to gain information about spin mobility, spin lattice relaxation times T_1 were measured in the same temperature range. The data for T_1 as a function of temperature for the one-layer and two-layer hydrates-corrected for the influence of constitutional hydroxyls and paramagnetic centres are shown in fig. 2.

For the two-layer hydrate, a minimum is observed around $+38°C$, beyond which T_1 levels off until it increases abruptly at $-70°C$. This behaviour can be explained by postulating the existence of a second minimum situated at around $-60°C$. This minimum would be the same as that observed for Na-montmorillonite by Touillaux et al.,[9] who proposed proton diffusion as the relaxation mechanism. In each case, an activation energy of 6 kcal is obtained. The minimum at high temperature could perhaps be attributed to a diffusion of the water molecules ($E_a = 4$ kcal). For the one-layer hydrate, these two minima also seem to be present, but they are less pronounced.

INFRA-RED ABSORPTION

One-layer and two-layer hydrates of Na-vermiculite were examined as oriented aggregates and the infra-red spectra were recorded for two angles of incidence, 0 and 40°. In the recorded spectra, the relatively sharp peaks at the higher frequencies which are produced by the structural OH groups of the silicate layers are easily distinguished from the very broad absorption bands at lower frequencies which are due to the interlayer water.

The spectra of the two-layer hydrate of Na-vermiculate (fig. 3) shows absorption at 3,675 cm^{-1} (due to structural OH groups), and a very broad band with a maximum

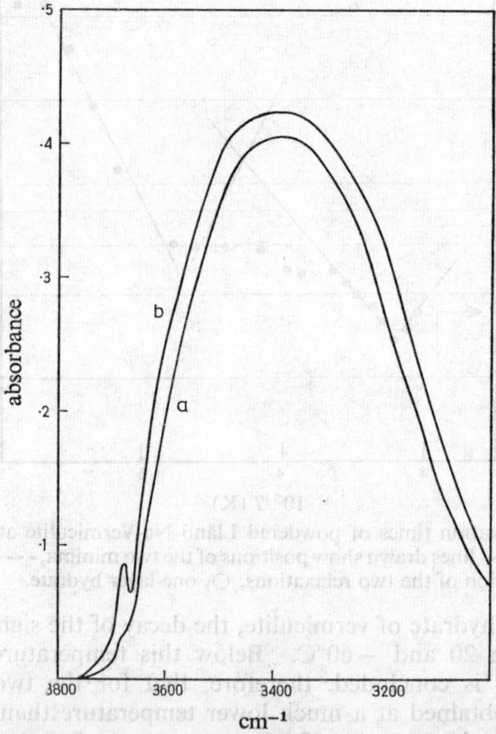

FIG. 3.—Infra-red absorption spectra of an oriented aggregate of the two-layer hydrate of Na-Vermiculite at two angles of incidence: (a) 0°; (b) 40°.

at 3,415 cm^{-1} which corresponds to the stretching vibrations of the interlayer water. There is no appreciable change in the intensity of this band with the angle of incidence, indicating that there is no preferential orientation of the interlayer water molecules.

For the one-layer hydrate (and at lower water contents of the clay) the maximum of the band corresponding to the stretching vibrations of water molecules moves to higher frequencies, i.e., 3,450 cm^{-1} (fig. 4).   The band associated with the deformation

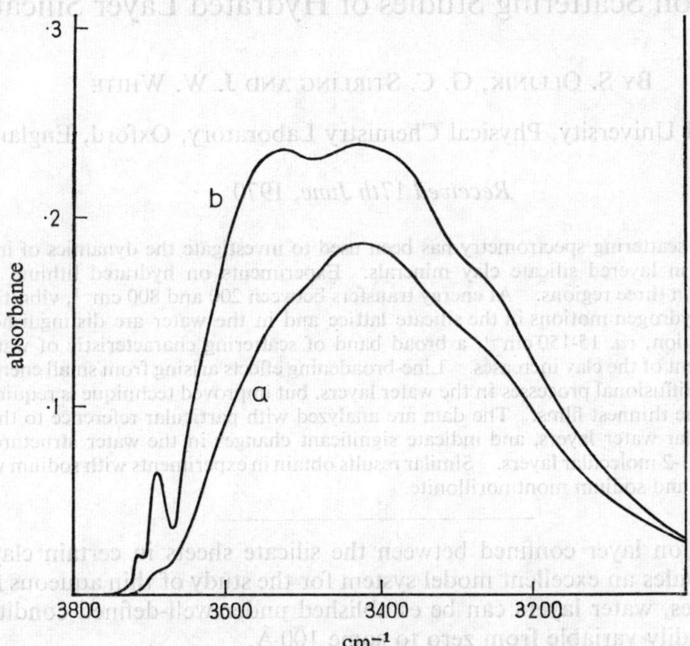

FIG. 4.—Infra-red absorption spectra of an oriented aggregate of the one-layer hydrate of Na-vermiculite at two angles of incidence : (a) 0° ; (b) 40°.

vibration, which appears at about 1,600 cm^{-1}, also shows a shift although in the direction of lower frequencies (not shown).   Both displacements indicate weakening of the hydrogen bonding of the water molecules and therefore that the association ion-dipole is the more prevalent.   For inclined incidence (fig. 4, curve (b)) a significant increase of intensity of the water band is observed, indicating that for these low water contents the molecules are not randomly disposed but do adopt a preferential orientation in the interlayer space.

The authors thank Mr. C. T. Deeds and Mr. W. K. Lumb for their assistance in the isotherm, X-ray, and density experiments, Dr. K. E. Manchester for measuring the heats of immersion, Prof. J. J. Fripiat for stimulating discussions on the n.m.r. work.   J. Hougardy is indebted to IRSIA for a Ph.D. grant.

[1] C. T. Deeds and H. van Olphen, *Adv. Chem. Series*, 1961, **33**, 332.
[2] D. M. Anderson and P. F. Low, *Proc. Soil Sci. Soc. Amer.*, 1958, **22**, 99.
[3] H. van Olphen, *J. Colloid Sci.*, 1965, **20**, 822.
[4] H. van Olphen, *Proc. Int. Clay Conf., Tokyo*, (Israel Universities Press, Jerusalem), 1969, **1**, 649.
[5] W. F. Bradley and J. M. Serratosa, *Clays Clay Minerals*, 1960, **7**, 260.
[6] J. Graham, G. F. Walker and C. W. West, *J. Chem. Phys.*, 1964, **40**, 540.
[7] M. Dupont, *Thèse de Doctorat* (Grenoble, 1965).
[8] D. E. Woessner and B. S. Snowden, Jr., *J. Chem. Phys.*, 1969, **50**, 1516.
[9] R. Touillaux, P. Salvador, C. Vandermeersch and J. J. Fripiat, *Israel J. Chem.*, 1968, **6**, 337.

Neutron Scattering Studies of Hydrated Layer Silicates

By S. Olejnik, G. C. Stirling and J. W. White

Oxford University, Physical Chemistry Laboratory, Oxford, England

Received 17th June, 1970

Slow-neutron scattering spectrometry has been used to investigate the dynamics of interlamellar water molecules in layered silicate clay minerals. Experiments on hydrated lithium vermiculite give information in three regions. At energy transfers between 200 and 800 cm^{-1}, vibrational bands associated with hydrogen motions in the silicate lattice and in the water are distinguished. In the low-frequency region, *ca.* 15-150 cm^{-1}, a broad band of scattering characteristic of water emerges as the water content of the clay increases. Line-broadening effects arising from small energy transfers are indicative of diffusional processes in the water layers, but improved technique is required to make this certain for the thinnest films. The data are analyzed with particular reference to the thickness of the interlamellar water layers, and indicate significant changes in the water structure for water thicknesses of *ca.* 1-2 molecular layers. Similar results obtain in experiments with sodium vermiculite, and with lithium and sodium montmorillonite.

The hydration layer confined between the silicate sheets in certain clay mineral structures provides an excellent model system for the study of thin aqueous films. In favourable cases, water layers can be established under well-defined conditions with thicknesses readily variable from zero to some 100 Å.

In vermiculite and montmorillonite the silicate sheets are separated by hydration layers containing exchangeable cations. The nature of the water in these layers has been the subject of considerable speculation.[1] It is generally conceded that the silicate surface confers a certain degree of order on the water structure, but the causes and range of this ordering are matters of some debate.[2, 3] Various techniques have been employed to study the dynamics of water molecules in the hydration layer. N.m.r. line-shape observations on vermiculite and montmorillonite indicate that proton mobility is dependent on the water content of the clay.[4] Pulsed n.m.r. on montmorillonite$+H_2O$ and $+D_2O$ systems has been used to show that the interfacial water is preferentially oriented.[5, 6] Self-diffusion of the water protons in hydrated vermiculite has been studied by pulsed spin-echo n.m.r.[7] Yet these and other experiments have not led to a full understanding of the interlamella water structure, and in particular the relative effects of the charged surfaces and dissolved cations in producing order have to be evaluated.

Neutron scattering spectrometry is well suited to study the dynamics of water in clays, since the silicate structure remains essentially " transparent " to the neutron beam owing to the low scattering cross-section of aluminosilicate compared to that of hydrogen. Additionally, the spectra record not only the frequency spectra of molecular modes but also the intermolecular and diffusive motions for included water molecules. The last are most strongly altered by structuring fields and can be studied here because of the high energy resolution available.

The neutron results can be used in two different ways. Relatively large energy transfers, up to about 800 cm^{-1}, arise from vibrational and rotational motions of hydrogen groups. In some respects this inelastic scattering is analogous to Raman spectroscopy, but has the advantage that hydrogen motions are selectively observed, and without the restrictions of optical selection rules.[8]   Secondly, very small energy transfers (*ca.* 0.1-10 cm^{-1}) can be measured with neutrons, and coupled with the angular dependence of scattering, they provide information on diffusional behaviour of the scattering nuclei, in this case the water protons.   We have shown that the neutron method can be used to determine diffusion coefficients in water and aqueous solutions with comparable accuracy to move conventional techniques.[9]

Naumann, Safford and Mumpton measured the neutron inelastic spectra for a range of layer silicate minerals,[10] with particular reference to OH motions in the 200-1,000 cm^{-1} region.   In some preliminary experiments with a line-source incident neutron beam, we investigated water motions in lithium vermiculites with relatively large water spacings, and showed how the self-diffusion coefficient of the water varied with the clay water content.[11]   In the present experiments we concentrate on thinner films, mainly in the region of 1-2 molecular layers, and extend the work to some other minerals.

THEORETICAL BACKGROUND

For an incident neutron spectrum with a narrow energy spread, the observed intensity distribution of the scattered neutrons from an incoherent scatterer like hydrogen as a function of the energy transfer ω and the solid angle of scatter Ω is identical to the differential scattering cross-section $\partial^2\sigma/\partial\Omega\partial\omega$:

$$\frac{\partial^2\sigma}{\partial\Omega\partial\omega} = b^2\frac{k}{k_0}\frac{1}{2\pi}\int\int \exp\left(i(\mathbf{Q}\cdot\mathbf{r}-\omega t)\right)\cdot G_s(\mathbf{r},\,t)\,\mathrm{d}\mathbf{r}\,\mathrm{d}t. \tag{1}$$

Here b is the scattering length of the nucleus; $\mathbf{k}$, $\mathbf{k}_0$ are the scattered and incident wave vectors of the neutron respectively, $\mathbf{Q} = \mathbf{k} - \mathbf{k}_0$ and is the momentum transfer in the collision, expressed in radians/s, and $G_s(\mathbf{r},t)$ is the space-time autocorrelation function describing the motion of the atom containing the scattering nucleus. $G_s(\mathbf{r},t)$ is the chance of finding a proton at $\mathbf{r}$ at time t if it was at the origin at time $t = 0$. It is defined by eqn (2):

$$G_s(\mathbf{r},\,t) = \frac{1}{N}\sum_n \langle\delta(\mathbf{r}+\mathbf{r}_n(0)-\mathbf{r}_n(t))\rangle. \tag{2}$$

Characteristics of $G_s(\mathbf{r},t)$ have been discussed by Egelstaff.[12]   The functional relationship between $\mathbf{r}$ and t may be simply sinusoidal as for a vibrator, or it may have a complicated form as for diffusive motions in liquids.   For liquids, a number of models have been proposed to describe the molecular motions and hence $G_s(\mathbf{r},t)$.[13] Using these it is possible to analyze the shape and angular dependence of the scattering to obtain the self-diffusion coefficients associated with a particular scattering centre. Models have been discussed, e.g., by Sjölander.[14]   For a molecule undergoing simple Fick's law diffusion the differential scattering cross-section is given by eqn (3),

$$\frac{\partial^2\sigma}{\partial\Omega\partial\omega} = \text{const.}\,\frac{DQ^2}{(DQ^2)^2+\omega^2}, \tag{3}$$

where D is the self-diffusion coefficient defined by Fick's Law.   This model is adequate for long-time behaviour (i.e., small values of $\mathbf{Q}$ and ω), and gives rise to an energy broadening which increases with the momentum transfer squared:

$$\Delta E = 2\hbar DQ^2. \tag{4}$$

EXPERIMENTAL

Young River (Western Australia) sodium vermiculite was treated, following the method of Posner and Quirk,[15] to produce lithium- and sodium-exchanged materials. After treatment, films approximately $4\,\text{cm} \times 5\,\text{cm}$ in area were prepared by sedimentation and drying of a carefully dialyzed suspension. They were equilibrated at known constant humidities to yield specimens of different water contents and hence water layer thicknesses. For neutron scattering experiments, samples were quickly sealed in aluminium casettes. Sample weights were measured before and after an experimental run which typically took about 12 h. During this time the sample was surrounded by an atmosphere of dry helium in the sample changer of the spectrometer. All samples were prepared to give a total scattering of between 7 and 9 % of the incident beam intensity.

Basal plane spacings were meaured by X-ray diffraction immediately after recording a neutron spectrum. The clay specimen was transferred to a glass slide and sealed under a thin Polythene film to prevent any change in moisture content. Spacings were measured with a Philips PW 1050 diffractometer using Ni-filtred Cu radiation. Samples were ignited at 310°C and weighed to give water contents.

Neutron spectra were measured on the 6H time-of-flight spectrometer on the reactor DIDO at A.E.R.E. Harwell.[16] Beryllium-filtered neutrons are monochromated by a phased chopper system and energy spectra of the scattered neutrons measured at nine angles between 18 and 90° to the incident beam direction. The incident neutron energy chosen for the present experiments was 4.60 meV (4.22 Å), with resolution 0.55 meV (full-width at half-maximum). All experiments were carried out at ambient temperature, $23 \pm 1°C$.

Experimental data are normalized against a vanadium standard run at the same time as the sample, and corrected for background and counter efficiency. The corrected data are displayed as " time-of-flight " spectra, where corrected counts are shown against the scattered neutron reciprocal velocity (μs m^{-1}). Inelastically scattered neutrons occur as an energy-gain spectrum (i.e., towards lower reciprocal velocities) owing to the low energy of the incident neutron beam.

To obtain ΔE, the energy broadening (taken to be Lorentzian following eqn (3)) and the incident spectrum (Gaussian) are deconvoluted using standard tables. Slopes of (ΔE, Q^2) plots may then be used to estimate D. For quasi-elastic scattering $k \approx k_0$ and

$$\mathbf{Q} = (4\pi/\lambda) \sin (\theta/2),$$

where λ is the neutron wavelength and θ is the angle of scatter. For water and aqueous solutions this is a reliable procedure provided slopes of ΔE against κ^2 are measured at sufficiently low $\mathbf{Q}$ values.[9]

RESULTS

Fig. 1 shows time-of-flight spectra measured at 90° scattering angle for neutrons scattered from (a) lithium vermiculite dried at 310°C and (b) liquid water. In these spectra the elastic peak is centred at 1,065 μs m^{-1}, with neutron energy-gain spectra deployed to the left. Energy transfers are indicated in cm^{-1}. The spectra can be conveniently described in three parts:

(i) MOLECULAR MODES.—These occur mainly in the region above 200 cm^{-1} energy transfer, and in the present experiments up to about 800 cm^{-1} (energy levels above this are not observed owing to the low populations of upper levels arising from the Boltzmann distribution). The scattering results from hydrogen motions, which in the dry clay reflects vibrations of OH groups in the clay structure. These bands correspond to those observed in infra-red and Raman spectroscopy, and have been discussed for neutron spectroscopy by Naumann et al.[10] A particular advantage of the neutron method is the selective scattering by hydrogen which can often facilitate the assignment of complex molecular spectra.

In the liquid water spectrum, the band at *ca.* 480 cm^{-1} arises from hindered torsional oscillations of the water molecules. The position and structure of the band is significantly affected by the presence of various ionic solutes,[17] thus providing information on the water structure in electrolyte solutions. The hydrogen-bond stretching frequency which occurs at 175 cm^{-1} in optical spectra has a low neutron scattering intensity, and was not resolved in the present experiments.

(ii) LOW-FREQUENCY INELASTIC SCATTERING.—The frequency region between approximately 15 and 150 cm^{-1} is moderately intense in the neutron spectrum of liquid water (fig. 1(*a*)). A broad spectrum of vibrations in this region is characteristic of liquids[18] and may arise from strongly damped intermolecular vibrations and torsions. For H_2O at 23°C this spectrum has a broad maximum at *ca.* 60 cm^{-1}. For D_2O measured under the same experimental conditions the maximum is at

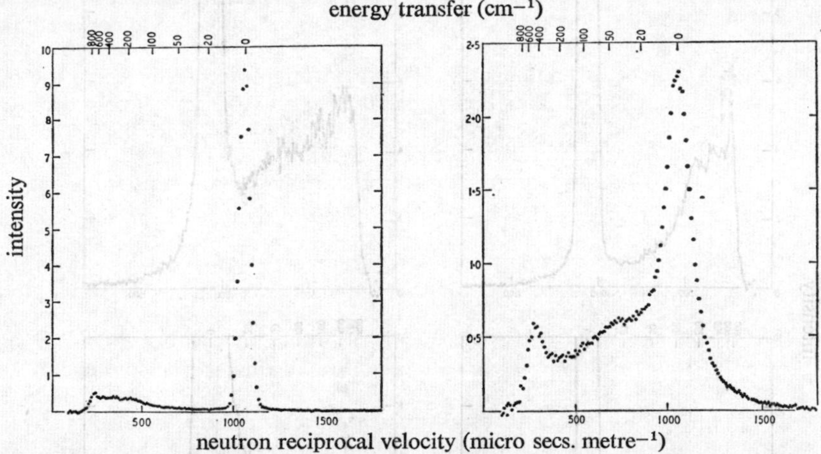

FIG. 1.—Neutron time-of-flight spectra, angle of scatter = 90°. (*a*) Lithium vermiculite dried at 310°C; (*b*) water.

ca. 40 cm^{-1}. The peak position in H_2O is almost independent of the scattering angle (and hence momentum transfer) but in D_2O a noticeable shift of order 10 cm^{-1} to lower frequencies occurs at high scattering angles.[19] The light water scatters incoherently and the spectrum is a density of states curve weighted by $(\mathbf{u}.\mathbf{Q})^2$, where $\mathbf{u}$ is the hydrogen amplitude. The D_2O spectrum is more akin to the " quasi-phonon" spectrum from polycrystalline and coherent scattering liquid metals.[20] It supports the analysis given to the hydrogen spectrum relating it to cooperative intermolecular modes.

(iii) QUASI-ELASTIC SCATTERING.—Small energy transfers arising from translational motions of the scattering nuclei lead to a broadening of the incident neutron energy spectrum as described earlier. In fig. 1(*a*) the elastic peak has the same energy width as the incident neutron beam, indicating that the scattering nuclei in the dry clay are fixed in the lattice. In fig. 1(*b*), for liquid water, there is a relatively large broadening, and it is from measurements in this quasi-elastic region that diffusional details can be obtained.

To exploit these characteristics we have obtained spectra for a series of lithium vermiculites containing water spacings of different thicknesses. A point of some concern has been the swelling characteristics shown by the clay specimens. We have found that vermiculite films equilibrated for some time in a constant humidity atmosphere yield homogeneous samples with basal spacings up to about 15 Å (interlamellar spacing about 5 Å, or two molecular layers of water). For larger spacings it is necessary to use direct wetting techniques. After wetting, specimens were sealed in the aluminium casettes used for the neutron measurements. However, subsequent measurements of total water contents and X-ray spacings showed that swelling behaviour could be erratic, and uniform swelling was not always achieved. This effect has been noted before,[21, 22] and requires that all specimens be carefully monitored both for water content and X-ray spacing. Because of the non-uniformity in swelling behaviour we do not report here any data on clay systems with large water contents.

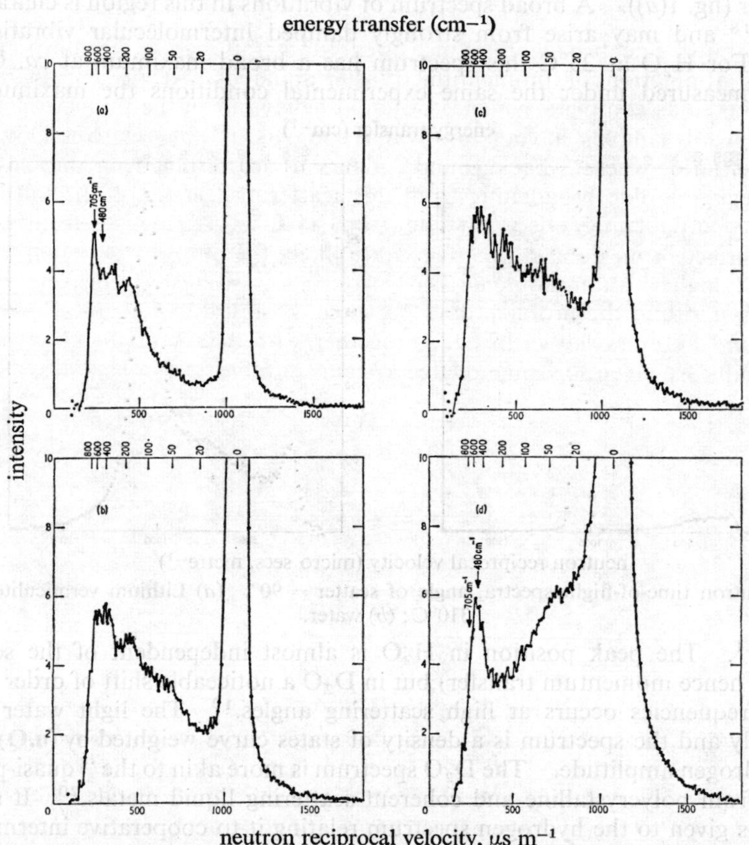

FIG. 2.—Inelastic neutron time-of-flight spectra, angle of scatter $= 90°$. (a) dried clay; (b) g H_2O/g clay $= 0.079$; (c) g H_2O/g clay $= 0.135$; (d) water.

In fig. 2, the inelastic spectra at 90° scattering angle for some homogeneous lithium vermiculites are shown, together with those of the dry clay, and water. Descriptions of these specimens are included in table 1. There is a continuous gradation in the spectra as the water content of the clay increases. Molecular modes associated with OH motions in the clay, which occur between 200 and 800 cm⁻¹ (fig. 2(a)), are progressively submerged as the water spectrum becomes more dominant at higher water contents. In the lower frequency region, 15-150 cm⁻¹, the characteristic scattering of liquid water also builds up significantly by the time the interlamellar layer is about two molecular layers of water thick. Further evidence for incipient water behaviour in these thin layers can be obtained from quasi-elastic line broadening measurements.

TABLE 1.—DIFFUSION DATA FOR LITHIUM VERMICULITE SAMPLES

sample	water content g H_2O/g clay	$d(001)$ Å	D cm^2 s^{-1} × 10^5
a	0	dry clay	0
b	0.079	12.75	0
c	0.114	13.3	0.05
d	0.135	14.05	0.11
e	—	water	2.14

Fig. 3 shows line broadenings measured for a number of lithium vermiculite samples as a function of momentum transfer squared. At present these results cannot be taken as complete evidence of liquid behaviour in the thinnest layers since a problem with samples having low water contents is the relatively low signal/ background ratio, where " background " refers to scattering from the clay lattice. There is negligible line broadening from this scattering (see fig. 3(a)), and this may well modify broadening effects arising from H_2O scattering. Self-diffusion co-efficients derived from data in this region are likely therefore to represent minimum values. At higher water contents the scattering is almost exclusively from the interlayer water, and the problem does not arise. Table 1 gives values of D derived directly from the plots shown in fig. 3, together with details of the clay specimens. Similar results are given for some related systems in table 2.

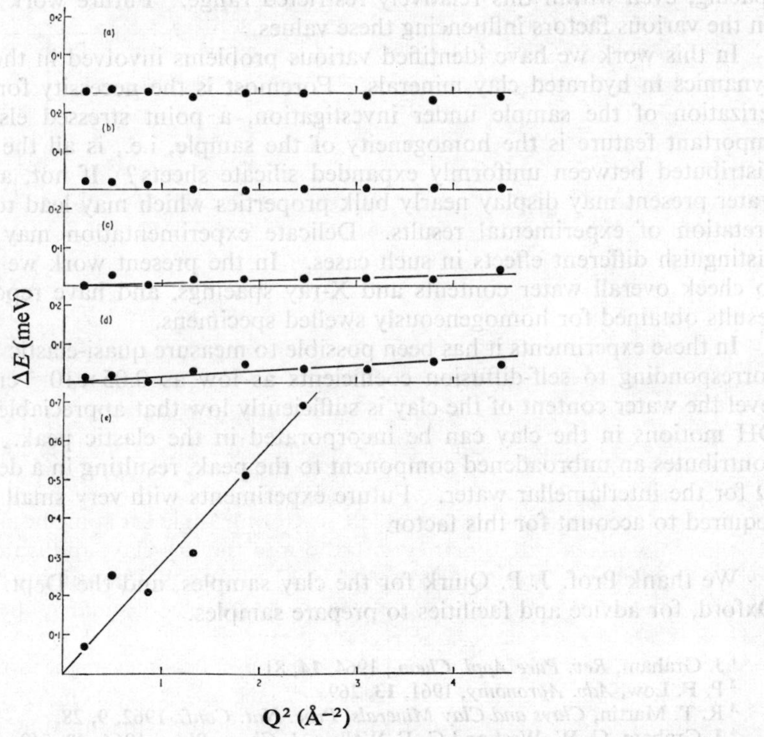

FIG. 3.—Observed line broadening against momentum transfer squared for lithium vermiculite samples. (a) dried clay; (b) g H_2O/g clay = 0.079; (c) g H_2O/g clay = 0.114; (d) g H_2O/g clay = 0.135; (e) water.

TABLE 2.—DIFFUSION DATA FOR VERMICULITE AND MONTMORILLONITE SAMPLES

material	water content g H_2O/g clay	d(001), Å	D cm^2 s$^{-1}\times 10^5$
Na-vermiculite	0.080	13.71	0.06
Na-vermiculite	0.111	13.92	0.06
Na-vermiculite	0.185	14.48	0.17
Li-montmorillonite	0.230	14.5	0.19
Na-montmorillonite	0.228	14.53	0.25

DISCUSSION

We have demonstrated some of the different ways in which neutron scattering can be applied in the study of H motions in clay minerals, with particular reference to water dynamics in the thin films in layered silicates. A unique feature of the neutron method is the time-scale of observation, *ca.* 10^{-12}-10^{-11} s, which means that self-diffusion coefficients can be estimated independently of macroscopic imperfections in the clay structure. In conventional " long-time " techniques (10^{-3}-10^{-2} s and longer), allowance has to be made for pores and cracks in the structure which can modify the diffusion coefficient.[7]

In films about 1-2 molecular layers of water thick, water self-diffusion coefficients are an order of magnitude lower than those in bulk water. Within the limits of the present experiments we are unable to detect significant differences between water in vermiculite and montmorillonite containing lithium or sodium as the exchangeable cations. Broadly, the self-diffusion coefficients appear to vary as the interlamellar spacing, even within this relatively restricted range. Future work will concentrate on the various factors influencing these values.

In this work we have identified various problems involved in the study of water dynamics in hydrated clay minerals. Foremost is the necessity for careful characterization of the sample under investigation, a point stressed elsewhere.[1, 3] An important feature is the homogeneity of the sample, i.e., is all the absorbed water distributed between uniformly expanded silicate sheets? If not, a fraction of the water present may display nearly bulk properties which may lead to spurious interpretation of experimental results. Delicate experimentation may be required to distinguish different effects in such cases. In the present work we have taken care to check overall water contents and X-ray spacings, and have reported only those results obtained for homogeneously swelled specimens.

In these experiments it has been possible to measure quasi-elastic line broadenings corresponding to self-diffusion coefficients as low as 0.05×10^{-5} cm^2 s^{-1}. At this level the water content of the clay is sufficiently low that appreciable scattering from OH motions in the clay can be incorporated in the elastic peak. This scattering contributes an unbroadened component to the peak, resulting in a decreased value of D for the interlamellar water. Future experiments with very small spacings will be required to account for this factor.

We thank Prof. J. P. Quirk for the clay samples, and the Dept. of Soil Science, Oxford, for advice and facilities to prepare samples.

[1] J. Graham, *Rev. Pure Appl. Chem.*, 1964, **14**, 81.
[2] P. F. Low, *Adv. Agronomy*, 1961, **13**, 269.
[3] R. T. Martin, *Clays and Clay Minerals, Proc. Nat. Conf.*, 1962, **9**, 28.
[4] J. Graham, G. W. West and G. F. Walker, *J. Chem. Phys.*, 1964, **40**, 540.
[5] D. E. Woessner and B. S. Snowden, *J. Colloid Interface Sci.*, 1969, **30**, 54.
[6] D. E. Woessner and B. S. Snowden, *J. Chem. Phys.*, 1969, **50**, 1516.
[7] B. D. Boss and E. O. Stejskal, *J. Colloid Interface Sci.*, 1968, **26**, 271.

[8] P. A. Reynolds and J. W. White, *Disc. Faraday Soc.*, 1970, **48**, 131.

[9] G. C. Stirling and J. W. White, unpublished.

[10] A. W. Naumann, G. J. Safford and F. A. Mumpton, *Clays and Clay Minerals, Proc. Nat. Conf.*, 1966, **14**, 367.

[11] R. J. Hunter, G. C. Stirling and J. W. White, unpublished.

[12] P. A. Egelstaff, *An Introduction to the Liquid State*, (Academic Press, London, 1967).

[13] G. H. Vineyard, *Phys. Rev.*, 1958, **110**, 999.

[14] A. Sjölander, *Thermal Neutron Scattering*, ed. P. A. Egelstaff (Academic Press, London, 1965), chap. 7.

[15] A. M. Posner and J. P. Quirk, *Proc. Roy. Soc. A*, 1964, **278**, 35.

[16] L. J. Bunce, D. H. C. Harris and G. C. Stirling, *U.K.A.E.A. Report* AERE-R 6246 (H.M.S.O., 1970).

[17] G. J. Safford, P. S. Leung, A. W. Naumann and P. C. Schaffer, *J. Chem. Phys.*, 1969, **50**, 4444.

[18] K.-E. Larsson, *Neutron Inelastic Scattering* (Copenhagen, 1968), (I.A.E.A., Vienna, 1968), vol. 1, p. 397.

[19] B. K. Aldred and J. W. White, unpublished.

[20] S. J. Cocking, *Neutron Inelastic Scattering* (Copenhagen, 1968), (I.A.E.A., Vienna, 1968), vol. 1, p. 463.

[21] W. G. Garrett and G. F. Walker, *Clays and Clay Minerals, Proc. Nat. Conf.*, 1962, **9**, 557.

[22] K. Norrish and J. A. Rausell-Colom, *Clays and Clay Minerals, Proc. Nat. Conf.*, 1963, **10**, 123.

GENERAL DISCUSSION

Prof. A. Watillon (*Université Libre de Bruxelles*) said: In the first part of his paper, Dukhin considers the presence of a stagnant layer at the solid-liquid interface. This idea is supported by Wijga [1] and implicitly by Mossman and Mason. [2] More recently at a glass/water interface, [3] based on the same assumption, we interpreted simultaneous measurements of ζ potentials and of surface conductance. For K^+, Ba^{2+} and La^{3+} ions used as counterions, we obtained for the stagnant layer a thickness in the range 60-80 Å. This fact suggests that the proposed double-layer model, for glass, seems to be reasonably valid.

In the last part of his paper, Dukhin proposes an interpretation of the relaxation of the dielectric constant of colloid suspensions, based on his theory of double-layer polarization combined with Maxwell-Wagner relaxation theory of heterogeneous systems. He considers that this approach could interpret experimental results such as those of Schwan [4] on polystyrene latex suspensions. In our laboratory, DeBacker [5] observed relaxation phenomena on Pyrex spheres (diam. = 40 μm) in very dilute electrolyte solutions. Taking into account their larger size and the lower ionic strength, this system can be compared with Schwan's polystyrene latices. DeBacker calculated on the IBM 7040 computer the complex dielectric properties of the system against frequency. Therefore he used the Maxwell equation [6] twice: first, to deduce the admittance of the particle embedded in its double layer, secondly to obtain the admittance when the particle surrounded with its atmosphere is embedded in the dispersion medium. The calculations fit the observed data well. Finally, we wonder if polarization effect is large enough to explain the major part of relaxation processes in colloid systems.

Prof. S. S. Dukhin (*Inst. Colloid and Water Chem., Kiev*) (communicated): In reply to Watillon, in the Dukhin-Shilov theory of concentration polarization of a thin double layer of spherical particles, allowance is made for the appearance of difference in ion concentration beyond the double layer due to tangential flows of the diffuse layer ions and ion exchange between the double layer and contiguous volume of electrolyte. The effective relaxation time of this polarization process is $\tau = a^2/2D$, since this time is required for uniformity of concentration through diffusion at a distance of the order of a. If the frequency is such that $\omega\tau$ is very great the concentration difference does not appear during a single cycle and the concentration mechanism of polarization is no longer the determining factor. With very large ω, O'Konski's theory holds true; in this, the effect of the double layer is taken into account through the surface conductivity. Although Watillon does not mention the frequency used by DeBacker it may be presumed that $\omega\tau$ was very high in his experiment since large particles with diameter $2a = 40$ μm were used, i.e., $\tau = a^2/2D \sim 1$ s was very large. These experiments cannot, therefore, be a test of our theory.

[1] P. W. O. Wijga, *Thesis*, (Utrecht, 1946).
[2] C. E. Mossman and S. G. Mason, *Can. J. Chem.*, 1959, **37**, 1153.
[3] A. Watillon and R. DeBacker, *J. Electronal. Chem.*, 1970, **25**, 181.
[4] P. H. Schwan, G. Schwarz, J. Maszuk and H. Pauly, *J. Phys. Chem.*, 1962, **68**, 2626.
[5] R. DeBacker and A. Watillon, to be published.
[6] J. C. Maxwell, *A Treatise on Electricity and Magnetism* (Constable and Co., 1891).
[7] C. T. O'Konski, *J. Phys. Chem.*, 1960, **64**, 605.

Prof. H. Pfeifer (*Karl-Marx Universität*) (*communicated*): The important statement in the paper by Clifford, Oakes and Tiddy that for the lamellar phase the average correlation time of water molecules at 8.3 Å will be about three orders of magnitude longer than in ordinary water depends only on one point in fig. 9. (The last circle on the right-hand side.) Since these relaxation rates (T_{1W}^{-1}) were calculated from measured rates which are not very different (fig. 8), the error may be very high and the decrease of T_{1W}^{-1} for a thickness of water layer less than 8.3 Å questionable.

Mr. J. Clifford (*Unilever Res., Port Sunlight*) said: In reply to Pfeifer in presenting our paper it was pointed out that subsequent experiments had shown that the errors involved in calculating the final point in fig. 9 were indeed large and that the figure of 2×10^{-9} s for the average correlation time for rotation of water molecules, for a layer of thickness of 8.3 Å, should be regarded as only very approximate. This does not affect the general conclusions in any way.

Dr. G. Peschel (*Universität Würzburg*) said: With regard to the paper by Clifford *et al.*, in a former paper dealing with nuclear magnetic resonance relaxation times of water protons in water layers near polyvinylacetate [1] the authors found a molecular long range effect for the water molecules which might be due to the highly porous structure of the polyvinylacetate used, since water molecules might be trapped in the pores and probably lead to a pattern of surface centres which can form hydrogen bonds with adjacent water layers but which do not fit into the water structure. In this way a highly disturbed surface zone may be produced. The polystyrene used in the present paper has a much smoother surface which should not give rise to a significant perturbation of adjacent water layers because the potential wells of the surface are less pronounced so that most of the vicinal water molecules can easily arrange in such a way that the mismatch between surface zone and bulk liquid is negligibly small. Would the authors' explanation tend in this direction?

Dr. A. L. Smith (*Liverpool Polytechnic*) said: The flocculation kinetic results cited in ref. (16) of the paper by Clifford, Oakes and Tiddy refer to narrow size distribution polystyrene latex dispersions of particle diameter 0.37 μm flocculated in excess electrolyte. The method of particle counting in a flow laser ultra-microscope was used. Some variation of derived second-order rate constant with initial particle concentration was found and the rates quoted were obtained by extrapolation to zero particle concentration where the Smoluchowski theory of rapid flocculation is most likely to be valid. At 25°C the particle counting method gave a rate 50 % of the Smoluchowski figure. The temperature coefficient of the second-order rate constant was found to be equal, within experimental uncertainty, to that predicted by the Smoluchowski expression $4\,kT/3\eta$ for the extrapolated zero particle concentration rates, but rose significantly at high particle concentrations. Sols not extensively dialyzed or passed through an ion-exchange column further showed a variation of flocculation rate with particle charge even in 100 mol m^{-3} MgSO$_4$ aq., with maximum rate near the zero point of charge. This effect is presumably due to the presence of residual adsorbed layers of surface-active material and can be reproduced by the addition of such material to dialyzed sols.

Prof. J. Th. G. Overbeek (*University of Utrecht*) said: At the Faraday Society Discussion at Nottingham in 1966, Deryaguin pointed out that Smoluchowski's

[1] G. A. Johnson, S. M. A. Lecchini, E. G. Smith, J. Clifford and B. A. Pethica, *Disc. Faraday Soc.*, 1966, **42**, 120.

theory does not take into account that the mobility of two particles when they are close together is smaller than when they are far apart. The remark has been worked out by Deryaguin and, with more precision by Honig (Philips, Eindhoven) and Roebersen and Wiersema (Utrecht). The effect is so large that the mobility becomes proportional to h, the distance of separation of the spherical particles, and therefore they would never touch ($h = 0$) unless the van der Waals attraction pulled them together. The final result, taking both the hydrodynamic and the van der Waals effect into account, is that for any acceptable van der Waals-Hamaker constant, rapid flocculation is slower than the Smoluchowski rate by about 10-50 %.

Dr. E. Willis (*Unilever Res., Port Sunlight*) (*communicated*) Mr. A. Lips and I have studied the flocculation of a 126 nm diam. polystyrene latex in excess electrolyte at 25°C by light scattering. The particle concentration was 1×10^9 ml^{-1}. Using a model which describes the scattering properties of aggregates in terms of optical interference between their constituent primary particles, we obtained agreement of better than 5 % between the theoretical and the measured (intensity, time) curves for a range of scattering angles between 30 and 135° for times up to the half-life. We found the second-order rate constant to be 68 % of the Smoluchowski value, in good agreement with Smith's result. It would appear from these results and from Overbeek's comments that bound water layers make little contribution to the colloid stability of polystyrene latices.

Dr. M. L. Hair (*Xerox Corporation, Rochester, U.S.A.*) said: In fig. 4, Hougardy *et al.* present an infra-red spectrum (*b*) which clearly shows the presence of two different types of hydroxyl species at 3,400 and 3,550 cm^{-1}. This spectrum is attributed to a one-layer hydrate of Na-vermiculite containing oriented water molecules. In fig. 3, the infra-red spectrum of the two-layer hydrate shows only a single band, at 3,400 cm^{-1}, and assigned to non-oriented water. I would ask the authors whether the two bands observed with the one-layer hydrate are in fact due to a mixture of oriented and non-oriented water, or is there an alternative explanation? The authors comment that the deformation band due to the water which appears at about 1,600 cm^{-1} is shifted to lower frequencies. Does this remain as a single band or are two bands observed due to two different species?

Dr. J. M. Serratosa, (*Inst. Edafología y Biología Vergetal C.S.I.C., Madrid*) said: In reply to Hair, the presence of two bands in the region of the OH stretching vibrations does not necessarily indicate the existence of two different species of water. In fact, a water molecule has three normal modes of vibration: symmetric stretching v_1, bending v_2 and asymmetric stretching v_3. In many cases the absorption bands are broad and the v_1 and v_3 frequencies cannot be differentiated. In the one layer complex we believe that only one kind of water molecule exists. The water molecule either maintains its symmetry, or alternatively the two OH bonds of the molecule are associated by hydrogen bonds of different energy. This alternative precludes one from reaching a definitive conclusion on the orientation of the molecules in the interlayer space. Only one band corresponding to the bending vibration v_2 is observed which indicates one species of water molecule present.

Prof. H. Pfeifer (*Karl-Marx Universität*) (*communicated*): It would be of interest to know how large is the influence of constitutional hydroxyls and paramagnetic centres on spin-lattice relaxation times T_1 and in what manner van Olphen, Stone, Hougardy and Serratosa were able to correct this influence (error of this method).

I would also ask if the authors have some opinion about the observed change in relaxation times in going from two-layer hydrate to one-layer hydrate (fig. 2). It seems that the interaction energy is lower (decreased intermolecular interaction) and that the correlation time is unchanged (no shift of the minima—which are said to be present in the one-layer hydrate—but less pronounced). The latter statement would be in contrast with the experimental results of the free induction decay measurements (" it is concluded, therefore, that for the two-layer hydrate the orientation effect is obtained at a much lower temperature than for the monolayer hydrate ") and of their infra-red absorption spectra.

Dr. W. Stone (*Inst. Sci. de la Terre*) said: In reply to Pfeifer, all T_1 measurements show a plateau at low temperature. This effect was considered as resulting from paramagnetic impurities. In these preliminary experiments, the low-temperature relaxation rate was then simply subtracted from the observed rates. The same procedure was utilized to correct for the constitutional protons. These corrective factors represent approximately 10% of the observed rates in the region of T_1 minimum. Concerning the slightly lower T_1 values observed in the two-layer hydrate case, this fact is not in contradiction with the new results obtained on the oriented flakes.

Preliminary results on oriented flakes of Na-vermiculite by pulsed n.m.r. show (i) that for the two-layer hydrate, an organization of the water molecules exists even at room temperature. This result agrees with the entropy calculations. The observed free induction decay is angle dependent (i.e., varies as the angle between the c-axis and the direction of the large magnetic field is changed). By examining this dependence it can be concluded that, for temperature above $0°C$, the H—H vector of the water molecule is axially symmetric about the c-axis. The angle θ between these two vectors is approximately $50°$. It is also found that at low temperature (around $-50°C$) the organization of the water molecules changes, the symmetry axis is no longer the c-axis. (ii) That for the one layer hydrate, the H—H vector of the water molecules is also axially symmetric about the c-axis (θ again about $50°$) and that this symmetry is retained whatever the temperature. (iii) That the spin-lattice relaxation times appear to be slightly angle-dependent.

Dr. R. G. W. Anderson and **Dr. J. W. White** (*Phys. Chem. Lab., Oxford*) said: To complement the studies of water diffusion near vermiculite surfaces [1] we present some neutron inelastic scattering measurements of water diffusion in aqueous sols made from fumed silica. This system is favourable for the neutron method [1] because the very low scattering from silica compared to water (scattering cross-section $SiO_2 = 12 \times 10^{-24}$ cm^2, $H_2O = 170 \times 10^{-24}$ cm^2) allows the water dynamics alone to be observed with high signal-to-background. For a 5 % mass/mass H_2O/SiO_2 dispersion the water scattering is still about 5 times that from silica. Six mass ratios from pure water down to 5 % have been studied. Values of the water diffusion coefficients as a function of coverage have been measured from the quasi-elastic neutron scattering and evidence for jump diffusion at low ratios H_2O/SiO_2 has been found. The inelastic scattering spectra reveal some changes in the vibration frequency spectrum of water near 200 cm^{-1} associated with the adsorption.

Measurements were made at $23°C$ with the 6H cold neutron spectrometer [2] on the DIDO reactor at A.E.R.E. Harwell. For all spectra the scattering was less than

[1] see S. Olejnik, G. C. Stirling and J. W. White, this Discussion.
[2] L. Bunce, D. H. C. Harris and G. C. Stirling, *U.K.A.E.A. Research Report.*, no. 26246, (H.M.S.O., 1970).

5 % to minimize multiple scattering events, and from the straightness of the Debye-Waller plots (fig. 2*b*), incoherent scattering from hydrogen was dominant.   The whole experiment was designed to estimate the maximum structure-forming effect of silica on water, and unannealed Cabosil HS-5 was used.   This material has a quoted particle size of *ca.* 70 Å (700 nm) and a B.E.T. surface area of $325 \pm 25 \; m^2 \; g^{-1}$. The results of Clifford and Lecchini [1] indicate that a large structuring effect could be expected for this type of unannealed material.   Samples were prepared by mixing weighed quantities of water and Cabosil for several hours in a ball mill, the resulting gel was transferred quickly to thin-walled aluminium sample containers which were then sealed to avoid changes in water content.   Samples with mass/mass H_2O/SiO_2 greater than about 60 % were seen to be efflorescent by allowing them to stand in a sealed glass vessel for several days when water distilled out.   The composition of all samples was checked by ignition and weighing.

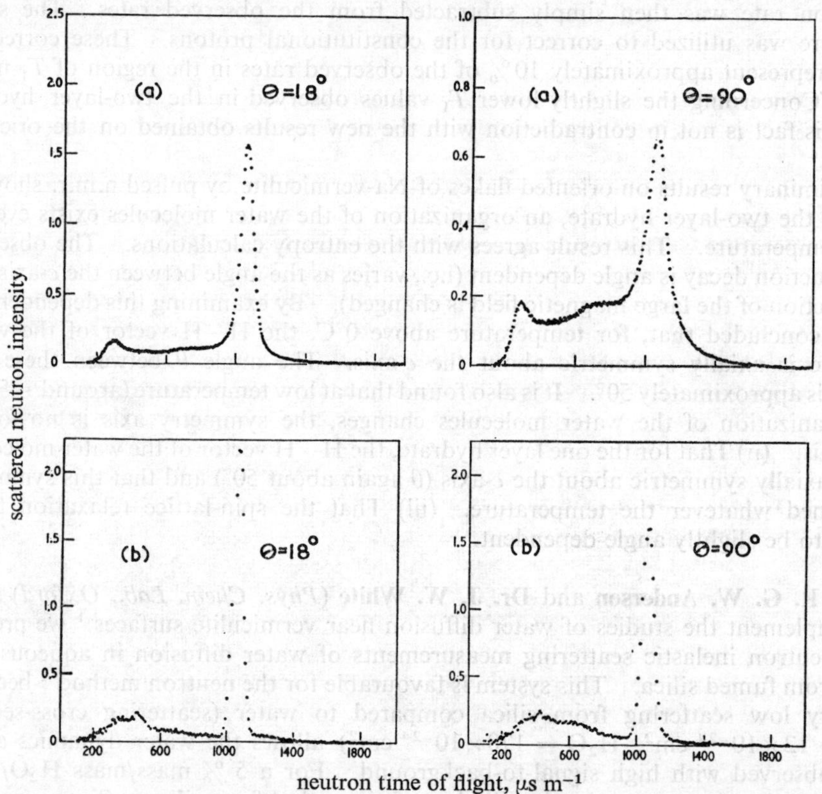

Fig. 1.—Neutron time of flight spectra for (*a*) mass/mass(H_2O/SiO_2) = 30 %; (*b*) Cabosil outgassed at 500°C.

Neutron time of flight spectra taken at scattering angles $\theta = 18$ and 90° to the incident beam direction are shown in fig. 1 for 30 % by mass water in Cabosil and for a Cabosil sample outgassed at 350°C and measured in an aluminium can.   The spectrum of the 30 % water gel resembles that of liquid water (ref. (1), fig. 1(*b*)). The water torsion band near 480 cm⁻¹ (*ca.* 300 μs m⁻¹) and the hindered translations near 60 cm⁻¹ appear strongly.   The smaller quasielastic broadening (peak near 1100 μs m⁻¹) for the gel is alone evident.   The intensities in the spectra for the gel

[1] J. Clifford and S. M. A. Lecchini, *Wetting*, (S.C.I., Monograph no. 25), 1967, p. 174.

and the silica are separately normalized to vanadium. To simulate the SiO_2 contribution to the gel spectrum the ordinate in fig. 1(*b*) should be divided by 30. The contribution is thus negligible. At the lowest water content studied (6 %), the torsion and translation peaks shifted to 460 ± 20 and 50 ± 10 cm^{-1} respectively from *ca.* 480 ± 20 and 58 ± 10 cm^{-1} in liquid H_2O. The hindered translations are now very broad.

In fig. 2 we show the angular dependences of the quasielastic scattering peak widths and intensities. The squared momentum transfer in the scattering collision $\mathbf{Q}^2$ is related to the scattering angle θ and neutron wavelength λ by

$$\mathbf{Q}^2 = (16\pi^2/\lambda^2)\sin^2(\theta/2) \tag{1}$$

and for molecules undergoing continuous diffusion obeying Fick's Law, the line broadening, ΔE is given by

$$\Delta E = 2\hbar D\mathbf{Q}^2, \tag{2}$$

where D is the diffusion coefficient of the scattering molecule. By increasing the scattering angle and hence $\mathbf{Q}$ the neutron is allowed to observe the diffusion on a progressively shortening time scale between 10^{-11} and 10^{-13} s. It can be seen from fig. 2(*a*) that at long observation times (small $\mathbf{Q}^2$) an almost linear relation between ΔE and $\mathbf{Q}^2$ is obtained as expected from eqn (2). For pure water, the value of D found on the longest neutron time scale agrees with the macroscopic value, from tracer and nuclear magnetic resonance spin echo studies.[1] For the gels the limiting D values are lower and depend on the water/silica ratio (see fig. 3). This behaviour is analogous to that of some dissolved ions such as Li$^+$, which reduce D for water.[1-3] Others (e.g., I$^-$) may increase it, but the diffusion constants measured macroscopically and by neutrons agree in both classes of solutions. We can therefore use the neutron measurement of D, at least empirically, as an indication of the extent of the water structure-forming tendency of a solute.

Fig. 2 and the concentration dependence of D (fig. 3) indicate a continuous decrease in water diffusion constant as a function of increased silica content in the gels. At the same time, the curvature at large $\mathbf{Q}$ values in fig. 2(*a*) shows that as the concentration of silica increases the water diffusion becomes less continuous and more like jump-diffusion. Again the water behaviour is analogous to water in strong salt solutions at low temperatures.[4]

A second measure of the restriction of water motions at high silica concentrations is provided by fig. 2(*b*). For an incoherent scattering solid, the elastic scattering intensity is related to the momentum transfer $\mathbf{Q}$, by the Debye-Waller factor

$$\exp[-2W] = \exp-[\mathbf{Q}^2\langle u^2\rangle], \tag{3}$$

where $\langle u^2\rangle$ is the mean square vibrational amplitude of the scattering atom. The log (peak area) is linear in $\mathbf{Q}^2$ with a slope proportional to this mean square vibrational amplitude. For a liquid where the atomic displacements are not bounded in time this formula is not correct, but Debye-Waller plots still give a qualitative measure of the net displacement on time scales with an upper bound defined approximately by the observation time of the neutron.

[1] G. C. Stirling and J. W. White, *Mol. Phys.*, to be published.
[2] G. J. Safford, P. S. Leung, A. W. Naumann and P. C. Schaffer, *J. Chem. Phys.*, 1969, **50**, 4444.
[3] J. W. White, *Deutsche Bunsen Gesellschaft Proc.*, (May, 1971).
[4] G. J. Safford, *Deutsche Bunsen Gesellschaft Proc.*, (May, 1971).

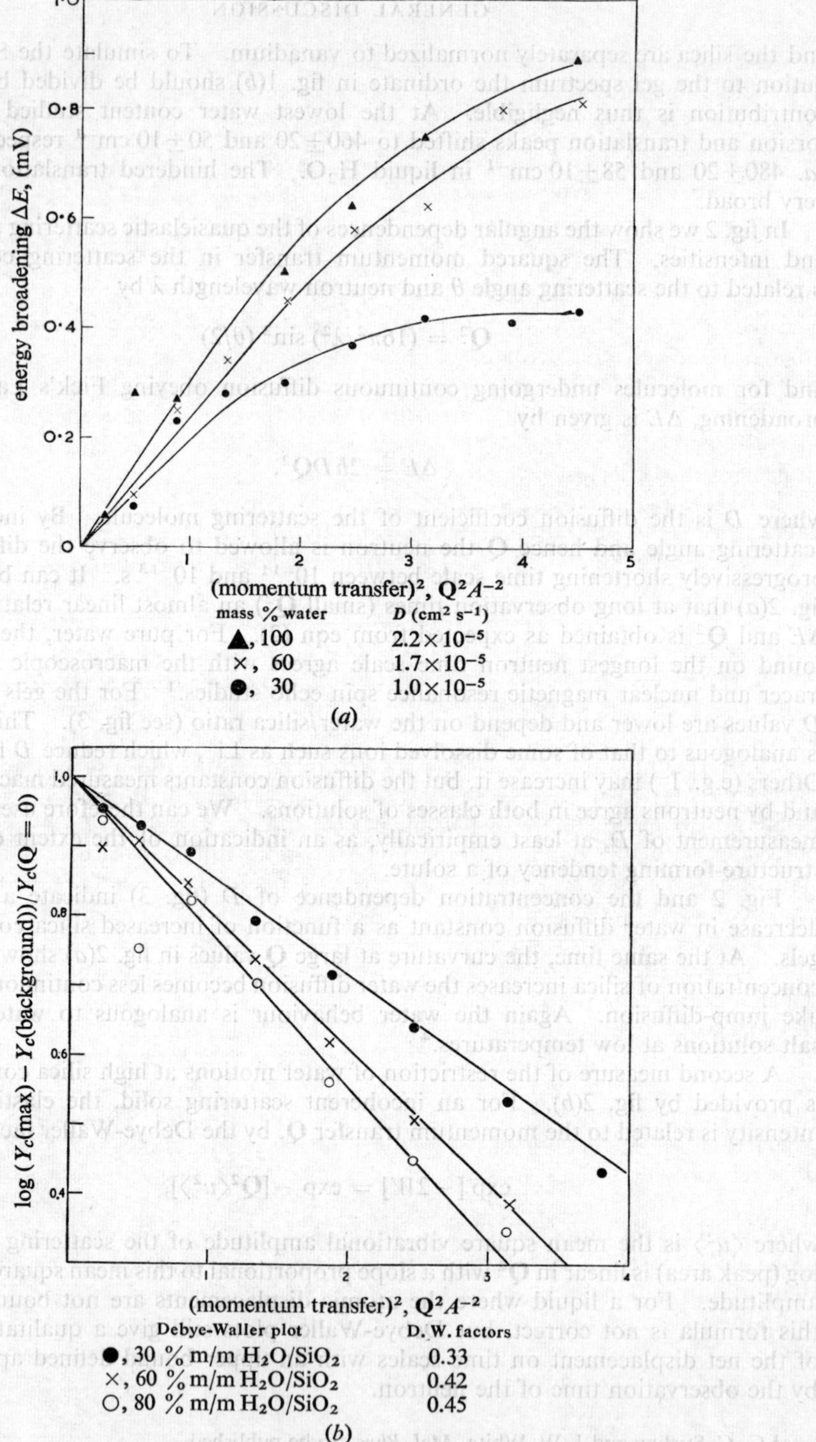

Fig. 2.—Quasi-elastic scattering dependence on squared momentum transfer and on water content: ▲, 100 % H_2O; m/m H_2O/SiO_2; ○, 80 %; ×, 60 %; ●, 30 %. (a) Energy broadening as a function of Q^2; (b) " Debye-Waller " plots of the quasi-elastic intensity.

Mathematically, the Debye-Waller factor for a solid is a frequency integral over the whole displacement (vibration and rotation) frequency spectrum from $\omega = 0$ to $\omega = \infty$,[1] eqn. (4),

$$\exp\left[-2W\right] = \exp\left[\frac{-\hbar Q^2}{6MN}\int_{\omega=0}^{\omega_{max}} \mathrm{d}\omega g(\omega)\frac{1}{\omega}\coth\left(\frac{\hbar\omega}{2kT}\right)\right], \tag{4}$$

where N is the number of atoms of mass M, and T is the absolute temperature of the sample. For neutron incoherent scattering from hydrogenous substances, the hydrogen contributions to eqn (4) usually dominate. Also in practical measurements, the lower limit of the integration in eqn 4 is set by the finite, and short, time of the neutron-molecule scattering event. This time τ may be calculated approximately for a solid from the neutron velocity and for a liquid by the method of Larsson[2] and lies between 10^{-11} and 10^{-14} s for thermal neutrons. The lower integration limit is of the order, $\omega_\tau = 2\pi/\tau$.

For most solids the frequency spectrum between $\omega = 0$ and, say, $\omega = 10^{11}$ radians s^{-1} is insignificant and an error in $2W$ produced by integrating from $\omega = 10^{11}$ rad s^{-1} to infinity is small. For a liquid, the Debye-Waller factor is given by the same formula as above but an appreciable fraction of the translational and rotational modes may lie at frequencies below ω_τ. The Debye-Waller factor is sensitive to solute effects in a different way to the quasi-elastic broadening since the frequency width *and* the area of the spectrum are both involved. Provided no modes are shifted from above ω_τ, the Debye-Waller can be used to compare the vibration amplitudes as for solids.

Close resemblance between the spectra in fig. 1 and those of water supports the evidence of fig. 2(b), that decreasing the water content of the gels from 80 to 30 % decreases the mean square displacements by about a factor of two. This behaviour correlates with the increased importance of jump diffusion rather than continuous diffusion in the system.

Unannealed Cabosil and dissolved ions both induce structure in water. Their effectiveness, and hence the range of the induced water structure, may be compared first, but considering the Cabosil as a normal solute and measuring the concentration dependence of the water diffusion coefficients for the two solutes. This reveals the mass effectiveness without regard to how many silica molecules are actually attached to water molecules. By using the specific surface area of the material (300 m^2 g^{-1}) and the area of a water molecule (9 Å^2)[3] it can be shown that at surface saturation there are nearly 3 silica molecules to one of water. The number of effective (surface) silica molecules was therefore chosen as one third the number of moles of silica present. Fig. 3 shows the dependence of the water diffusion coefficient on (a) total mol ratio of SiO_2/H_2O; (b) effective mol ratio of SiO_2/H_2O (one third (a)); and (c) mol ratio $LiCl/H_2O$. When account is taken of the number of silica molecules exposed to the water, silica has a stronger structure making effect on water than lithium chloride in accordance with the suggestion of Pethica. Unless some mechanism for proton transport, other than centre of mass diffusion is operating here, these data suggest, by analogy with lithium chloride solutions, that water-ordering occurs near silica surfaces out to several molecular layers, but probably not for many 100 Å.

[1] I. I. Gurevich and L. V. Tarasov, *Low Energy Neutron Physics* (North Holland, 1968), p. 145.
[2] K. E. Larson, *Thermal Neutron Scattering*, ed. P. A. Egelstaff, (Academic Press, 1965), chap. 8, p. 348.
[3] C. K. Hersh, *Molecular Sieves*, (Reinhold, 1961), p. 49.

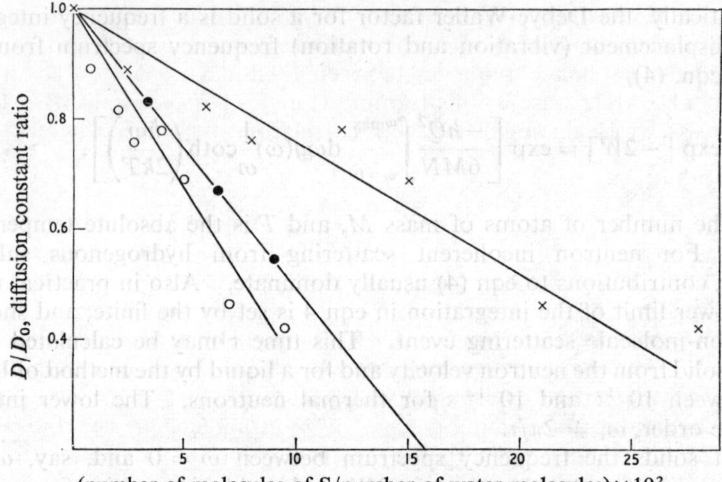

(number of molecules of S/number of water molecules) $\times 10^2$

(a) $\times$, S = total silica molecules; (b) $\bigcirc$, S = (total silica molecules)/3; (c) $\bullet$, S = total lithium chloride (Stirling and White [4])

FIG. 3.—Comparison of the effect of SiO_2 and LiCl on the diffusion coefficient of water: (a) total SiO_2; (b) surface SiO_2 in contact with water; (c) lithium chloride.

Dr. A. M. Hecht and **Dr. E. Geissler** (*Lab. de Spectrométric Physique,*[*] *Grenoble Gare*) said: With regard to the paper by Olejnik *et al.*, although there exists a large body of work devoted to clay+water systems, apart from the results of neutron scattering experiments we have little detailed knowledge of the structure and motions of water molecules adsorbed on clay interfaces. We therefore undertook a series of pulsed and continuous wave measurements of proton magnetic relaxation times of water adsorbed in oriented films of synthetic fluorine-substituted sodium montmorillonite. The synthetic clay has the double advantage of not possessing structural protons since the hydroxyls are replaced by fluorines, and of containing a small ($\lesssim 10$ p.p.m.) concentration of paramagnetic ions. Two samples of different hydration were examined: (I) containing a monomolecular layer, and (II) containing a bimolecular layer of water between the clay platelets (hydrated respectively at 20°C with relative humidity 0.4, and at 27°C with relative humidity 0.8). The n.m.r. lineshape between 220°K and room temperature consists of a narrow central line flanked by a doublet. The latter is generated by the water molecules in a rotational mode around the *c*-crystal axis of the clay plates. With increasing temperature the central line increases in intensity at the expense of the doublet, the variation being described by the activation energies 6.9 kcal/mol (sample I) and 9-10 kcal/mol (sample II). The central line is Lorentzian, characterized by a phase coherence time τ_2 for the spins which remains approximately constant at 600 μs in the temperature range 220-300 K. The shape of the line indicates the presence of motional narrowing $\Delta\omega$ of the static magnetic interaction between protons. The characteristic frequency of such a movement would be

$$1/\tau_c = \Delta\omega^2\tau_2 \gtrsim 10^7 \text{ s}^{-1}.$$

In order to identify the various kinds of movement prevailing in the water layer, we measured the proton spin-lattice relaxation time $\tau_1(T)$. The relaxation curves

* Laboratoire associé au C.N.R.S.
[1] A. M. Hecht and E. Geissler, *J. Colloid Interface Sci.*, to be published.

show a single τ_1 at each temperature which is practically independent of the orientation of the sample in the magnetic field. The measurement of τ_1 was also carried out as a function of the resonance frequency between 2.0 and 11.3 MHz. The minimum values of τ_1 at 11.3 MHz were 9.5 ms (sample I) and 9.0 ms (sample II). Different activation processes were distinguished, with correlation times as follows:

SAMPLE I
$$\tau_{c_1} = 2 \times 10^{-14} \exp (6.9/RT) \text{ s, and}$$
$$\tau_{c_3} = 5 \times 10^{-13} \exp (4.9/RT) \text{ s;}$$
SAMPLE II
$$\tau_{c_1} = 2 \times 10^{-14} \exp (6.6/RT) \text{ s.}$$

In both samples the relaxation time τ_1 shows a magnetic field dependence not yet understood. A lower energy activation process was also observed in sample II, but its parameters have not been determined with accuracy.

The pre-exponential time factor in τ_{c_2} is in good agreement with the rotational frequency of about 60 cm^{-1} seen in the neutron spectra of Olejnik, Stirling and White, and is not much different from the rotational period of the free water molecule. We associate this mode with the rotation responsible for the doublet in the c.w. spectrum.

The activation energy in τ_{c_1} is too weak for a normal molecular diffusion process, and the pre-exponential factor is short for a rotational motion. As vibrations hardly contribute to the spin-lattice relaxation, we consider proton jumps between molecules as a possible mechanism. In such a movement the dipolar coupling time-averages to zero, resulting in a narrow central line. It is notable that the activation energy for the intensity of the central line is identical to that of τ_{c_1} for sample I (though not for sample II, where more complex movements are expected to exist). If the central line is generated by such proton jumps the shortness of the pre-exponential time factor suggests that the movement is coherent over an area involving several hundreds of protons.

Dr. B. A. Pethica (*Unilever Res., Port Sunlight*) said: The neutron scattering data presented in the paper by Olejnik, Stirling and White relate to low levels of hydration of vermiculite, and show strong localization of the inter-lamellar water. The behaviour of water in larger amounts is of even greater interest in relation to the behaviour of liquid layers near a solid wall. The extra data provided in discussion concerning the properties of water in Aerosil+water mixtures is consequently important. It showed that the effect per mol of SiO_2 in reducing water diffusion is much less than that of LiCl. However, if allowance is made for the fact that only some few percent of the SiO_2 groups are in contact with the water, the effect of the Aerosil in restricting water diffusion is seen to be very great. Probably, the Aerosil was un-annealed, and it would be expected, as mentioned in the introductory paper, that annealed Aerosil would have a much smaller effect on the water for the same surface area in contact. If this is the case, it would provide important evidence in judging data obtained on silica surfaces in general.

Dr. M. M. Breuer (*Unilever Res. Lab., Isleworth*) said: Some experimental results obtained by Dr. C. B. Baddiel, Mrs. S. G. Clode and myself, also suggest that solid surfaces affect the nature of liquid layers in their near vicinity, and that these effects do not extend beyond a few molecular layers. We studied the molecular structure of adsorbed H_2O and D_2O on oriented films of synthetic polypeptides, using polarized infra-red radiation. In particular, we measured the dichroism of the i.-r. absorption band at $v = 3,500 \text{ cm}^{-1}$ of H_2O and $2,500 \text{ cm}^{-1}$ of D_2O molecules adsorbed on poly-L-alanine films in which the polypeptide molecules were in the α-helical conformation and the helices were aligned in oriented parallel arrays. As the only polar groups present in the system capable of binding H_2O (D_2O) molecules were the peptide linkages, which in turn were aligned almost parallel to the helix axes, it was possible that the H_2O (D_2O) molecules would exhibit considerable dichroism. In fact, we found at low relative humidities (r.h.) (0.30 %), i.e., within the monolayer domain, that the band at $3,500 \text{ cm}^{-1}$ $(2,500 \text{ cm}^{-1})$ showed some dichroism (1.3⊥). At higher r.h. values, however, the dichroic ratio decreases which may be attributed to an increase in the water adsorbed (the overall H_2O (D_2O) intensity increased) whilst the proportion of ordered molecules remained the same. It is difficult to judge accurately the central position of these broad bands but there appears to be little or no change in the absorption band frequency $(\pm 10 \text{ cm}^{-1})$ compared to that of liquid H_2O (D_2O). These results suggest that the structure of the first layer of H_2O (D_2O) is affected by the orientation of the peptide groups at the polypeptide surface, most probably through interactions of the dipole moments of the peptide bonds with those of the water molecules, i.e., C=O and N—H with H_2O. On the other hand, the ordering effect of the polypeptide surface on subsequent layers of H_2O (D_2O) appears to diminish rapidly with distance from the interface. Finally, I should mention that simultaneously with our work, at the University of Edinburgh, Dr. B. Malcolm also obtained similar results (except he also claims his spectral data indicated more than one species of H_2O) with oriented polypeptide films which were, however, prepared by a different technique and were of much lower dichroic ratio.[1]

<hr>

[1] B. Malcolm, *Nature*, 1970, **227**, 1358.

Measurement of Viscosity of Liquids in Quartz Capillaries

By N. V. Churayev, V. D. Sobolev and Z. M. Zorin

Dept. of Surface Penomena, Institute of Physical Chemistry,
The Academy of Sciences of the U.S.S.R., Moscow.

Received 30*th April*, 1970

A method is developed for measuring radii of microcapillaries and the viscosity of liquids in them. The viscosity of water, benzene, and carbon tetrachloride was measured in quartz capillaries of radius $r = 0.5$-0.04 μm. The viscosity of water in such capillaries is elevated (by 40 % in capillaries 0.04 μm radius), but the viscosity of non-polar CCl_4 and benzene remains normal. The temperature dependence of the increased viscosity of water is studied; the viscosity becomes normal at $t = 60$-$70°$C When water is drawn into a capillary with " dry " walls, the wetting angle differs from zero. In these cases the contact angle is not constant, but depends on the rate of entry of the water.

Owing to the action of surface forces the structure of liquids in fine capillaries may differ from their structure in bulk. Investigation of these structural changes yields information, not only concerning the range of action of surface forces, but also about the structure of the liquids. We studied the viscosity of water and non-polar liquids in capillaries from 1 to 0.05 μm in radius, drawn from quartz glass containing over 99.99 % SiO_2. This material completely eliminated the effect of leaching and dissolving of the capillary walls, and interpretation of the experiments was not complicated by pore geometry.

The first task when using fine capillaries is to determine their radius to a sufficient accuracy. Microscopic methods are useless for radii below 1 μm owing to wave limitations. Electron microscopy may give satisfactory results,[1] but this method involves rigid requirements as to quality of the end-cut, because the capillary hole is difficult to distinguish from surface irregularities.

The radius can be determined in simple manner from the capillary pressure as the pressure of air compressed by a liquid is sucked into the capillary which is sealed at one end.[2] However, since the liquid advances along an unwetted surface, the contact angle θ and the radius r are unknown. More reliable results can be obtained by calibration of the capillary by thermal expansion of the liquid in a communicating cavity.[3] However, this method is laborious and the accuracy is not better than 10-15 %.

The method used here for measuring capillary radii was that of measuring the pressure of nitrogen P_0 needed to form and detach a bubble from the capillary end immersed in a liquid. Then the radius r_0 is found from the equation:

$$r_0 = 2\sigma/P_0. \tag{1}$$

The major advantage of the method is that tabulated values of surface tension σ can be used in the calculations, since bubble formation occurs in the bulk of the liquid, outside the sphere of influence of surface forces. The detaching pressure recorded corresponds to the minimum curvature radius of the meniscus, thus eliminating the difficulty—determined value cos θ from the calculations. The capillary radius measured by this method differs from the actual r by the thickness

h of the film coating the capillary surface: $r = r_0 + h$. However, a similar correction must be introduced when other methods are used (except for microscopic ones).

Fig. 1 is a schematic diagram of the experimental apparatus. It was designed not only for determining capillary diameters, but also for measuring the viscosity and surface tension of the liquids contained in the capillaries. An empty sealed quartz capillary (1) is cemented with an epoxy resin into the stopper of a high-pressure chamber (3). After cementing, the capillary ends are trimmed, one of them being inserted in the ampoule (2) containing the liquid under study, and the other into the high-pressure chamber. The high-pressure chamber communicates with a pneumatic system by means of which the gas pressure in the chamber can be changed rapidly within a range of up to 100 atm. Pressure is measured by a set of standard (interchangeable for different ranges) gauges (4) of 0.3 % class accuracy. The accuracy of the pressure reading when working within the latter two-thirds of the gauge scales was not worse than 1 %.

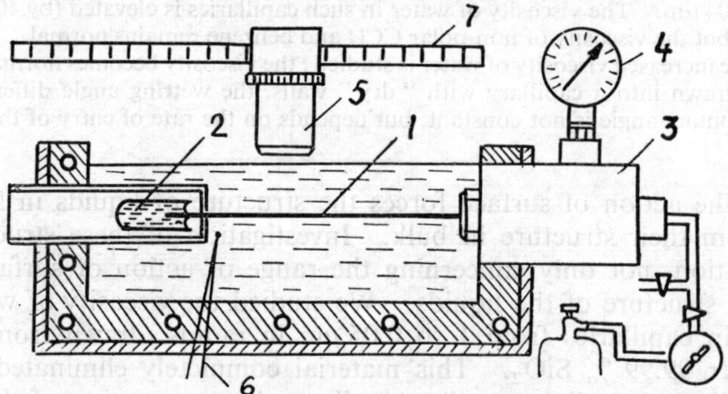

FIG. 1.—Diagram of unit for investigating liquids in microcapillaries.

The movements of the meniscus of the liquid filling the capillary were observed with a long-focus microscope (100×) (5) by the dark background method. A micrometer eyepiece was used for measuring meniscus travel (to ± 1 μm). The time of travel was measured with a stopwatch. The capillary and the ampoule containing the liquid under study were controlled thermostattically at the assigned temperature to within $\pm 0.1°$. The temperature of the liquid was measured with a differential thermocouple (6), the tip of which was located near the capillary. The entire unit was arranged on the movable stage of a reading microscope, by means of which the position of the meniscus in the capillary could be read off the scale (7) of a micrometric screw to within ± 10 μm.

When the free end of the capillary comes in contact with the liquid in the ampoule, the liquid is sucked into the capillary if the gas pressure in the capillary $P < P_c$ (P_c being the capillary pressure) or is forced out if $P > P_c$; when $P = P_c$ the meniscus stops. The capillary pressure P_c was registered on approach to equilibrium from both directions (at $P > P_c$ and $P < P_c$), the advancing meniscus always moving along a wetting film preformed by the movement of the receding meniscus. Equality of the measured values of P_c showed that under these conditions there was no wetting hysteresis. The effect of the limiting shear stress of the bulk water ($\tau \approx 10^{-2}$ dyne/cm²)[4] was much smaller than the possible error of pressure measurement and therefore undetectable.

At a pressure slightly exceeding the capillary pressure, the liquid is forced out of the capillary slowly so that the formation and detachment of a gas bubble from the capillary end can be observed and the minimum pressure P_0 needed for this measured.

With P_0 constant the following events were observed: as the bubble size increased the pressure inside the bubble decreased, the rate of its growth being determined by the resistance of the capillary to gas flow. At the moment of bubble detachment this pressure is smaller than P_c, and therefore immediately after detachment the liquid is drawn into the capillary. However, since, the pressure P_0 in the system is kept constant, the liquid column is gradually forced out of the capillary, and the events are repeated. When working with a capillary of $r<0.1$ μm the bubbles are difficult to see, but the periodic intake and outflow of the liquid from the capillary is readily observed, and this occurs only when the pressure rises to $P = P_0$. Hence, the applicability of the method is restricted only by the possibility of observing the meniscus in the capillary. The results of measurement of P_c and P_0 in the same capillaries for different liquids are presented in table 1.

TABLE 1.—MEASURED VALUES OF CAPILLARY PRESSURE P_c, PRESSURE OF AIR BUBBLE DETACHMENT P_0, AND CAPILLARY RADIUS r_0

experiment no.	liquid	P_c atm	P_0 atm	t °C	σ dyne/cm	r_0 μm	$\Delta r_0/r_0$ %
1	water	2.82	2.82	19.5	72.7	0.525	0.77
	CCl$_4$	1.06	1.06	19.5	26.8	0.521	
2	water	6.05	—	21.8	72.6	0.247	0.81
	CCl$_4$	2.23	—	21.8	26.7	0.245	
3	water	8.21	8.21	18.9	72.9	0.181	
	CCl$_4$	2.99	2.99	19.0	26.8	0.182	0.55
	benzene	3.24	3.24	19.0	28.95	0.182	
	mercury	36.4	—	19.1	323*	0.181	
4	water	37.5	—	20.4	72.72	0.0394	
	CCl$_4$	13.14	13.14	20.5	26.75	0.0416	1.7
	benzene	14.06	14.06	20.8	28.85	0.0414	
5	water	35.4	—	22	72.6	0.0418	
	CCl$_4$	13.14	—	22	26.7	0.0415	0.72
	benzene	14.04	—	22	28.9	0.0418	

* $\sigma \cos \theta$

Since P_c was measured by the receding meniscus, i.e., under conditions of complete wetting [5,6] the data given are evidence of equality (to the accuracy of measurement) of the surface tension of the liquids in the capillaries (of the radii indicated in table 1) and in the bulk. The capillary radii were calculated by eqn (1) from the pressures P_0 measured for different liquids and the bulk values of σ. Experiments with different liquids gave closely agreeing values for r_0, the average deviation not usually exceeding 1 %.

With fine capillaries ($r<0.1$ μm) allowance must be made for a systematic error $\sigma = (h/r_0)$ 100 %, where h is the thickness of the adsorption film. If, taking into account undersaturation due to concavity of the meniscus, the adsorption film thickness is taken to be 10-15 Å,[7] the possible error of determination of the capillary radius, even for the finest of the capillaries studied, is not more than 3-4 %. If the capillary has a slight taper, the " profile " of the channel $r(x)$ can be found, or constancy of r over the length of the capillary can be verified by measuring the pressure P_c and the corresponding coordinate of the meniscus.

Poiseille's law for a liquid column of length l, moving at a velocity v in a cylindrical capillary of radius r is

$$v = r^2(P_c - P_1)/8\eta l, \qquad (2)$$

where η is the viscosity of the liquid, P_c is the capillary pressure, and P_1 is the air pressure near the meniscus. The pressure P_1 is not equal to the pressure recorded by the gauge because there is a pressure drop due to movement of the gas column in the capillary. To determine P_1 we make use of the equation of gas movement in a capillary for Knudsen numbers $K = 0.001\text{-}0.1$ [8] :

$$v = r^2(P-P_1)(1+4\xi/r)/8\eta^*l^*, \tag{3}$$

where P is the gas pressure in the chamber, η^* is the viscosity of the gas, l^* is the length of the gas column, and $\xi = 1.38\lambda$ is the coefficient of slip [8] (λ being the free path of the gas molecules). Eqn (3) is applicable under the conditions of our experiments, since for capillaries from 10 to 0.01 μm in radius, $K = 0.003\text{-}0.03$. Simultaneous solution of eqn (2) and (3) gives

$$l_1 v = r^2(P_c - P)/8\eta, \tag{4}$$

where l_1 is the effective length of the liquid column :

$$l_1/l = 1 + \eta^*l^*/\eta l(1+4\xi/r). \tag{5}$$

Usually $l > l^*/3$. Then the difference between l_1 and l is not more than a few per cent. Since the second term of eqn (5) is small compared to 1, and as the viscosity of air depends little on the pressure, the dependence of vl_1 on P (at $P_c = $ const.) should be practically linear. All our experimental data obtained are in good accord with this assumption.*

Fig. 2 illustrate one of the (vl_1, P) dependencies for water (curve 2), and fig. 3 for CCl_4 (the graphs are plotted as v against P at $l = $ const). The intersection of the graphs with the abscissa axis corresponds to the capillary pressure P_c, which coincides with P_0, since the measurements are made with a receding meniscus or with a meniscus advancing on a part of the capillary coated with a wetting film. The microscope is set up a definite distance l from the end of the capillary. The travel of the meniscus (receding at first, and then advancing along the previously traversed section) are observed in the field of view of the fixed microscope at different pressures P. The length of the section of observation was 0.4 mm for large v and 0.02 mm for the smallest velocities of movement. Then the same measurements were repeated at a different value of l. The fact that the (v, P) graphs intersect at one point (fig. 3) is evidence that the radius was constant over the length of the capillary. Using eqn (4), the coefficient of viscosity η of the liquids in the capillaries could be determined from the angular coefficient of the (v, P) graphs. The P_c values were found from the points of intersection of the graphs with the abscissa axis. These values were also used for determining the capillary radius.

The results of measurement of the viscosity of water (points 1), CCl_4 (points 2) and benzene (points 3) at $t \approx 20°C$ are presented in fig. 4. In this graph the ordinates are relative viscosity η/η_0 (η_0 being the viscosity of the liquid in bulk), and the abscissae are capillary radii. It is evident from the data obtained that no changes in viscosity occur for non-polar CCl_4 and benzene with decrease of r down to 0.05 μm. For polar water, the supermolecular structure of which varies under the action of surface forces, the viscosity rises with decreasing capillary radius. The difference of the viscosity from the bulk value becomes perceptible at $r \leqslant 0.5$ μm and reaches ca. 35-40 % when r is decreased to 0.05 μm.

It is noted that the data obtained can be interpreted not only as increase of the viscosity of water in fine capillaries. If it is assumed that the properties of the liquid

---

* Except for travel of the advancing meniscus of polar liquids along a dry surface (fig. 2, curve 1).

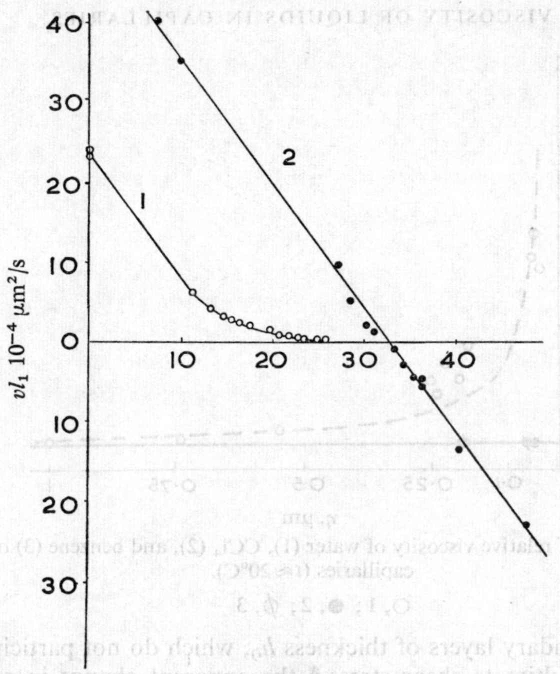

FIG. 2.—Dependence of product vl_1 on gas pressure in chamber P for movement of a water column along dry (1), and prewetted (2), surfaces of a quartz capillary ($r = 0.045$ μm, $t = 22.2°C$, $l = 11.26$ mm).

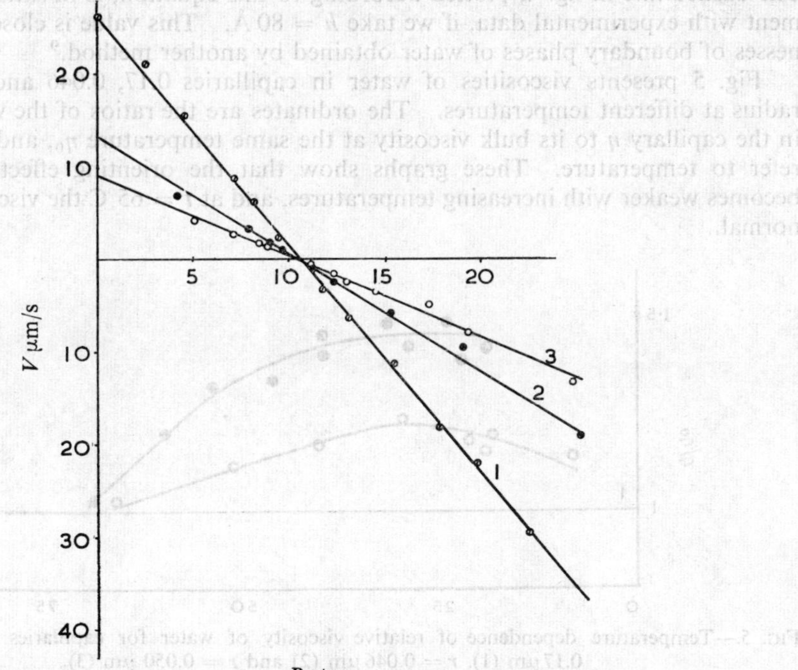

FIG. 3.—Dependence of velocity of movement v of a CCl_4 column in a quartz capillary ($r = 0.051$ μm, $t = 19.5°C$) on gas pressure in chamber P at $l = 11.55$ mm (1), 25.11 mm (2) and 37.52 mm (3).

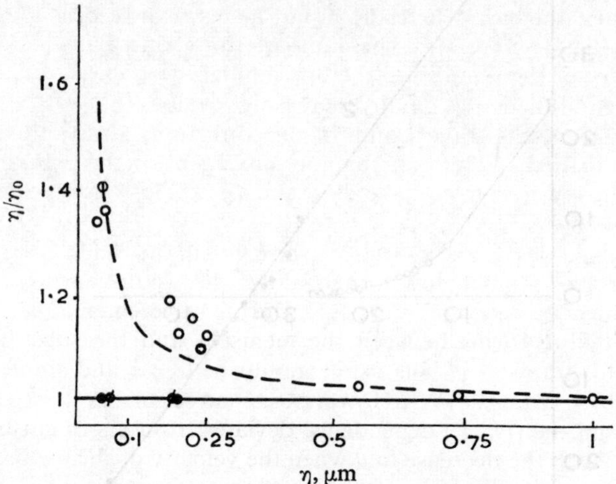

FIG. 4.—Dependence of relative viscosity of water (1), CCl_4 (2), and benzene (3) on radius of quartz capillaries ($t \approx 20°C$).

O, 1; $\bullet$, 2; ϕ, 3

change only in boundary layers of thickness h_0, which do not participate in the flow owing to the high ultimate shear stress,* the apparent change in relative viscosity amounts to

$$\eta/\eta_0 = [r/(r - h_0)]^2. \qquad (6)$$

The dashed line in fig. 4, plotted according to this equation, is in satisfactory agreement with experimental data, if we take $h = 80$ Å. This value is close to the thicknesses of boundary phases of water obtained by another method.[9]

Fig. 5 presents viscosities of water in capillaries 0.17, 0.046 and 0.050 μm in radius at different temperatures. The ordinates are the ratios of the water viscosity in the capillary η to its bulk viscosity at the same temperature η_0, and the abscissae refer to temperature. These graphs show that the orienting effect of the walls becomes weaker with increasing temperatures, and at $t = 65°C$ the viscosity becomes normal.

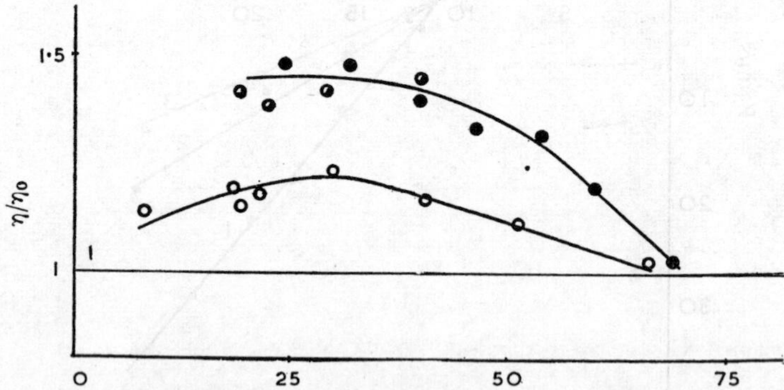

FIG. 5.—Temperature dependence of relative viscosity of water for capillaries for radius $r = 0.17\,\mu$m (1), $r = 0.046\,\mu$m (2) and $r = 0.050\,\mu$m (3).

$\bullet$, 1; $\ominus$, 2; O, 3

* For boundary layers of water $\tau_0 = 100$ dyne/cm².[4] Calculations show that the gradients of pressure used in our experiments were insufficient to put them in motion.

We now consider the case where the liquid meniscus advances along the surface of a dry capillary (fig. 2, curve 1). These observations were made as follows. First, the velocity of entry of the liquid into the dry capillary was measured at atmospheric pressure ($P_0 = 0$), and then at gradually increasing P values (but with $P < P_c$). Under all conditions the meniscus travels only in one direction, always advancing on the unwetted surface of the capillary. After the movement of the column stopped, the pressure was again lowered to atmospheric and the cycle was repeated on another section of the capillary.

Comparison of curves 1 and 2 (fig. 2) shows that the velocity of the meniscus along the dry surface is much lower than along the wetted surface. This can be attributed to the effect of wetting hysteresis. During travel of an advancing meniscus, a finite contact angle θ forms between the meniscus and the substrate surface.[4, 6] For $\theta = $ const., the graph of vl_1 against P should be linear and should intersect the abscissa axis at $P_c = (2\sigma \cos \theta)/r$. However, as shown in fig. 2, at small velocities of the liquid column the (vl_1, P) dependence deviates from a straight line. This can be associated only with the decrease in θ when the velocity of the meniscus is lowered.

Fig. 6 shows the values of cos θ calculated from eqn (4) under the assumption that $P_c = (2\,\delta \cos \theta)/r$, plotted against the meniscus velocity v for water (curves 1-4) and CCl_4 (curve 6). With CCl_4, as investigations have shown, $\theta = 0$ for travel along a surface both dry or wet. For water at $v > 5\ \mu m/s$ the contact angle differs from 0 and remains practically constant in value. At flow velocities $v < 5\ \mu m/s$, cos $\theta \to 1$ with $v \to 0$, this effect being more pronounced in fine capillaries.

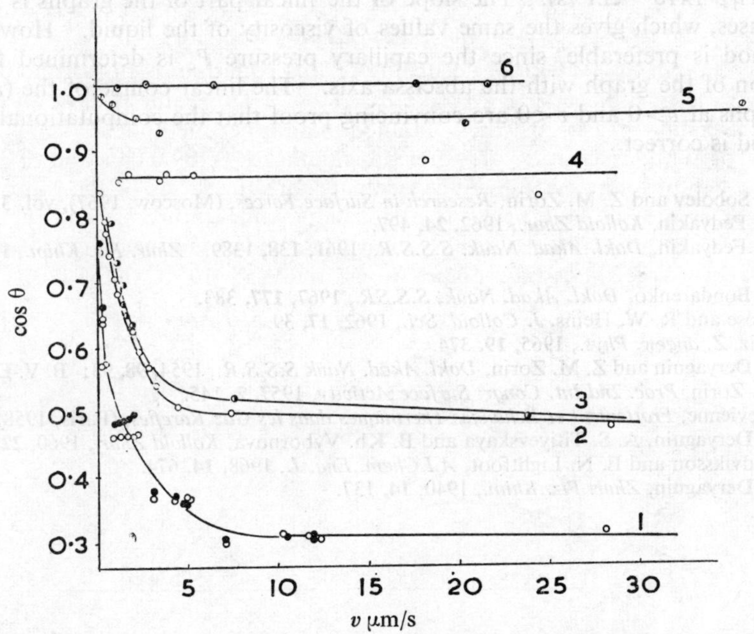

FIG. 6.—Cos θ against veloctiy v of meniscus advancing along unwetted capillary surface ($t = 20°C$). Water : $r = 0.181\ \mu m$ (1), $r = 0.049\ \mu m$ (2), $r = 0.054\ \mu m$ (3), $r = 0.5\ \mu m$ (4). Water along adsorbed film, $r = 0.22\ \mu m$ (5); carbon tetrachloride : $r = 0.051\ \mu m$ (6).

The dependence of θ on v cannot be attributed to any known purely hydrodynamic causes,[6, 10] because the latter change the shape of meniscus perceptibly only at velocities of $v \geqslant 2$-3 cm/s, i.e., much greater than those used in our experiments. In all probability, this phenomenon is related to the conditions of formation of an absorbed film near the advancing meniscus, and to specific meniscus-film interactions (formation of a transitional zone [11] and heat effects of film wetting).

To verify this statement we measured $\cos\theta$ on suction of water into a capillary whose walls had preliminarily been coated with an adsorbed film of water. The film was formed as follows. A water column, 10-15 mm in length, was drawn into one end of capillary 50-70 mm long, after which the capillary was sealed at both ends and placed in a thermostat. The decrease in length of the column in time was observed with a comparator until equilibrium was established. If it is assumed that at equilibrium the water evaporating from the column coats the free walls of the capillary with a uniform film, the thickness of the film can easily be calculated, from the capillary radius.[1] After cementing this capillary in the unit, the part filled with water was broken off and the free end was brought in contact with the ampoule (2) (fig. 1). Measurements were made by the method described above. Curve 5 of fig. 6 is a plot of $\cos\theta$ against velocity v in a capillary of radius $r = 0.22\ \mu m$, coated with an adsorbed film 15 Å thick. It is evident from the graph that in this case the wetting angle θ is independent of the velocity, and $\cos\theta = 0.96$. Further investigation is necessary.

The results of the above experiments show that reliable measurements of the viscosity of liquids in capillaries can be made only if $\cos\theta = \text{const}$. It is evident from fig. 2 that this can be accomplished either with the meniscus moving along a wetted surface (curve 2, $\cos\theta = 1$), or along a dry one, but only at high velocities (curve 1, $vl_1 \geqslant 7\times10^{-4}\ cm^2/s$). The slope of the linear part of the graphs is the same in both cases, which gives the same values of viscosity of the liquid. However, the first method is preferable, since the capillary pressure P_c is determined from the intersection of the graph with the abscissa axis. The linear course of the (vl_1, P) or (v, P) graphs at $v>0$ and $v<0$ are convincing proof that the computational basis of the method is correct.

[1] V. D. Sobolev and Z. M. Zorin, *Research in Surface Forces*, (Moscow, 1967), vol. 3, p. 36.

[2] N. N. Fedyakin, *Kolloid Zhur.*, 1962, **24**, 497.

[3] N. N. Fedyakin, *Dokl. Akad. Nauk. S.S.S.R.*, 1961, **138**, 1389. *Zhur. Fiz. Khim.*, 1962, **36**, 1450.

[4] N. F. Bondarenko, *Dokl. Akad. Nauk. S.S.SR.*, 1967, **177**, 383.

[5] W. Rose and R. W. Heins, *J. Colloid. Sci.*, 1962, **17**, 39.

[6] G. Friz, *Z. angew. Phys.*, 1965, **19**, 374.

[7] B. V. Deryaguin and Z. M. Zorin, *Dokl. Akad. Nauk S.S.S.R.*, 1954, **98**, 93; B. V. Deryaguin, Z. M. Zorin, *Proc. 2nd Int. Congr. Surface Activity*, 1957, **2**, 145.

[8] M. Devienne, *Frottement et Échanges Thermiques dans les Gaz Rarefies*, (Paris, 1958).

[9] B. V. Deryaguin, A. S. Titiyevskaya and B. Kh. Vybornova, *Kolloid Zhur.*, 1960, **22**, 398.

[10] V. Ludviksson and E. N. Lightfoot, *A.I.Chem. Eng. J.*, 1968, **14**, 674.

[11] B. V. Deryaguin, *Zhur. Fiz. Khim.*, 1940, **14**, 137.

Preliminary Studies of Thick Surface Films

By A. J. Smith and A. Cameron

Lubrication Laboratory, Mechanical Engineering Department, Imperial College of
Science and Technology, Exhibition Road, London, S.W.7, England

Received 9th April, 1970

A new technique is described for studying the existence of a thick viscous lubricant film adjacent
to the surface of a metal. It uses mercury instead of a solid plate to displace a hydrocarbon from the
metal surface, thereby avoiding spurious effects due to dirt or surface asperities. The capacitance
between the solid metal and the mercury is used to indicate the film thickness. A special circuit
enables the potential difference applied to the surfaces to be reduced to the order of microvolts. A
thick film is formed when there is a surfactant present in the hydrocarbon which reacts chemically
with the metal. The soap so formed appears to enmesh the hydrocarbon near the surface, forming
a grease layer some 10^3-10^4 Å thick. Preliminary results show the effect of carrier and surfactant
matching.

The addition of polar molecules to non-polar lubricants improves their frictional
qualities. Since the work of Hardy[1] and Bowden[2] it has been assumed that the
action of the additives is due to the formation of a close-packed layer one molecule
thick on the lubricated surface. There has, however, been a continuing series of
reports that these mixtures can form a "thick" film of semi-fluid material adjacent to
the metal surface.

Many workers[3-10] have investigated this problem by studying the behaviour of
liquids squeezed out from between plane surfaces, and some have variously detected
gross departures from Newtonian flow at film thicknesses ranging from 200 Å to
10,000 Å. Studies of flow through capillaries[10, 11] have yielded similar results with
the upper limit of anomalous film formation reported as high as 3×10^6 Å. Although
some of these studies have been conducted with scrupulous care, and have been of a
convincing nature, others have been successfully criticized on the grounds that solid
dust or colloidal particles or asperities on the approaching surfaces may have affected
the experimental results. Notwithstanding the fact that other experiments conducted
on one surface only[12-18] have tended to support the findings of these squeeze-film
experiments, the squeeze-film technique generally has become discredited on the grounds
that it is fundamentally susceptible to spurious interference due to dirt or asperities.

In this report the possibility of dirt or surface asperities keeping the squeeze
surfaces apart has been avoided by displacing the fluid by mercury, which might be
expected to flow around any such protruberances. The film thickness was deduced
from measurements of capacitance.

EXPERIMENTAL

Two forms of the same basic apparatus have been used: these will be designated the ball
apparatus and the plate apparatus. The plate apparatus (fig. 1) consists of a vertical glass
plate A on whose surface are deposited by vacuum evaporation a number of rectangular
areas of metal B. Holes bored through the plate and plugged with a conducting epoxy

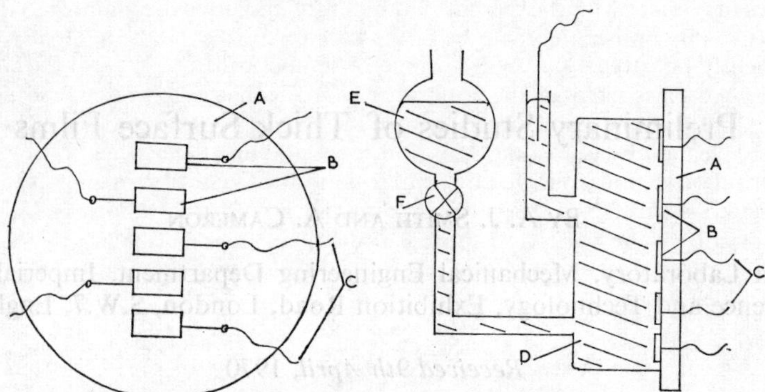

FIG. 1.—Plate apparatus. A, glass plate; B, metallized areas; C, electrical connections; D, glass chamber; E, mercury reservoir; F, glass cock.

resin permit electrical connections C to be made to the metal film from the reverse side of the plate. The metallized side of the plate covers the open face of a glass chamber D connected by flexible PTFE tube to a mercury reservoir E. By opening the cock F and raising or lowering the reservoir, the level of the mercury may be controlled. A wire passing into the reservoir makes electrical connection with the mercury. The apparatus was used in the following manner. The mercury in the chamber was raised to cover the metallized surface and the electrical circuit from the mercury to the metal film checked. The mercury was lowered, exposing some of the plates, and sufficient lubricant was added to the chamber through an opening in its top to cover them. The oil was left in contact with the metal for a measured period of time. The mercury level in the chamber was then raised to cover the metal plates, and the capacitance between a plate and the mercury was measured as a function of time.

In the ball apparatus (fig. 2), the metallized glass surface is functionally replaced by a steel ball A suspended by an insulated steel wire from a stopper B. A guard coil of steel

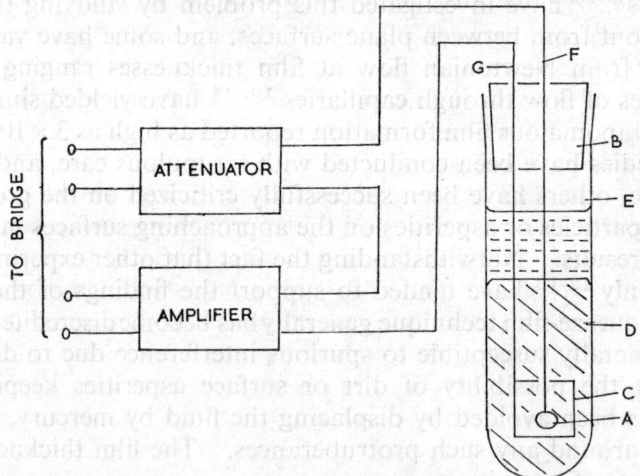

FIG. 2.—Ball apparatus. A, steel ball; B, stopper; C, Guard coil; D, mercury pool; E, test tube; F, G, electrical connections.

wire C encircles the ball and makes contact with the mercury pool D contained in a test tube E. This apparatus was used in a similar manner to the plate apparatus, the stopper being raised or lowered to effect the relative motion of ball and mercury. In order to ensure that the electrical measurements made at the terminal pair F, G would not be affected by the presence of a film of lubricant on the guard coil, care was taken that the lower end of the coil was at all times in contact with the mercury both before and after the addition of the lubricant to the apparatus.   By immersing the tube of the ball apparatus in a heated bath, experiments could conveniently be carried out over a range of temperatures.

CLEANING.—All glass parts of the apparatus were cleaned in chromic acid and rinsed in distilled water and acetone immediately before use.   Metal parts, including the metallized glass plate were cleaned by ultrasonic agitation in an alkaline detergent solution (ARDROX 1618) for a period of 15 min. followed by similar periods of agitation in an emulsifier solution (Imperial Smelters ISCEON 113S) and in pure tetra-chloro-tetrafluoro-ethane (Imperial Smelters, Fluorisol).   The parts were rinsed in distilled water between cleaning stages.   In all the cleaning procedures, spreading of a film of distilled water on the surface was used as the criterion of cleanliness.

MATERIALS

The mercury used in the experiments was trebly distilled *in vacuo* and stored in polythene bottles prior to the tests.   The hexadecane and benzene which acted as solvents (i.e., the test lubricants) were supplied by Koch-Light (*Purissimus* grade) and Hopkins and Williams (A.R.) respectively.   The solvents were purified by passing them through glass columns 700 mm long by 20 mm diam. packed with activated silica gel.   The solvents were left in contact with the adsorbant for a minimum period of 24 h.   The organic acids and amines used were also supplied by Koch-Light (*Purissimus*).

CAPACITANCE MEASUREMENTS

The electrical capacitance of the mercury-oil film-metal surface system was measured in order to deduce the thickness of the oil film, which acted as the dielectric of capacitor.   A modified Wayne-Kerr Universal Bridge type B221 in conjunction with the Auto-balance Adaptor AA221 was used in making these measurements.   This bridge operates on the transformer ratio arm principle [19] and, unmodified, presented to the test capacitor over the range of capacitance of interest, a peak voltage of 70 mV.   Under the experimental conditions this voltage was sufficient to cause dielectric breakdown of the oil film, thereby invalidating the measurements.   The voltage applied to the oil film was therefore reduced by inserting an attenuator between the E terminals of the bridge and the capacitor, and an amplifier between the capacitor and the I terminals of the bridge.   Fig. 3 and 4 are much simplified

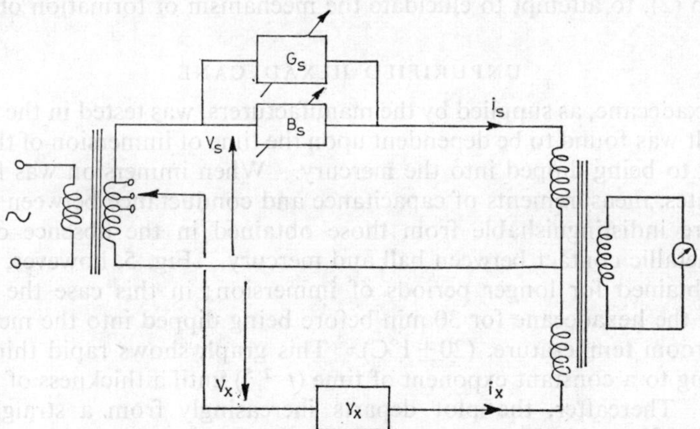

FIG. 3.—Unmodified bridge.

circuit diagrams of the unmodified and modified bridges respectively.   The original bridge applied a voltage V_x to the unknown admittance, through which passed a current i_x.   Adjusting the standards G_s and B_s brought it into balance with i_x.   By inserting the attenuator R_1 and R_2, the voltage across Y_x was reduced to V_x/A.   The resulting current i_x/A was boosted by the amplifier to give the output current i_x B/A. By setting B equal to A, the bridge could be used in the same way as before modification.

In practice, A and B were each set at 1,000, thereby reducing the peak voltage applied to the oil film to 70 μV. This voltage was sufficiently small to avoid breakdown of oil films above 1,000 Å thick, but for measurements on thinner films an additional attenuator could be inserted in the circuit in order to reduce the voltage by a further factor of 100 (i.e., to 700 nV). The system was calibrated to give capacitance measurements of an accuracy of ± 2 %.

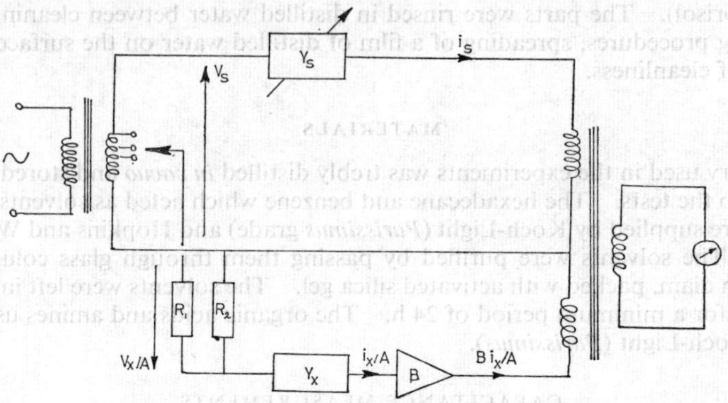

FIG. 4.—Modified bridge.

RESULTS AND DISCUSSION

This study of thick surface films has been in three main steps : these are : (1) an attempt to demonstrate the existence of a statically stable film of thickness greater than about 100 Å (this approaches the practical limit of resolution of the technique) : (2) to identify a group of liquid solvents which when pure would not give rise to a film, in order that the effect of adding small concentrations of impurities might be studied ; (3) by observing under varying conditions the film forming properties of the solutions mentioned in (2), to attempt to elucidate the mechanism of formation of the films.

UNPURIFIED HEXADECANE

When hexadecane, as supplied by the manufacturers, was tested in the ball apparatus, the result was found to be dependent upon the time of immersion of the ball in the alkane prior to being dipped into the mercury. When immersion was for less than several minutes, measurements of capacitance and conductance between the ball and mercury were indistinguishable from those obtained in the absence of lubricant, indicating metallic contact between ball and mercury. Fig. 5, however, is typical of the result obtained for longer periods of immersion; in this case the ball was in contact with the hexadecane for 30 min before being dipped into the mercury. The test was at room temperature, $(20 \pm 1°\text{C})$. This graph shows rapid thinning of the film according to a constant exponent of time $(t^{-2 \cdot 5})$ until a thickness of 4,000 Å was approached. Thereafter, the plot departs increasingly from a straight line and eventually the film stabilized at a thickness of 1,800 Å. The thickness curve was derived from the capacitance measurements according to the equation,

$$h = A \varepsilon \varepsilon_0 / C,$$

where A is the surface area of the ball, ε is the relative permittivity of the dielectric, ε_0 is the rationalized permittivity of free space, C is the capacitance of the system.

The value used for ε was that of bulk liquid hexadecane (2.29).   The uncertain validity of this value in this situation is an unfortunate feature of this method of thickness measurement.   The minimum possible value of ε, however, is unity, when the corresponding minimum film thickness would be 900 Å, a value far in excess of that expec-

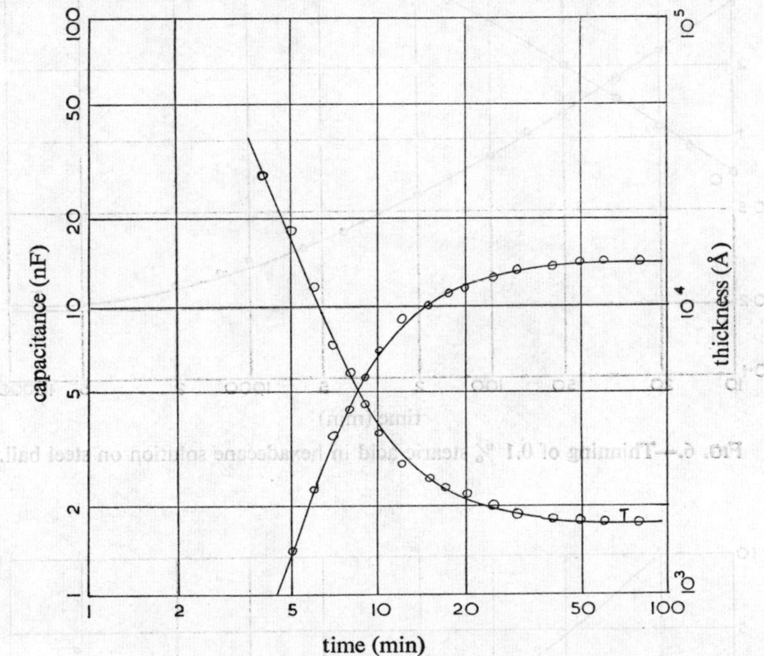

FIG. 5.—Thinning of unpurified cetane on steel ball.

ted from the normal theory of adsorption referred to earlier.   There is no corresponding maximum limiting value of ε but there is little reason to expect that the value of ε for the film would be far from that which has been assumed.

PURIFIED HEXADECANE

When the alkane was purified by adsorption on silica gel, no film was obtained from hexadecane in either of the two forms of the apparatus.   This was true even when the period of immersion of the steel ball or evaporated chromium plate was increased to 90 min.   The criterion by which the absence of a film was judged was that the capacitance and conductance of the system measured in the presence of an oil were similar to those measured without an oil.   A.R. benzene, purified in a similar way, gave a similar negative result, as did untreated A.R. acetone.

STEARIC ACID IN HEXADECANE SOLUTION

A solution of 0.1 % w/w of stearic acid in purified hexadecane, when tested in the ball apparatus, yielded a film whose thinning is depicted in fig. 6.   The growth time of the film was 15 min at $20 \pm 1°C$, the experimental temperature.

The calculated minimum value of film thickness was 2,100 Å.   This is based on a value of $\varepsilon = 2.2$.   (ε for solid stearic acid is 2.21, ε for hexadecane is 2.29.)   Fig 5.

SP1—H

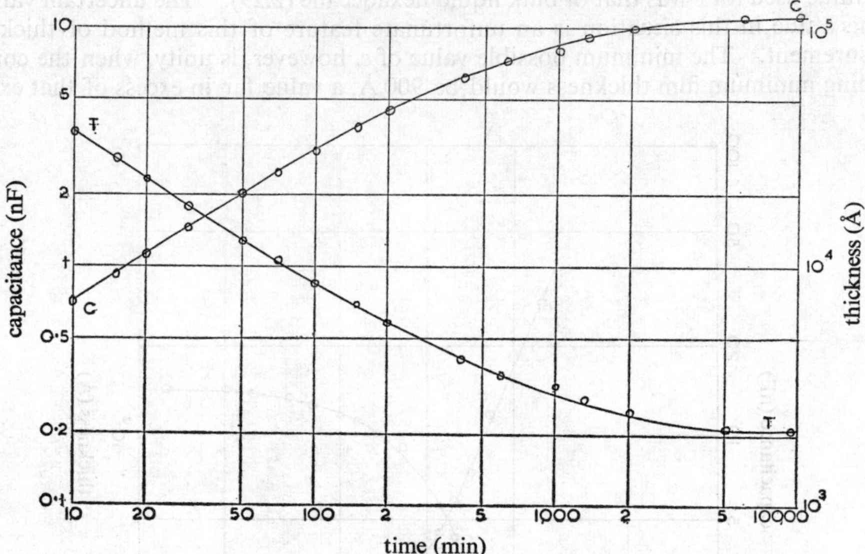

FIG. 6.—Thinning of 0.1 % stearic acid in hexadecane solution on steel ball.

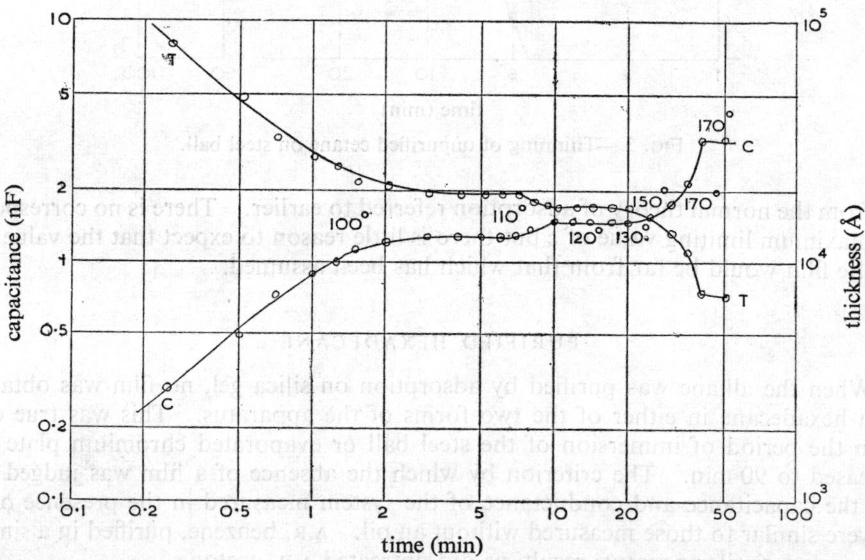

FIG. 7.—Thinning of 0.1 % stearic acid in hexadecane solution at varying temperatures.

represents the thinning of a film formed in the ball apparatus at a temperature of 100°C, the period of growth having been 15 min. The temperature was maintained at 100°C until the film had stabilized and was then raised in steps of 10°C. At a temperature of 100°C the film thinned more rapidly than a similar film formed and compressed at 20°C, but stabilized at a greater thickness (19×10^3 Å). Slight further thinning occurred on heating to 110°C, but over the temperature range 110-140°C the film thickness was constant.   Between 150 and 170°C the film thickness again fell rapidly, but stabilized when the temperature was fixed at 170°C, the maximum attainable by the oil bath used.   Similar transition temperatures have been reported for greases.[20]

The thinning of a stearic acid + hexadecane film which was both grown and compressed on a steel ball at 170°C is depicted in fig. 8.   The final thickness here was slightly greater (1.2×10^4 Å) than that of the film grown at 100°C and later heated to 170°C.

Similar experiments were run with 0.1 % cetyl amine solution in hexadecane, using the ball apparatus.   Immersion time was for 1 h at temperatures of 20°C and

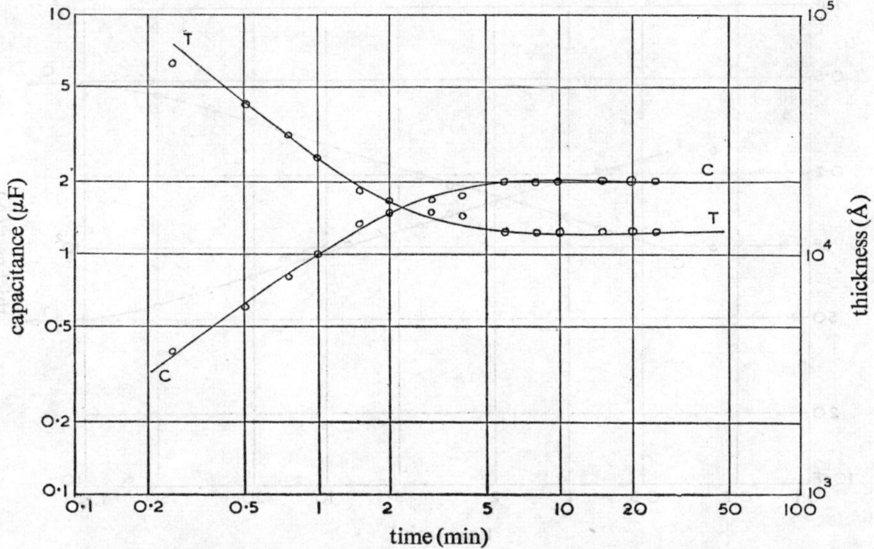

FIG. 8.—Thinning of 0.1 % stearic acid in hexadecane solution at 170°C.

95°C.   In neither case was a film found.   These results strongly suggest that the thick film might be caused by chemical means.   A possible mechanism being the formation of soap fibrils which entrain the solvent molecules, forming a two-phase soap thickened grease near the metal surface.

NOBLE METALS

A gold surface, tested in the plate apparatus with a 0.1 % solution of stearic acid after a period of 15 min immersion failed to yield a film.   The gold visibly amalgamated during the test.   It was expected that a 0.1 % solution of cetylamine in hexadecane would protect a copper surface against amalgamation, but this was not so, even after 2 h of immersion in the solution prior to the test.   There were apparently sufficient defects in the amine film to permit the mercury to penetrate it.

The ball apparatus was modified by replacing the steel ball and wires by a rectangular slip of platinum foil and platinum wires.   A 0.1 % solution of stearic acid in hexadecane failed to produce a film after immersion of the platinum in the lubricant for 30 min.   A similar negative result was obtained with 0.1 % hexadecylamine solution and platinum.

A valuable feature of these results is that they demonstrate that the mercury surface has no measurable effect upon the film under observation. This is presumably due either to the absence of a thick film on the mercury surface or to the breaking of a fresh mercury surface as the solid surface moves under the mercury, and to the movement of the mercury with the film as the film is squeezed during thinning,

ORGANIC ACIDS IN BENZENE SOLUTIONS

Fig. 9, 10, and 11 represent the thinning at room temperature of films formed from 0.1 % solutions of sebacic acid, stearic acid and oleic acid respectively in benzene.

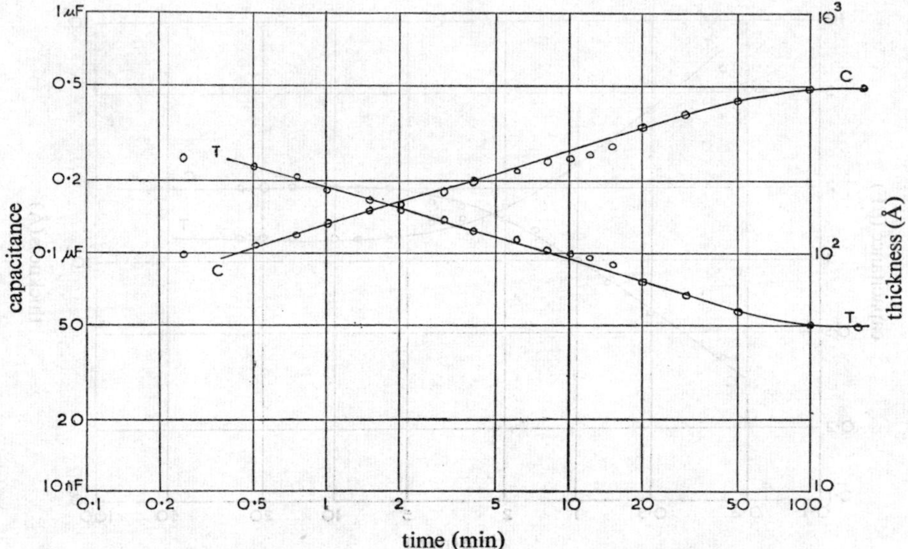

FIG. 9.—Thinning of 0.1 % sebacic acid solution in benzene.

The capacitance measurements have been interpreted as before, in each case the value of ε used has been that of the bulk solid acid. A striking feature of these results is that in all cases the capacitance values are at least three orders of magnitude greater than those obtained with stearic acid in hexadecane solution, and indicate values of film thickness in the order of small numbers of molecular lengths. The sebacic acid result is interesting in this respect in that the calculated film thickness for this material (fig. 9) is of the same order as those of the other two acids despite the fact that an adsorbed monolayer of sebacic acid is approximately one tenth of the thickness of that of a monobasic fatty acid of comparable carbon chain length.

Clearly, the calculated values of film thickness plotted in fig. 9, 10, and 11 must be regarded as approximate only. With films of this thickness, the effect of the surface micro-relief of the steel ball would be to increase the effective surface area considerably over that calculated from the macroscopic topology of the ball. The uncertainty of the value of ε used in the calculation of thickness, discussed earlier, is another possible source of error. The fact that stearic acid gives a far thinner film with benzene than it does with hexadecane is a striking demonstration of the effect, noticed by previous workers [5-7, 9] of the solvent on the film forming properties of this type of solution.

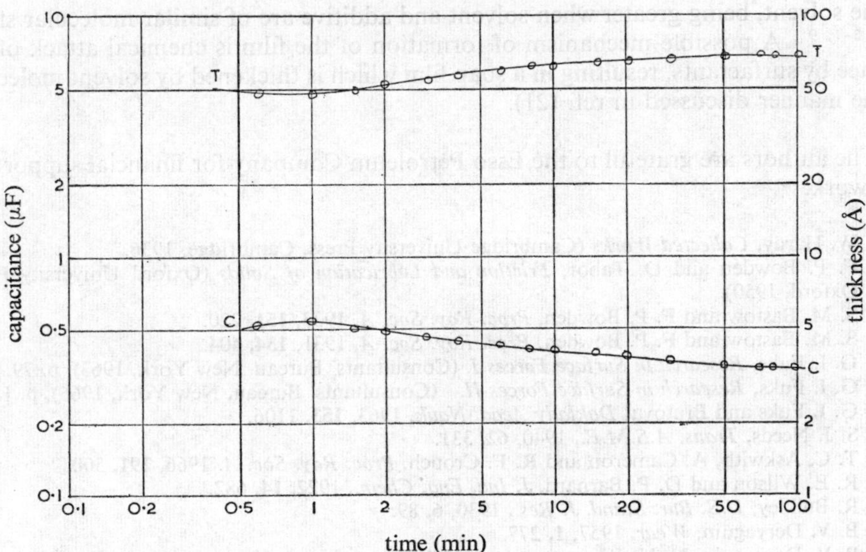

FIG. 10—Thinning 0.1 % stearic acid solution in benzene.

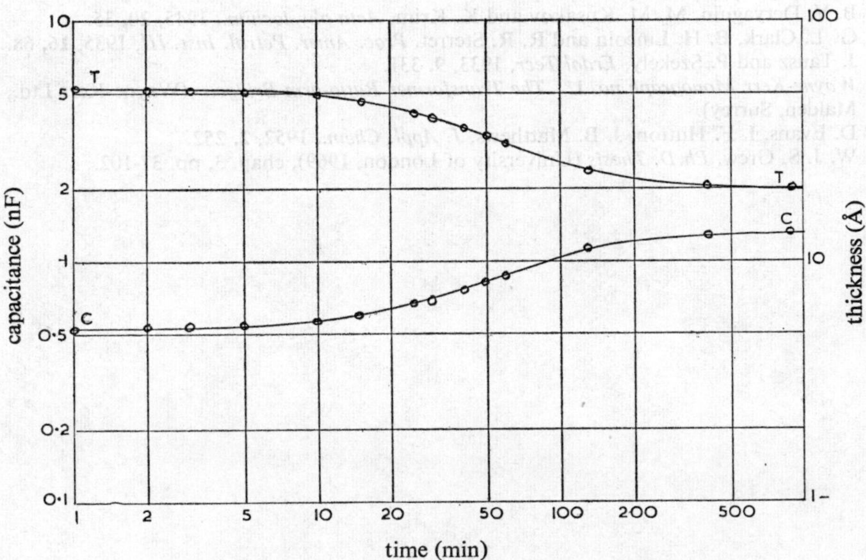

FIG. 11.—Thinning of 0.1 % oleic acid solution in benzene.

CONCLUSION

The most important conclusion of this preliminary study is that the thick surface film found by many previous workers does exist. The process of formation would appear to be due to chemical reaction rather than long-range surface physical forces. This is supported by their failure to form on noble metals, or with hexadecyl amine on steel, and by the similarity of the results obtained from stearic acid and sebacic acid solution in benzene. The thickness of the film is strongly dependent upon the nature of the solvent, being greater when solvent and additive are of similar molecular structure.[5-7, 9] A possible mechanism of formation of the film is chemical attack of the surface by surfactants, resulting in a soap film which is thickened by solvent molecules in the manner discussed in ref. (21).

The authors are grateful to the Esso Petroleum Company for financial support for this work.

[1] W. Hardy, *Collected Works* (Cambridge University Press, Cambridge, 1936.
[2] F. P. Bowden and D. Tabor, *Friction and Lubrication of Solids* (Oxford University Press, Oxford, 1950).
[3] S. M. Bastow and F. P. Bowden, *Proc. Roy. Soc. A*, 1933, **151**, 220.
[4] S. M. Bastow and F. P. Bowden, *Proc. Roy. Soc. A*, 1931, **134**, 404.
[5] G. I. Fuks, *Research in Surface Forces I.* (Consultants' Bureau, New York, 1963), p. 79.
[6] G. I. Fuks, *Research in Surface Forces II.* (Consultants' Bureau, New York, 1966), p. 159.
[7] G. I. Fuks and Bratova, *Doklady Acad. Nauk*, 1963, **153**, 1106.
[8] S. J. Needs, *Trans. A.S.M.E.*, 1940, **62**, 331.
[9] T. C. Askwith, A. Cameron and R. F. Crouch, *Proc. Roy. Soc. A*, 1966, **291**, 500.
[10] R. E. Wilson and D. P. Barnard, *J. Ind. Eng. Chem.*, 1922, **14**, 682.
[11] R. Bulkley, *U.S. Bur. Stand. J. Res.*, 1930, **6**, 89.
[12] B. V. Deryaguin, *Wear*, 1957, **1**, 277.
[13] B. V. Deryaguin, N. N. Kharaeva, A. M. Khomutov and S. V. Andreev, *Research in Surface Forces, II.* (Consultants' Bureau, N.Y., 1966), p. 156.
[14] A. M. Taylor and A. King, *J. Opt. Soc. Amer.*, 1933, **23**, 308.
[15] R. S. Bradley, *Z. Krist.*, 1936, **96**, 499.
[16] B. V. Deryaguin, M. M. Kusakov and K. Krim, *Acta physiochim.*, 1945, **20**, 35.
[17] G. L. Clark, B. H. Lincoln and R. R. Sterret, *Proc. Amer. Petrol. Inst. III*, 1935, **16**, 68.
[18] J. Tausz and P. Szekely, *Erdol Teer*, 1933, **9**, 331.
[19] *Wayne-Kerr Monograph no. 1. The Transformer Ratio-arm Bridge.* (Wayne-Kerr Ltd., New Malden, Surrey).
[20] D. Evans, I. F. Hutton, J. B. Matthews, *J. Appl. Chem.*, 1952, **2**, 252.
[21] W. J. S. Grew, *Ph.D. Thesis* (University of London, 1969), chap. 3, pp. 37-102.

Thickness of Very Thin Films in Elastohydrodynamic Lubrication

By A. Dyson

Shell Research Ltd., Thornton Research Centre, P.O. Box 1,
Chester CH1 3SH, England

Received 9th April, 1970

Films of lubricant were formed between two loaded rolling steel discs, and the thicknesses of the films were estimated from measurements of the electrical capacitance between the discs. Two mineral oils and one synthetic diester were used as lubricants. The minimum film thicknesses ranged down to approximately 10 nm (100 Å), and agreed approximately with the predictions of an isothermal Newtonian theory in which the viscosity of the lubricant, measured in bulk, was used. Of the four sets of results examined, only one, that for the di-ester, gave measured film thicknesses greater than the theoretical ones. The discrepancy of 13 % is probably within the systematic errors of the experiment and of the theory. With this exception, there was therefore no evidence that the viscosity of the lubricant was affected by the close proximity of the metal surfaces. The maximum shear rate in these experiments was of the order of 10^7 s^{-1}, i.e., several orders of magnitude greater than those obtaining in other experiments reported in the literature, in which the viscosity of a fluid was increased by the presence of the surfaces.

There have been reports from many fields that the flow properties of a liquid may be modified by the close proximity of a solid surface. The modifications range from a small increase in viscosity to rigidity, and the depth of the affected zone ranges from 1 nm to 10 μm (10 to 10^5 Å). A review of work up to 1949 by Henniker[1] gives references in support of the existence of such effects. In a later review, Hayward and Isdale[2] concluded that rheological abnormalities in pure liquids do not extend more than a few molecular diameters from a solid boundary, although they admit that there is still some controversy about liquids containing surface-active materials. One remarkable result is that of Askwith, Cameron and Crouch,[3] who report that there is a " plastic layer ", of a rigidity sufficient to arrest completely the descent of a flat metal plate, at a depth of ca 1.9 μm ($\frac{3}{4} \times 10^{-4}$ in.), even in pure cetane.

Such zones of enhanced viscosity would have important effects in the lubrication of practical machinery. Ordinary mineral lubricating oils contain various polar compounds, many of them surface-active, in a non-polar hydrocarbon medium, and thus belong to the class of liquids for which the evidence of the effect is strongest. Machine elements such as gear teeth, cams and tappets, and rolling contact bearings, are lubricated by elastohydrodynamic lubrication, and the minimum thickness of the film of lubricant in such devices is 0.1-1.0 μm, i.e., within the reported range of action of the effect discussed. The rheological abnormalities could therefore have a considerable effect on the minimum film thickness, which, in turn, strongly influences wear and fatigue failure in service.

The work reported here was designed to investigate the practical importance of such effects. Fundamental work with real machinery is difficult, but most of the essential elements are present in a disc machine, which is considerably simpler both experimentally and theoretically. In such a machine the minimum thickness of the lubricant film may be estimated from measurements of the electrical capacitance between the discs and may be compared with the predictions of a suitable theory, based on the assumption that the viscosity is not influenced by the close proximity of the surfaces. If the viscosity were increased by the effect considered, then the experimental estimates of film thickness would be greater than the theoretical ones. The fact that, with one possible exception, no such increase was detected implies that there were no significant increases in viscosity arising from the close proximity of the metal surfaces in the conditions of the experiments. Some estimates are given of the depth and degree of viscosity enhancement which it would have been possible to detect.

EXPERIMENTAL

The equipment used and the interpretation of the capacitance measurements in terms of minimum film thicknesses have been discussed previously,[4] and only an outline of the essential details will be given here.

Two discs each of case-hardened En34 steel, 76.2 mm (3 in.) diam and 25.4 mm (1 in.) width, were loaded together in rolling contact with a force of 2.45 kN (550 lbf). The edges of one disc were rounded to give an effective contact width of 22.2 mm (0.875 in.). The surfaces were ground to an eccentricity of less than 2.5 μm (0.000 1 in.) and to a surface roughness of 0.037 5-0.05 μm (1.5-2 microinch) c.l.a. Polishing with diamond paste improved the finish to approximately 0.02 μm (0.8 microinch) c.l.a. These surface-finish measurements were made by traversing a stylus instrument in an axial direction, the datum line being generated by a skid. The bulk temperatures were measured by thermocouples embedded in the discs. Lubricant was supplied by a jet to the inlet to the contact and to the sides of the discs. The disc temperature was controlled by the temperature and rate of the oil supply.

The capacitance between the discs was measured by an r.f. bridge at a frequency of 19 kHz. For the interpretation of the capacitance measurements in terms of film thicknesses, it was assumed that the discs had the same shape as in the Hertzian case of dry contact, with the addition of a constant separation h_m. The dielectric constants of the lubricants were measured at various temperatures, at atmospheric pressure and at a pressure of 345 MN m^{-2} (50,000 lbf in^{-2}), approximately equal to the mean Hertzian pressure in the contact, 357 MN m^{-2} (51,800 lbf in^{-2}). The lubricants used were two mineral oils and one synthetic oil (a di-ester). Their properties are given in table 1, reproduced from an earlier paper.[4] The

TABLE 1.—PROPERTIES OF LUBRICANTS

code		A	B	L
description		HVI mineral oil	MVI mineral oil	di(2-ethylhexyl) sebacate
kinematic viscosity/cS	37.8°C (100°F)	175.3	83.0	12.58
	98.9°C (210°F)	15.36	8.8	3.31
absolute viscosity/P, at	30°C	2.50	1.22	0.149
atmospheric pressure	60°C	0.505	0.263	0.062 1
	100°C	0.126	0.073	0.028 2
absolute viscosity/P, at	30°C	5.90	3.10	0.249
gauge pressure 34.5 MN	60°C	1.05	0.555	0.097
m^{-2} (5,000 lbf in^{-2})	100°C	0.232	0.135	0.042

di-ester was a commercial product, but it was stipulated that it should contain no unneutralized or partly neutralized sebacic acid. No special precautions were taken to avoid contamination. When the lubricant was changed, the discs were cleaned by being rotated under no load while tissues soaked with a light paraffinic solvent were pressed against them.

RESULTS

The results are given in fig. 1 and 2 in the form of graphs of the experimental film thickness h_e against the theoretical film thickness h_t, both quantities being expressed in nm. The theoretical estimates are based on the expression given by Dowson, Higginson and Whitaker [5]

$$h_t = 1.6\,(\bar{u}\eta_0)^{0.7}\alpha^{0.6}(E')^{0.03}R^{0.43}w^{-0.13},$$

where $\bar{u}$ is the peripheral velocity of either disc; η_0 is the absolute viscosity of the

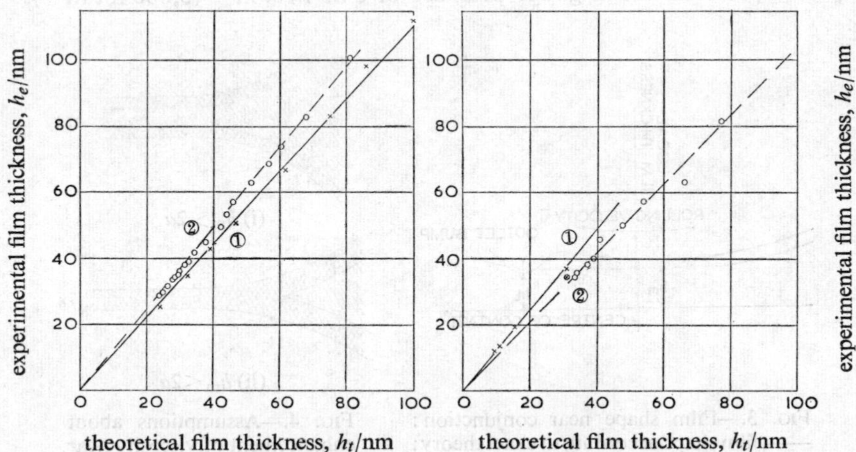

FIG. 1.—Comparison of experimental and theoretical film thickness for lubricants A and L : 1, lubricant A; 2, lubricant L.

FIG. 2.—Comparison of experimental and theoretical film thickness for lubricant B : 1, constant temperature; 2, constant speed.

lubricant at atmospheric pressure, at the mean temperature indicated by the embedded thermocouples; α is the pressure coefficient of viscosity; E' is the reduced elastic modulus, given by $E/(1-v^2)$, where E is Young's modulus and v is Poisson's ratio for the material of the discs; R is the radius of relative curvature, and w is the load per unit width of contact.

The theory is an isothermal Newtonian one, and the viscosity used is that of the fluid in bulk. There is some evidence [4] that thermal effects, probably caused by viscous shear in the inlet zone, are important for minimum film thicknesses of 1 μm or greater, while some order-of-magnitude calculations suggested that such thermal effects would be negligible for film thicknesses of 0.1 μm or less. The results used in the analyses were therefore restricted to a minimum film thickness of 0.1 μm or less.

The numerical results are given in appendix 1. They were analyzed statistically as a regression of h_e on h_t, all the error being assumed to arise in h_e. There was no evidence of non-linearity, or that the regressions did not pass through the origin. The slopes of the lines were taken as the coefficients of the regression, constrained to pass through the origin.

The theory gives the *minimum* film thickness under the outlet bump (fig. 3), while the experiment gives the thickness of an assumed parallel film under the Hertzian contact zone which would give the observed electrical capacitance. The difference between these two estimates of film thickness is sketched in fig. 3, and in the example given by Dowson *et al.*[5] the theoretical minimum film thickness h_t is approximately

85 % of the mean thickness in the Hertzian zone.   It has been assumed that the same ratio applies to the present results.

A further difficulty arises from the fact that the theory assumes a relation between viscosity and pressure of the form, $\eta = \eta_0 \exp(\alpha p)$.   This is only an approximation to the properties of real fluids.   The slope of the curve of the logarithm of viscosity against pressure usually decreases with increasing pressure.   For lubricants A and B, this curve was known only up to pressures of 103.5 MN m^{-2} (15,000 lbf in^{-2}), and for lubricant L up to approximately 0.828 GN m^{-2} (120,000 lbf in^{-2}).   In the theoretical estimates, the value of α was taken to be the mean slope of the above curve between atmospheric pressure and a gauge pressure of 34.5 MN m^{-2} (5,000 lbf in^{-2}).   This

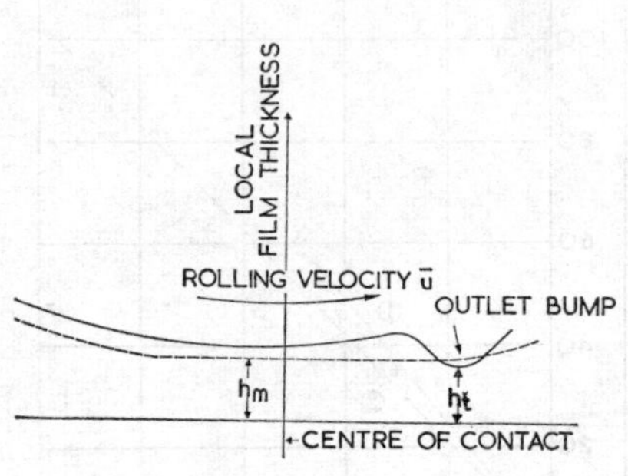

FIG. 3.—Film shape near conjunction: ——, film shape according to exact theory; - - - -, film shape assumed for purposes of estimation of film thickness from measurements of capacitance.   (*Sketch only, not to scale.*)

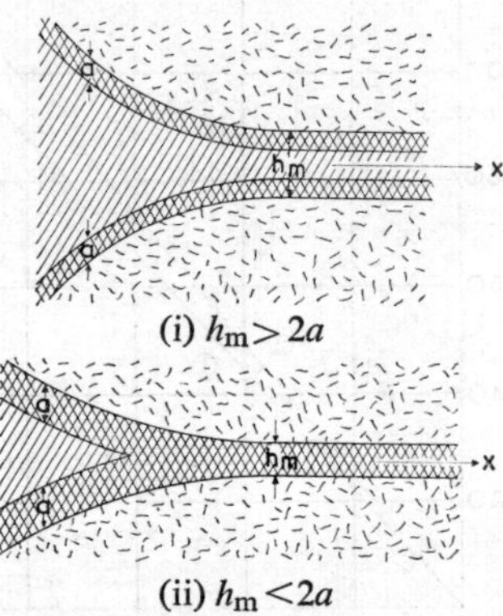

FIG. 4.—Assumptions about enhancement of viscosity near surfaces.

Film of lubricant with viscosity, $\eta = \eta_0 \exp(\alpha p)$, equal to viscosity of the fluid in bulk

Film of lubricant with enhanced viscosity, $k\eta = k\eta_0 \exp(\alpha p)$

Material of disc

procedure over-estimates [4] the effective value of α by approximately 8 % for lubricant A, by 3 % for B, and by 16 % for L.   The estimate of the error is much more reliable for lubricant L than for lubricants A and B.

The results of the statistical analyses are given in table 2.   The crude regression coefficients have been corrected for the expected 15 % discrepancy between the theoretical and experimental film thicknesses, and for the error in the estimates of α.   The reference to 95 % confidence limits refers to errors of a random nature only, and does not take into account any possible systematic errors.

DISCUSSION

The observed results are now compared with those to be expected if there were any significant degree of enhancement of viscosity.   Suppose that the viscosity is increased by a factor k compared with the bulk viscosity over a film of thickness a, adjacent to each surface, as illustrated in fig. 4.   The film thickness h_m at the inlet edge of the Hertzian contact zone has been estimated by the method of Grubin,[6] in which boundary conditions on the distribution of hydrodynamic pressure, $p = p(x)$, are imposed:

$$p \to \infty, \quad dp/dx = 0 \text{ at } h = h_m.$$

The shape of the gap between the two surfaces is assumed to be the Hertzian deformed shape, with the addition of a constant separation h_m. This procedure gives a good approximation [7] to the results both of more refined theoretical solutions of the elastohydrodynamic problem and of experiment. An approximation due to Crook [8] for the Hertzian deformed shape near the edge of the contact zone has been used. The resulting value of h_m, with viscosity enhancement assumed, has been compared with the corresponding value, $h_m(0)$, with no viscosity enhancement, i.e., with $k = 1$, $a = 0$. The detailed analysis is given in appendix 2, and the results are shown in fig. 5 in the form of curves of $h_m/2a$ against $h_m(0)/2a$. There are two linear asymptotes, one through the origin with a slope of $k^{\frac{3}{4}}$ for $h_m(0)/2a \ll (1-k^{-1})(k^{\frac{3}{4}}-1)^{-1}$, and the other with unit slope and an intercept of $(1-k^{-1})$ on the positive axis of $h_m/2a$, for $h_m(0)/2a \gg (1-k^{-1})(k^{\frac{3}{4}}-1)^{-1}$. These asymptotes are independent of the assumptions made in the analysis except that the index of k may vary between 2/3 and 3/4. The results at intermediate values are so dependent.

TABLE 2.—RESULTS OF STATISTICAL ANALYSES OF COMPARISON OF EXPERIMENTAL AND THEORETICAL FILM THICKNESSES

lubricant	A	B	B	L
disc temperature/°C	75-104	135	84-135	30-95
rotational speed/rev min^{-1}	11.5-182	20-108	103-108	106-107
number of points	10*	4	10	25
estimate of residual variance/nm^2				
about regression line	2.02	0.17	6.15	1.19
intercept/nm, which could have				
been detected at 5 % significance	3.2	1.4	5.9	1.1
level				
crude regression coefficient	1.111	1.222	1.035	1.219
corrected regression coefficient	0.99	1.06	0.90	1.13
95 % confidence limits of				
regression coefficient	±0.022	±0.035	±0.037	±0.011

* Two points, with film thicknesses slightly greater than 100 nm and not shown in fig. 1, have been included in the statistical analysis.

In fig. 1 and 2, h_e is regarded as an experimental estimate of h_m, and h_t as a theoretical estimate of $h_m(0)$, subject to certain corrections discussed in the results section. The quantity a is unknown, but since the relations between h_e and h_t are all linear over nearly a decade of variation, then the results must lie in one of the two asymptotic regions. Since there is no evidence of an intercept greater than about 1 nm in the linear relation between h_e and h_t, then if there is an effect, the results must lie on the asymptote passing through the origin.

Of the four sets of results examined, only one, that for di(2-ethylhexyl) sebacate (lubricant L), shows evidence of a corrected slope significantly greater than unity, while the mean of the four slopes is 1.02. If the slope of 1.13 for lubricant L is accepted, the corresponding value of k would be approximately 1.18, while the depth a of the zone of enhanced viscosity would be large compared with 100 nm. This does not seem to be a likely situation, particularly as the mineral oils A and B correspond more closely than does the di-ester L to the class of lubricants for which the effect would be expected to be greatest, i.e., a solution of polar materials in a non-polar solvent.

With the possible exception of lubricant L then, there is no evidence of any enhancement of the viscosity near a surface. The cause of the differences in the corrected slopes given in table 2, and in particular the difference between the two sets of results for lubricant B, is not known.   Random errors in the slopes are comparatively small, and the errors are mainly systematic.   One possible source of error is a deviation of the real shape of the gap between the two discs from that assumed in the interpretation of the capacitance measurements in terms of film thicknesses, but it is difficult to give a quantitative estimate.   The difference between the points obtained for lubricant B at constant speed and at constant temperature may reflect an inadequacy in the theory. A similar discrepancy has been noted by Greenwood.[9]   It seems probable that the 13 % discrepancy in the slope for lubricant L is a result of systematic errors in the experiment and in the theory.

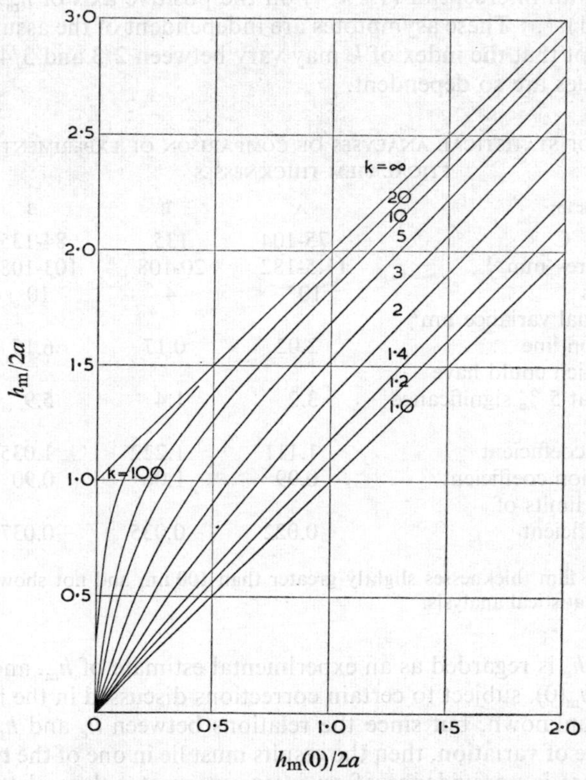

FIG. 5.—Theoretical effect of viscosity enhancement on film thickness.

If there were a zone of liquid of enhanced viscosity adhering to either surface, and if the lubricant in this zone had a dielectric constant higher than that of the bulk liquid, then the effect would escape detection.   But it is unlikely that the compensation would be exact over any wide range of film thicknesses, since the calculation of the capacitance depends on $\int \dfrac{dx}{h-2a[1-(k')^{-1}]}$, where k' is the ratio of the dielectric constant of the layer near the surface to that of the bulk fluid.   This integral is different in form from that in eqn (2,10), appendix 2, which governs the hydrodynamic behaviour.

Another possible source of error is that smooth surfaces are assumed, both in the hydrodynamic theory and in the interpretation of the capacitance measurements in terms of film thicknesses. In practice, the surfaces are rough, with a scale of roughness comparable with the minimum film thickness. But some evidence of nonlinearity in the relations between experimental and theoretical film thicknesses would be expected if this roughness were an important source of error.

In these experiments the lubricant is subject to a shear rate [10] in the inlet zone which varies from zero up to a maximum of $3\bar{u}/2h_{\mathrm{m}}$, i.e., of the order of 10^7 s^{-1}. The shear stress is more difficult to estimate, but the maximum must be much larger than 100 kN m^{-2}. These figures for the shear stress and the shear rate are several orders of magnitude higher than in those experiments reported in the literature, in which positive evidence of a much greater degree of enhancement of viscosity was reported. On the other hand, Cameron and Gohar [11] found evidence of a layer 50-100 nm thick, with a viscosity 3 to 4 times the bulk value. Their experiments were similar to those reported here, but they used a steel ball sliding against a glass plate, and measured the film thickness by an optical method. If such a layer had been present in the work reported here, it would almost certainly have been detected.

The author thanks Mr. A. R. Wilson and Mr. W. J. Cairney for making the experimental measurements of film thickness, and Mrs. A. C. Rowlands and Mr. A. Prothero for the computation of the numerical integrals in appendix 2.

[1] J. C. Henniker, *Rev. Mod. Phys.*, 1949, **21**, 322.
[2] A. T. J. Hayward and J. D. Isdale, *Brit. J. Appl. Phys. (J. Phys. D)*, 1969, **2**, 251.
[3] T. C. Askwith, A. Cameron and R. F. Crouch, *Proc. Roy. Soc. A*, 1966, **291**, 500.
[4] A. Dyson, H. Naylor and A. R. Wilson, *Proc. Inst. Mech. Eng.*, 1965-6, **180 (3B)**, 119.
[5] D. Dowson, G. R. Higginson and A. V. Whitaker, *J. Mech. Eng. Sci.*, 1962, **4**, 121.
[6] A. N. Grubin, *Fundamentals of the Hydrodynamic Theory of Heavily Loaded Cylindrical Surfaces* in *Symposium. Investigation into the Contact of Machine Components* (Central Scientific Institute for Technology and Mechanical Engineering, Moscow, 1969, book no. 30) (D.S.I.R. *Trans.* no. 377).
[7] D. Dowson and G. R. Higginson, *Elastohydrodynamic Lubrication* (Pergamon Press, Oxford, London, Edinburgh, New York, Toronto, Paris, Braunschweig, 1st ed., 1966), chap. 6.
[8] A. W. Crook, *Phil. Trans. A*, 1961, **254**, 223.
[9] J. A. Greenwood, *A Re-examination of Elastohydrodynamic Film Thickness Results* (University of Salford, Department of Mechanical Engineering Report, November 1969).
[10] A. Dyson and A. R. Wilson, *Proc. Inst. Mech. Eng.*, 1965-6, **180**(3K), 97.
[11] A. Cameron and R. Gohar, *Proc. Roy. Soc. A*, 1966, **291**, 520.

APPENDIX 1.—EXPERIMENTAL RESULTS

lubricant	speed/rev min^{-1}	temp./ °C	h_e/nm	h_t/nm	lubricant	speed/rev min^{-1}	temp./ °C	h_e/nm	h_t/nm
A	41	76	82.8	74.9		107	95	28.5	23.6
	30	75	66.6	61.9			91	29.5	24.7
	20	75	51.4	47.0			87	31.4	26.1
	15	74	42.8	39.0			83	33.7	27.6
	11.5	75	34.6	32.4			82	34.2	28.2
	11.5	85	25.2	24.2			81	34.2	28.6
	50	71	112.4	100.6			79	35.0	29.5
	45	73	98.0	86.0			77.5	36.2	30.0
	155	104	106.7	95.2			75	38.1	31.4
	182	104	118.0	106.7	L		73	39.2	32.6
							70	41.9	34.5
B	108	135	37.1	30.6			64	44.9	37.9
	40	135	19.4	15.4		106	58	49.5	42.5
	24	135	13.1	10.9			56	53.3	44.1
	20	135	11.8	9.5			54	57.1	46.1
							48	62.8	51.6
B	103	84	81.5	77.4			44	68.5	57.0
	103	90	62.8	66.3			41	73.9	60.6
	104	100	57.1	54.0			36	82.9	68.1
	106	107	50.0	47.8			30	101.0	81.0
	106	114	45.7	41.0					
	106	118	40.0	39.0					
	106	122	38.1	37.1					
	106	128	35.6	34.2					
	107	129	34.2	33.3					
	108	135	34.5	30.8					

APPENDIX 2

EFFECT ON FILM THICKNESS IN PURE ROLLING OF ENHANCEMENT OF LUBRICANT VISCOSITY NEAR SURFACES

MODIFICATION TO REYNOLDS' EQUATION

Let the local thickness of the whole film of lubricant be h and the bulk viscosity of the lubricant be η. Let the viscosity be increased to $(k\eta)$ within two regions, each of thickness a, and each bounded on one side by a solid surface, as in fig. 6. It is first assumed that $h \geqslant 2a$. Under the usual assumptions of the theory of hydrodynamic lubrication by thin films, the hydrostatic pressure p in the film is a function only of the coordinate x in the direction of motion, while the shear stress τ is a function only of the coordinate y measured in a direction normal to that of the surfaces. These quantities are related by the momentum equation

$$dp/dx = d\tau/dy. \tag{2.1}$$

For a Newtonian lubricant

$$\tau = \eta(du/dy), \tag{2.2}$$

where η is the viscosity and u is the velocity of a particle of fluid in the x direction. Substitution of eqn (2.2) in eqn (2.1) and integration twice with respect to y gives the velocity distribution:

$$u = (1/2\eta)(dp/dx)y^2 + Ay + B,$$

where A and B are constants of integration, and may be expected to differ in the two regions

$$0 \leqslant |y| < (h/2 - a); \quad (h/2 - a) < |y| \leqslant (h/2).$$

By symmetry, $A = 0$ and the velocity distributions in the two regions become

$$0 \leqslant |y| < (h/2-a),\ u = (2\eta)^{-1}(\mathrm{d}p/\mathrm{d}x)y^2 + B_1,$$

$$(h/2-a) < |y| \leqslant (h/2),\ u = (2k\eta)^{-1}(\mathrm{d}p/\mathrm{d}x)y^2 + B_2.$$

At $y = \pm h/2$, the velocity U must equal the peripheral velocity $\bar{u}$ of the surfaces bounding the film, and this gives a value for B_2:

$$B_2 = \bar{u} - (k\eta)^{-1}(\mathrm{d}p/\mathrm{d}x)(h^2/8).$$

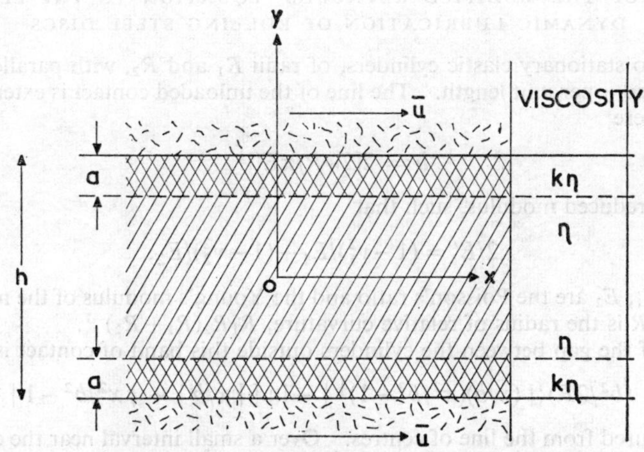

FIG. 6.—Co-ordinate system used in analysis.

The velocity of the fluid at $y = \pm(h/2-a)$ is given by

$$u_1 = (2k\eta)^{-1}(\mathrm{d}p/\mathrm{d}x)(h/2-a)^2 + B_2 = \bar{u} - (2k\eta)^{-1}(\mathrm{d}p/\mathrm{d}x)a(h-a).$$

This must be equal to the velocity at the same point, derived from the velocity distribution in the other region of the film, $0 \leqslant |y| < (h/2-a)$.   Therefore

$$(2\eta)^{-1}(\mathrm{d}p/\mathrm{d}x)(h/2-a)^2 + B_1 = \bar{u} - (2k\eta)^{-1}(\mathrm{d}p/\mathrm{d}x)a(h-a),$$

or

$$B_1 = \bar{u} - (2k\eta)^{-1}(\mathrm{d}p/\mathrm{d}x)a(h-a) - (2\eta)^{-1}(\mathrm{d}p/\mathrm{d}x)(h/2-a)^2.$$

The flow rate in the x direction, per unit transverse width, is $Q = 2\displaystyle\int_0^{h/2} u\,\mathrm{d}y$

$$= 2\int_0^{h/2-a}\left[(2\eta)^{-1}\frac{\mathrm{d}p}{\mathrm{d}x}y^2 + B_1\right]\mathrm{d}y + 2\int_{h/2-a}^{h/2}\left[(2k\eta)^{-1}\frac{\mathrm{d}p}{\mathrm{d}x}y^2 + B_2\right]\mathrm{d}y$$

$$= \bar{u}h - (12\eta)^{-1}(\mathrm{d}p/\mathrm{d}x)\{h^3 - (1-k^{-1})[h^3 - (h-2a)^3]\}. \tag{2.3}$$

If the lubricant is incompressible and if there is no side leakage, this flow rate must remain constant as h varies.   In particular, it must equal the value at that position at which $\mathrm{d}p/\mathrm{d}x = 0$.   If the film thickness at this point is h_m, the flow rate is $\bar{u}h_\mathrm{m}$, and if this is equated to the flow rate at the general position, given by eqn (2.3), an expression for the pressure gradient is obtained:

$$\frac{\mathrm{d}p}{\mathrm{d}x} = \frac{12\eta\bar{u}(h-h_\mathrm{m})}{h^3 - (1-k^{-1})[h^3 - (h-2a)^3]}. \tag{2.4a}$$

This is valid for $h \geqslant 2a$. If $h \leqslant 2a$, then the viscosity is $k\eta$ over the entire thickness of the film, and the conventional version of Reynolds' equation gives

$$\mathrm{d}p/\mathrm{d}x = 12k\eta u(h-h_\mathrm{m})/h^3 \tag{2.4b}$$

For $k = 1$ or $a = 0$, eqn (2.4a) and (2.4b) reduce to the conventional integrated form of Reynolds' equation:

$$\mathrm{d}p/\mathrm{d}x = 12\eta\bar{u}(h-h_\mathrm{m})/h^3.$$

APPLICATION OF THE MODIFIED REYNOLDS' EQUATION TO THE ELASTOHYDRODYNAMIC LUBRICATION OF ROLLING STEEL DISCS

Consider two stationary elastic cylinders, of radii R_1 and R_2, with parallel axes, loaded together by a load w per unit length. The line of the unloaded contact is extended to a band of width $2b$, where

$$b = 2[2wR/\pi E']^{\frac{1}{2}},$$

where E' is the reduced modulus, such that

$$2/E' = (1-v_1^2)/E_1 + (1-v_2^2)/E_2,$$

and v_1, v_2 and E_1, E_2 are the Poisson's ratio and the Young's modulus of the materials of the discs 1 and 2; R is the radius of relative curvature, $R_1 R_2 (R_1+R_2)^{-1}$.

The shape of the gap between the cylinders outside this band of contact is given by [6]

$$h_1 = (b^2/2R)\{|\,(x/b)(x^2/b^2-1)^{\frac{1}{2}}\,| -\ln[\,|\,x/b\,| + |\,x^2/b^2-1\,|^{\frac{1}{2}}]\},$$

where x is measured from the line of centres. Over a small interval near the edge of the flat region of contact, this expression may be expanded into the form [8]

$$h_2 = 2^{\frac{3}{2}}b^2(3R)^{-1}\varepsilon^{\frac{3}{2}} + \text{higher powers of } \varepsilon \tag{2.5}$$

where

$$|\,x\,| = b(1+\varepsilon), \qquad 0 \leqslant \varepsilon \ll 1.$$

In the elastohydrodynamic lubrication of rolling steel discs, it is assumed that the shape of the gap between the discs on the inlet side is given by the Hertzian gap with the addition of a constant separation h_m,

$$h = h_2 + h_\mathrm{m}. \tag{2.6}$$

The viscosity is assumed to vary exponentially with pressure;

$$\eta = \eta_0 \exp(\alpha p), \tag{2.7}$$

and the boundary conditions for the pressure distribution are assumed to be

$$p = 0 \text{ at } x = -\infty,$$

$$p \to \infty, \ \mathrm{d}p/\mathrm{d}x = 0 \text{ at } x = -b. \tag{2.8}$$

The above procedure gives a good approximation to the true film thicknesses, both those obtained by a more refined theory, including the simultaneous solution of the governing elastic and hydrodynamic equations, and also to those determined experimentally.

INTEGRATION OF REYNOLDS' EQUATION

Suppose first that $h_\mathrm{m} \geqslant 2a$. Eqn (2.4a) and (2.7) give

$$\exp(-\alpha p)\frac{\mathrm{d}p}{\mathrm{d}x} = \frac{12\eta_0\bar{u}(h-h_\mathrm{m})}{h^3-(1-k^{-1})[h^3-(h-2a)^3]}. \tag{2.9}$$

Integration of eqn (2.9), with the boundary conditions (2.8) gives

$$\alpha^{-1} = 12\eta_0\bar{u}\int_{-\infty}^{-b}\frac{(h-h_m)\,dx}{h^3-(1-k^{-1})[h^3-(h-2a)^3]}$$

In eqn (2.10), h is given by eqn (2.5) and (2.6). A convenient substitution is

$$b^2(2\varepsilon)^{\frac{1}{2}}(3Rh_m)^{-1} = \tan^2\theta.$$

Eqn (2.6) and (2.5) then give

$$h = h_m\sec^2\theta. \tag{2.11}$$

Eqn (2.10) becomes

$$(12\eta_0\bar{u}\alpha)^{-1} = (2/3)^{1/3}(2Rh_m/b^2)^{2/3}(b/h_m^2)I_1(k,0)$$

where

$$I_1(k,\theta) = \int_\theta^{\pi/2}\frac{(\sin\theta)^{7/3}(\cos\theta)^{5/3}\,d\theta}{1-(1-k^{-1})[1-(1-\beta\cos^2\theta)^3]}, \tag{2.12}$$

and

$$\beta = 2a/h_m. \tag{2.13}$$

Thus, for constant β, h_m is proportional to $[I_1(k,0)]^{\frac{3}{4}}$. If there is no increase in viscosity near the surface, then $k = 1$, and $I_1(k,0)$ in eqn (2.11) becomes $I_2(\pi/2)$, where

$$I_2(\theta) = \int_0^\theta(\sin\theta)^{7/3}(\cos\theta)^{5/3}\,d\theta.$$

By a standard result in the theory of gamma functions,

$$\int_0^{\pi/2}(\cos\theta)^{m-1}(\sin\theta)^{n-1}\,d\theta = \tfrac{1}{2}\frac{\Gamma(m/2)\Gamma(n/2)}{\Gamma[(m+n)/2]} \text{ if } m,n>1.$$

By the use of other standard results:

$$\Gamma(z+1) = z\Gamma(z),$$

$$\Gamma(z)\Gamma(1-z) = \pi/\sin(\pi z),$$

$I_2(\pi/2)$ may be evaluated as $8\pi/9\sqrt{3}$.

The ratio of the film thickness h_m that takes into account the enhancement of the viscosity, to the value $h_m(0)$ in the absence of such enhancement is therefore

$$r = h_m/h_m(0) = [(9\sqrt{3}/8\pi)I_1(k,0)]^{\frac{3}{4}}. \tag{2.14}$$

Eqn (2.14) is valid only if $h_m \geqslant 2a$. If $h_m \leqslant 2a$, eqn (2.4a) is valid only from $h = \infty$ to $h = 2a$, or, by eqn (2.11) and (2.13),

$$\cos\theta = \beta^{-\frac{1}{2}}$$

For films thinner than this, eqn (2.4b) must be used, and $I_1(k,0)$ in eqn (2.14) must be replaced by

$$kI_2(\theta_1)+I_1(k,\theta_1),$$

where $\theta_1 = \cos^{-1}(\beta^{-\frac{1}{2}})$. Eqn (2.14) therefore becomes

$$r = h_m/h_m(0) = \{(9\sqrt{3}/8\pi)[kI_2(\theta_1)+I_1(k,\theta_1)]\}^{\frac{3}{4}} \tag{2.15}$$

ASYMPTOTIC FORMS

The most convenient way to present the results is in the form of a plot of

$$h_{\mathrm{m}}/2a = \beta^{-1} \text{ against } h_{\mathrm{m}}(0)/2a = (r\beta)^{-1}.$$

The asymptotes $\beta \to 0$, $\beta \to \infty$ are of interest. If $\beta \ll 1$, eqn (2.12) may be expanded as a power series in β,

$$I_1(k,0) = \int_0^{\pi/2} (\sin\theta)^{7/3}(\cos\theta)^{5/3}\mathrm{d}\theta + 3(1-k^{-1})\beta \int_0^{\pi/2} (\sin\theta)^{7/3}(\cos\theta)^{11/3}\mathrm{d}\theta + \cdots$$

$$= \tfrac{1}{2}[\Gamma(5/3)\Gamma(4/3)/\Gamma(3)] + 3(1-k^{-1})\beta(\tfrac{1}{2})[\Gamma(5/3)\Gamma(7/3)/\Gamma(4)] + \cdots$$

$$= 8\pi/(9\sqrt{3})[1 + (4/3)(1-k^{-1})\beta + \cdots].$$

Eqn (2.14) then gives

$$r = 1 + (1-k^{-1})\beta + \cdots.$$

Neglect of terms of the second and higher orders in β, and division by $r\beta^2$ then gives a quadratic equation for β^{-1} in terms of $(r\beta)^{-1}$:

$$(\beta^{-1})^2 - \beta^{-1}(r\beta)^{-1} - (1-k^{-1})(r\beta)^{-1} = 0.$$

The solution is

$$\beta^{-1} = \tfrac{1}{2}\{(r\beta)^{-1} \pm [(r\beta)^{-2} + 4(1-k^{-1})(r\beta)^{-1}]\}^{\frac{1}{2}}.$$

If k tends to unity, β^{-1} must tend to $(r\beta)^{-1}$, so the positive sign of the square root must be taken. If $(r\beta)^{-1} \gg 1$, the solution becomes

$$\beta^{-1} = (r\beta)^{-1} + (1-k^{-1})$$

or

$$h_{\mathrm{m}}/2a = h_{\mathrm{m}}(0)/2a + (1-k^{-1}).$$

If $\beta \gg 1$, then θ_1 in eqn (2.15) tends to $\pi/2$, $I_1(k,\theta_1)$ tends to zero, and $I_2(\theta_1)$ tends to $8\pi/9\sqrt{3}$.

Thus, $$r = h_{\mathrm{m}}/h_{\mathrm{m}}(0) \to k^{\frac{3}{4}}$$

or $$h_{\mathrm{m}}/2a \to k^{\frac{3}{4}}[h_{\mathrm{m}}(0)/2a].$$

These two asymptotes may be derived from first principles and are independent of any assumptions about the nature of the lubrication or about the shape of the gap. Results at intermediate values of $h_{\mathrm{m}}(0)/2a$ depend on such assumptions. The two asymptotes intersect at

$$h_{\mathrm{m}}(0)/2a = (1-k^{-1})(k^{\frac{3}{4}}-1)^{-1};$$

$$h_{\mathrm{m}}/2a = k^{\frac{3}{4}}(1-k^{-1})(k^{\frac{3}{4}}-1)^{-1}.$$

Results for intermediate values of β were obtained by numerical integration of eqn (2.14) or (2.15).

Mechanical Properties of Very Thin Surface Films

By A. D. Roberts and D. Tabor

Surface Physics, Cavendish Laboratory, Cambridge, England.

Received 2nd April, 1970

This paper comprises two parts : the first briefly describes some earlier work on the determination of the shear strength of a calcium stearate bimolecular layer deposited between two mica surfaces. The second part deals with some more recent studies on the mechanical properties of thin liquid films sandwiched between a rubber and glass surface. The rubber surface was soft and optically smooth so that certain liquids squeezed between it and polished glass formed a very thin film of uniform thickness. Squeeze film studies made with this system revealed the importance of repulsive forces exerted by electrically-charged double layers residing on the solid surfaces. These forces were able to support the externally applied pressure so that an equilibrium film some 200 Å thick was obtained.

When rubber and glass surfaces were sheared in the presence of an electrolyte solution, double-layer repulsion helped to support the normal load and so protect the surfaces from serious abrasion. The presence of a soap in the solution led to further protection by an oriented monolayer on the rubber surface. Measurements suggested that these two factors operated to provide effective lubrication—the electrical repulsive forces to keep surfaces apart and the SDS monolayer for increased protection at points of intimate contact where the separating fluid had been locally penetrated.

The mechanical properties of very thin surface films sandwiched between bearing surfaces are of great interest in lubrication practice. In order to study these properties, methods are required by which the true area of contact between the surfaces and the thickness of the films between them may be determined precisely. In addition, the effect of surface asperities projecting through films must be eliminated. It is this last requirement that presents a great problem because, however carefully surfaces are prepared, asperities remain which are vast when considered on a molecular scale. In this paper, two ways are described which overcome the problem.

One of these is to sandwich thin films between molecularly smooth surfaces, the other between highly deformable surfaces. Both allow films of unbroken uniform thickness to be maintained between them over relatively large areas. The surfaces are mica and rubber. Some time ago it was found that mica could be cleaved to yield areas up to 80 cm² which were step free on both sides of the sheet and molecularly smooth. More recently, optically smooth rubber surfaces have been prepared and when these are pressed against glass they conform well to the micro-irregularities upon it. Certain liquids sandwiched between the two form, in effect, very thin films of uniform thickness.

Optical interference techniques have been applied to both types of surface to determine precisely the film contact area and thickness. An outline is given first of work carried out 15 years ago on the shearing of a calcium stearate bimolecular layer between mica surfaces followed by a description of recent work on the squeezing of very thin liquid films between rubber and glass.

SHEAR STRENGTH OF A CALCIUM STEARATE BIMOLECULAR LAYER

The surfaces produced by the cleavage of mica are so smooth that the true area of contact between them is identical with the geometric area. Clean mica adheres readily to itself and to many other surfaces, its sliding friction is large and surface damage severe. These phenomena reflect the strong action of short-range surface forces. An investigation into the nature of the contact between molecularly smooth mica surfaces was carried out by Bailey and Courtney-Pratt [1] and the determination of the shear strength of a calcium stearate bimolecular layer formed an integral part of their investigation.

Two cleaved smooth surfaces of mica were bent into cylindrical sheets supported with their axes at right angles to one another (fig. 1a). The rear surface of each sheet was silvered. The cylinders could then be brought together and the contact region between them studied by multiple-beam interferometry. By viewing the contact in monochromatic light the fringe pattern obtained was a system of concentric rings surrounding an area of uniform tint. The boundary of the contact area was determined as the position within the first ring where the first visible change in light intensity occurred (fig. 1b). As a further aid to contact assessment, the specimens were illuminated with white light and contact area fringes of equal chromatic order (FECO) examined through a spectroscope (fig. 1c). With this system separations between mica surfaces along any selected line across the area of contact may be determined to an accuracy of ± 4 Å. In addition the length of the straight portion of the fringes may be measured and, for a particular cross-section of the contact, the boundary of the region of intimate contact thus known with certainty.

A monolayer of calcium stearate was applied to each mica surface by the Langmuir-Blodgett technique. After deposition the mica cylinders were brought together under a normal load and the contact region found to consist of a circular sandwich containing a bimolecular layer of stearate. Interference measurements indicated a total bi-film thickness of about 45 Å. A tangential force was then applied by a spring device to one of the mica shells. At any fixed value of normal load when the tangential force was increased a maximum value was reached at which the surfaces began to slide smoothly one over the other. At the point of slip if the area of contact corresponding to this value be measured accurately from interference data, then the ratio of the tangential force to contact area gives a value of shear strength.

Measurements made yielded an average value for the shear strength of the bi-film of 250 ± 10 g/mm², compared with a value of 10,000 g/mm² obtained for *clean* mica surfaces. A few experiments with three monolayers present on each mica surface indicated even lower values than 250 g/mm². In these experiments the mean pressure over the bi-film during shearing was only about 10 atm which is some 500 times less than pressures encountered at the points of asperity contact between sliding metal surfaces. There is some evidence which suggests that the shear strength of monolayers is roughly proportional to pressure but direct experimental confirmation is lacking.

We now turn to the squeezing of very thin liquid films between rubber and glass. In most of these experiments the mean pressures developed over films during squeezing were only of the order of 1 atm.

THIN LIQUID FILMS SQUEEZED BETWEEN RUBBER AND GLASS

Recently we have been able to prepare optically smooth rubber surfaces by hot curing rubber compounds against optically smooth formers of metal and glass.[2] When spherical rubber surfaces are prepared they show excellent Newton's rings fringe-patterns against a glass plate. Scanning electron microscopy reveals that the ultimate smoothness of the rubber depends very much upon the surface finish of the metal or glass former. Under optimum conditions the rubber will form an exact replica of them.

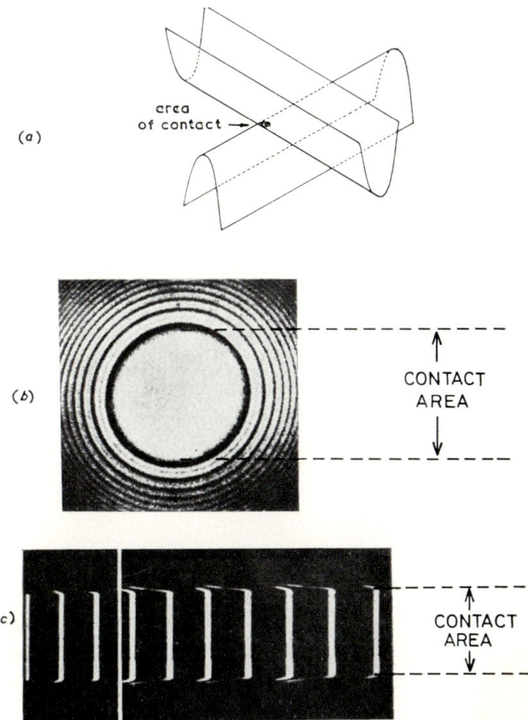

Fig. 1.—(*a*) Crossed cylinders of mica in contact (top). (*b*) Interferogram of contact area between mica cylinders. The central light region is of uniform density and over all this region the specimens are in molecular contact. The area of contact extends as far as the first visible change in intensity; magnification 40× (middle). (*c*) The appearance of fringes of equal chromatic order along a selected line across the area of contact. The contact shown is between clean sheets of mica; magnification 17×(bottom).

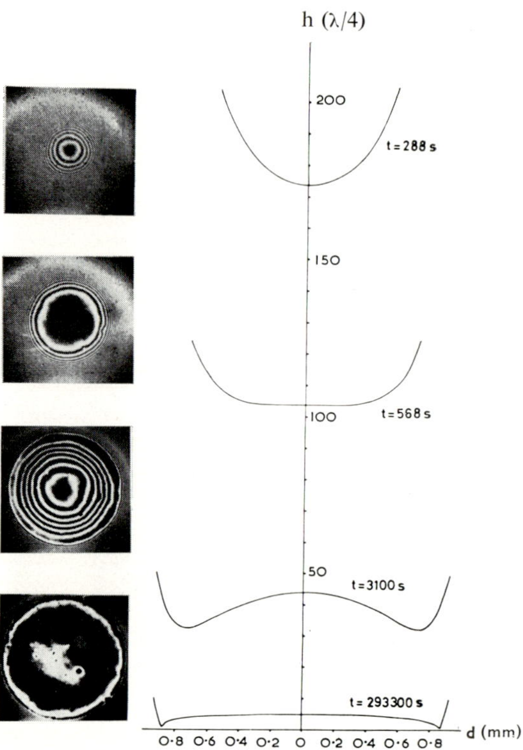

FIG. 2.—Interferograms produced between a spherical rubber surface and glass plate during the normal approach under a constant load (5 g) of one surface toward the other through a viscous oil. The oil was a dimethyl silicone (type MS 200, Midland Silicones Ltd.) of viscosity 10^6 cSt. The elastic modulus of the rubber surface was 6×10^6 dyn/cm^{-2} and it had a radius of curvature of 2 cm. The pictures together with deduced profiles show the formation and collapse with time t of a " bell " entrapment of liquid. $\lambda = 5461$ Å.

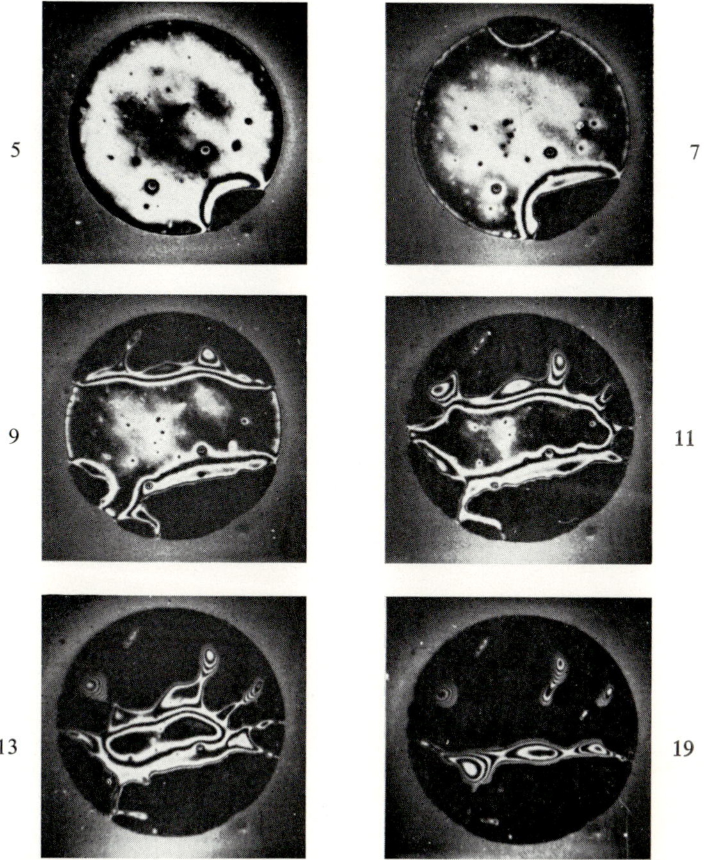

FIG. 3.—Interferograms of the final stages of collapse of a liquid " bell " entrapment. The rubber surface appears to be attracted towards the glass at its nearest point of approach (initially at about " 5 o'clock " on the contact periphery) and adjacent parts wrinkle under local stress. With time (indicated in days by each frame) the seal spreads over the whole contact zone so trapping small "islands" of liquid. The contact load, oil specification and rubber surface constants are the same as those in fig. 2. Magnification $14\times$.

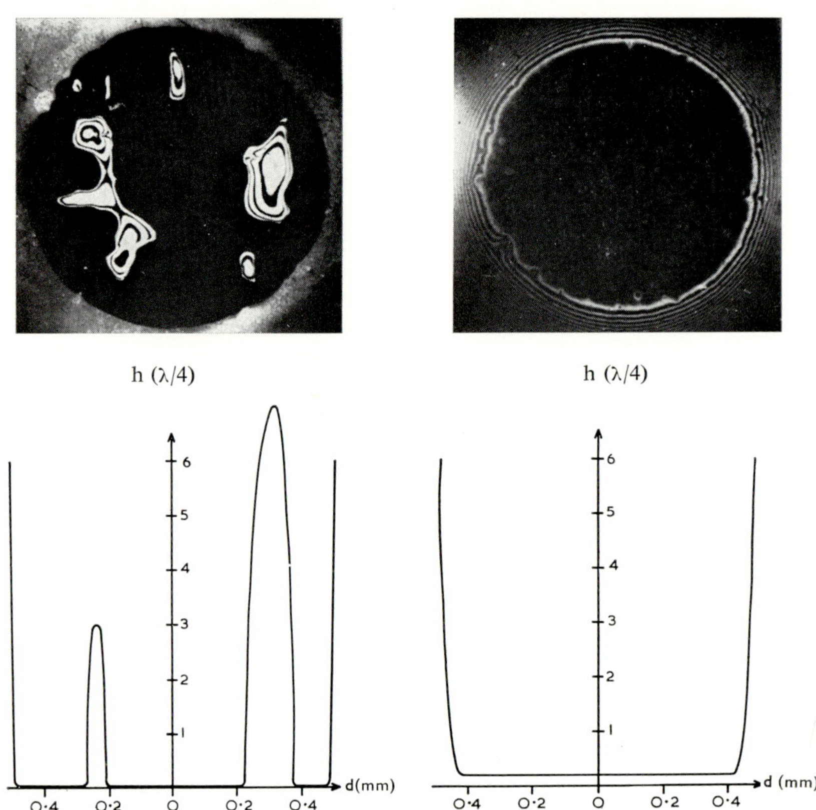

FIG. 4.—Interferograms and deduced profiles showing the marked difference in contact between distilled water (left) and water containing 0.3 % SDS (right).   With distilled water trapped pockets of water are seen, but for the SDS solution a thin equilibrium film of about 200 Å uniform thickness is obtained. $\lambda = 5\,461$ Å.

Although these surfaces are only as smooth as their optically polished formers, which are covered with asperities hundreds of Ångstroms high, they can be brought into molecular contact with glass or any other surface provided it is fairly smooth. This is for two reasons. First, the rubber is very soft so that asperities upon it may be easily deformed. Secondly, the size of asperities is relatively small so their displacement has only to be slight to conform with opposing surface asperities. Strong adhesion to itself and other materials, a high dry sliding friction and severe surface damage accompanying this, all of which recalls the behaviour of mica, reflect the extensive action of surface forces between optical rubber and other surfaces placed in contact.

The optical interference between a spherical rubber surface and glass plate has been examined during the normal approach under a constant load of one surface toward the other through a viscous oil. This approach traps a " bell " of liquid in the centre of the contact zone. Interferograms of the entrapment using a very viscous dimethyl silicone oil are shown in fig. 2. They were produced without silvering the rubber or glass surfaces, good fringe contrast following because both surfaces had the same refractive index (~ 1.51) and good fringe visibility because scattered background light had been eliminated.

With time most of the liquid becomes extruded from the contact area and the " bell " begins to collapse. The final stages of collapse are shown in fig. 3. It appears that the rubber surface is attracted towards the glass at its nearest point of approach, which is the peripheral " lip ", and seals. So strong is this sealing that adjacent parts of the rubber surface wrinkle under local stress. Eventually the seal spreads over the whole contact zone, cutting off and trapping small " islands " of liquid. These islands would seem to be permanently trapped for they do not shrink in size after many days. However, if the experiment be repeated with less viscous silicones, similarly trapped islands do disappear. This suggests that the rubber and glass are not in intimate contact but separated by a monolayer or two of silicone.

With distilled water the bell formation and collapse is rapid and difficult to observe. Once again the rubber seals to the glass but islands of trapped liquid escape after a few hours. Presumably a few monolayers of water remain between the rubber and glass. But if the experiment be repeated with water containing only 0.3 % of sodium dodecyl sulphate (SDS) a drastically different result ensues. The surfaces no longer appear to seal and trap islands of liquid. The intense black hue of sealing is not seen, but instead a grey-black tint extending over the entire contact area (fig. 4). The grey-black tint indicates the presence of a very thin liquid film. This film is far too thin to show interference fringes, but its thickness can be measured photometrically. If dry intimate contact gives " perfectly " black interference of zero light intensity—there is a phase change of 180° at the air-rubber interface when the contact is viewed through the glass*—and a gap of optical thickness $\lambda/4$ gives bright interference of intensity I_0, any gap of intermediate optical thickness nt will give an intensity $I = I_0 \sin^2 (2\pi nt/\lambda)$ at normal incidence, where n is the refractive index of the material in the gap and t its geometric thickness. A mean value of $n = 1.33$ is assumed for thin aqueous films and no corrections made for slight phase changes that might be incurred at intermediate structural interfaces in the liquid gap. Photometric measurements indicate film

---

* It is not easy to obtain precise determinations of the phase change. Some ellipsometric measurements carried out by Dr. R. J. King at N. P. L. suggests that the phase change at the rubber surface may be less than 180 by 1 or 2°. This would lead to an error of 12-24 Å in the thickness. However, our own measurements show that, for dry contact, the light reflected from the interface has within the sensitivity of our apparatus zero intensity. This implies that the effective phase change between the rubber and the glass is, in fact, 180°.

thicknesses of about 200 Å for these SDS films which do not collapse with time so long as the liquid does not evaporate.

Aqueous solutions of SDS are surface-active so it seems reasonable to assume that the SDS molecules become adsorbed on to the rubber surface as an orientated close-packed monolayer with their negative polar end-groups in the water. It is possible that some SDS molecules may become similarly oriented on the glass surface but in any case the glass itself exposes a negative charge of OH^- ions adsorbed on the surface from the aqueous solution. The negatively-charged groups may be expected to attract positive ions in the water so establishing a double-layer of charge at either surface. The concentration of positive ions in solution would be greatest at a surface and decrease with distance away from it. The double-layers of charge at either surface would mutually repel and tend to keep the surfaces apart. Thus, if the charge density at either surface remains constant then an increase in pressure on the equilibrium film should thin it. Also, if the number of ions in solution be increased the equilibrium film should thin again because the distribution of positive ion charges in solution becomes more constricted toward the surfaces owing to a type of screening effect and results in the electrical repulsion forces operating over a smaller distance. These two parameters, pressure and ionic strength, were chosen as variables for studying the properties of the electrical double-layers residing on the rubber and glass surfaces.

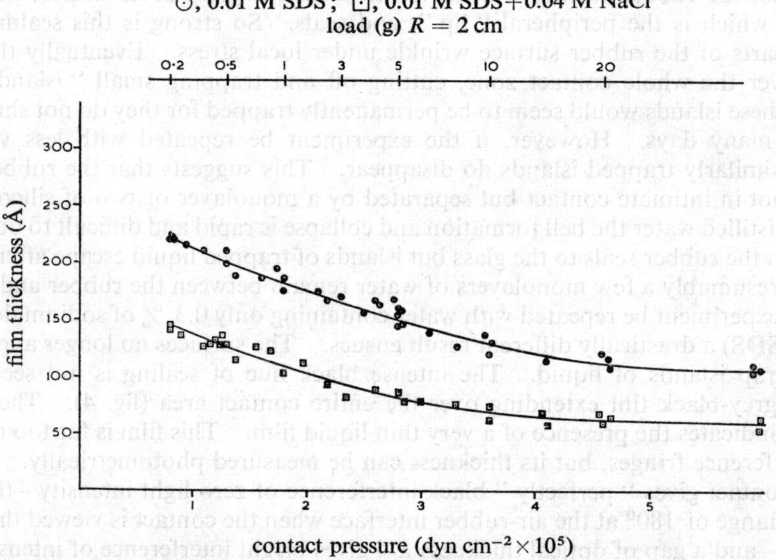

Fig. 5.—Variation in the equilibrium thickness of SDS films sandwiched between a rubber sphere and glass plate with contact pressure and with solution ionic strength. Photometric measurements were made at the centre of the contact zone and the pressures quoted are the Hertzian pressures at this point: they are 1.5 times the mean contact pressure as found from the load and area of contact.

The results of measurements made upon thin sandwiched films of SDS are shown in fig. 5. Equilibrium films attained decreased in thickness both for an increase in pressure and ionic strength. Mysels and Jones [3] have measured the thickness of " first black " soap films under compressive stresses up to about 10^6 dyn cm^{-2} by applying air pressure to the film. First black soap films are of variable thickness determined by a balance of van der Waals attraction and contact pressure against double-layer repulsion. Their results compare favourably with ours (fig. 6) which suggests that the charge distribution is similar in both cases. In making this comparison it is necessary to consider the magnitude of the van der Waals attractive forces between the rubber and glass. Calculations show that these forces are small compared with the applied pressure so that the true mean pressure on the film between rubber and glass is the applied load divided by the contact area.

With our apparatus it is possible to set the glass surface in motion in a tangential direction and measure the frictional force experienced by the rubber hemisphere. When such experiments were carried out at different speeds of continuous sliding under a constant load, it was found that at all speeds the frictional force was minute. The results (table 1) show this. At all speeds the effective liquid viscosity calculated from these data is less than 10 cP, whether the film be 200 or 2,000 Å thick.

—, Results for suspended soap films ; ⊙, results for films between rubber and glass ; - - - , theoretical results.

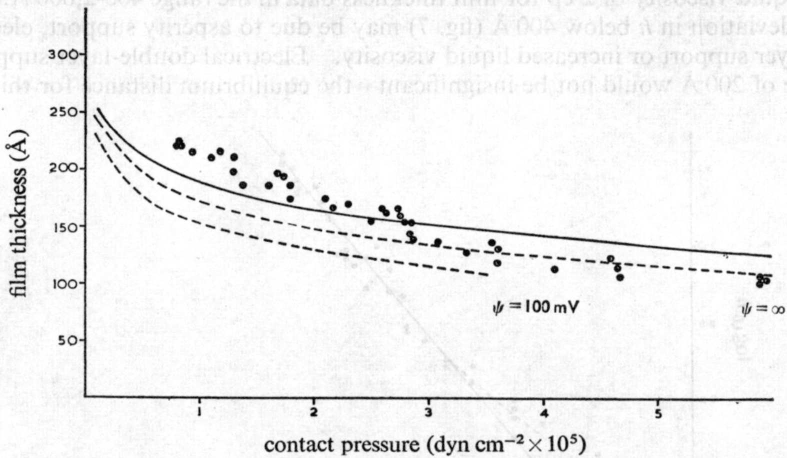

FIG. 6.—Comparison of SDS results for suspended soap films and for films sandwiched between rubber and glass. ψ = surface potential.

TABLE 1.—CONTINUOUS SHEAR OF THIN LIQUID FILMS

shear speed, v (cm/s)	film thickness, h (Å)	frictional force, F (g)	effective liquid viscosity, η (cP)
0.010	270	0.05	8
0.027	370	0.03	3
0.044	460	0.03	2
0.078	570	0.03	1
0.096	630	0.04	1
0.156	910	0.04	1
0.242	1,100	0.04	1
0.311	1,400	0.04	1
0.426	1,720	0.04	1

Load 50 g, rubber sphere constants : radius of curvature 2 cm, elastic modulus 6×10^6 dyn cm⁻². Lubricant 0.3 % aqueous SDS.

It is always difficult to decide what fraction of the frictional force is due to viscous shear alone and what is due to surface asperity interactions. In our particular case some insight into the matter is gained by examining the change in film thickness h

with velocity v of sliding.   It was found (see fig. 7) that h was proportional to $v^{0.6}$ for film thicknesses between 400 and 2,000 Å.   Herreburgh [4] has derived an equation $h = 1.2\, \eta^{0.6}\, v^{0.6}\, R^{0.6}\, w^{-0.2}\, E^{-0.4}$ for predicting the minimum film thickness in an isoviscous elastohydrodynamic *line* contact for rubber-like materials, where η is the lubricant viscosity under normal pressures, R the radius of curvature of the rubber-bearing surface of elastic modulus E, and w the contact load.   His result is in good agreement with the theoretical and experimental line contact results of Roberts and Swales [5] for films about 20,000 Å thick.   Therefore, it is most interesting to see that our experimental results for a *point* contact in the film thickness range 400-2,000 Å also show the same variation in h with v as Herreburgh predicts.   Thus, it seems reasonable to calculate the liquid viscosity of our very thin films using his equation. It is only an approximate calculation and assumes a 40 % *drop* in h for a " point " contact as compared with a " line " contact due to increased side leakage of lubricant. According to Dowson and Higginson [6] this is a fair assumption.   The calculation yields a liquid viscosity of 2 cp for film thickness data in the range 400-2,000 Å.   The " high " deviation in h below 400 Å (fig. 7) may be due to asperity support, electrical double-layer support or increased liquid viscosity.   Electrical double-layer support at a distance of 200 Å would not be insignificant—the equilibrium distance for this load

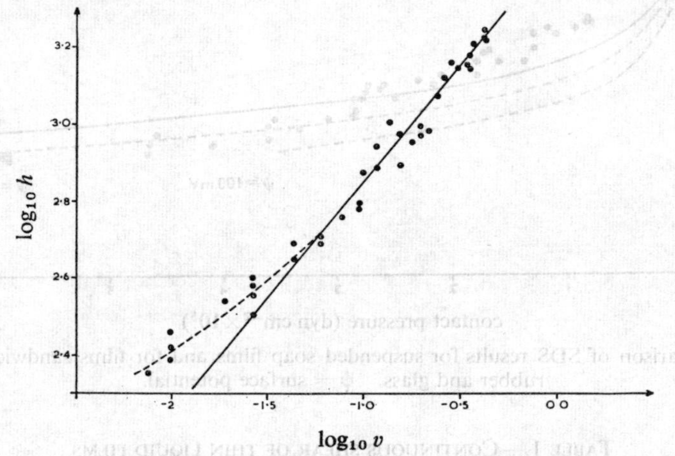

FIG. 7.—Variation in dynamic minimum film thickness h with sliding speed v for thin (200-2 000 Å) aqueous films of SDS sandwiched between a rubber sphere and glass plate.   The continuous line has a slope of 0.60.

is 110 Å—so the " high " values in h may be mainly due to this factor.   Thus, it may be that the viscosity of water in films only 200-300 Å thick sandwiched between solid surfaces is no greater than the bulk viscosity.

The preceding sliding experiments were carried out at arbitrary values of the film thickness determined by the sliding velocity.   A further series of experiments were carried out in which the surfaces were first allowed to reach their static equilibrium separation, determined by the balance between normal load and repulsive double-layer forces.   The surfaces were then set in relative motion.   The frictional force increased rapidly as the glass surface was set in uniform motion, reached a maximum and then fell as a relatively thick film formed between the surfaces.   The peak value corresponds to the force required to shear the equilibrium film.   These results are shown in table 2.   For films thicker than 80 Å the friction is low and increases steadily as the film thickness diminishes.   This is probably due to viscous shear of the liquid film alone. For thinner films the friction is appreciably higher. Presumably, shear of the SDS monolayer is involved or there may even be some rubber-glass contact. As regards the liquid viscosity of these extremely thin films, the instantaneous frictional force is complicated by viscoelastic relaxations of the rubber surface, so no estimate may be made of viscosity. Only continuous shear experiments, in which dynamic equilibrium has been established, may be used for viscosity determination. However, the instantaneous results are useful for *relative* assessment.

TABLE 2.—INSTANTANEOUS SHEAR OF EQUILIBRIUM FILMS

molar conc. of SDS	molar conc. of NaCl	equilibrium film thickness Å	peak value of coefficient of friction
0.01	none	150	0.15
0.01	0.01	120	0.18
0.01	0.02	95	0.27
0.01	0.04	80	0.35
0.01	0.08	70	0.60
0.01	0.16	60	0.90
0.01	0.30	55	1.16
none	0.1	40	4
none	0.01	40	5
none	0.001	90	4
distilled water		40	5
dry contact		" zero "	9

Load 5 g, contact pressure 3×10^5 dyn/cm^{-2}

With distilled water alone, the instantaneous friction is very high and only a little lower than for dry surfaces. A few experiments have been carried out with small amounts of strong electrolytes added to water with *no* SDS present. Unstable equilibrium films have been obtained with these solutions but they do not provide very effective lubrication. Films of strong electrolytes are very sensitive to contaminants and easily collapse leaving only a few monolayers of liquid between surfaces. Our thickness measurements at this stage are not very accurate but these films appear to be about 40 Å thick. Collapsed films of distilled water also appear to be about the same thickness. Films of this thickness are termed " second black " and tend to be of constant thickness ; the nature of the forces determining their structure is uncertain.

We conclude that in the presence of sodium dodecyl sulphate solution double-layer electrical repulsive forces occur between rubber and glass surfaces which are able to support pressures of the order of 10^4-10^6 dyn cm^{-2} (i.e., 10^{-2}-1 atm). An oriented monolayer is formed on the rubber surface and perhaps on the glass too. Friction measurements in the presence of the SDS solution suggest that under pressures of about 1 atm two factors operate to provide effective lubrication—the electrical repulsive forces which tend to keep the surfaces apart and the SDS monolayer itself which protects the surfaces from intimate contact if the separating liquid is locally penetrated. The viscosity of the liquid (0.3 % SDS in water) sandwiched between the monolayers of SDS appears to be constant approximately (the bulk viscosity) whether it be 200 or 2,000 Å thick.

USE OF MICA AND RUBBER FOR THIN FILM STUDIES

In the two investigations that have been described two different types of surface were used to trap the thin film under examination. In many ways, mica is the ideal surface. It is molecularly smooth and simple to prepare. Films sandwiched between two of these surfaces may be resolved to ± 3 Å, though the optics for doing this is cumbersome. Rubber cannot be made molecularly smooth, the preparation of an optically smooth surface is technically difficult and films squeezed between it may only be resolved to about 40 Å. Yet it has been found useful because when placed in contact with another material it provides a parallel contouring surface over relatively large areas. In practical manipulation, optical rubber has advantages over mica. It is not fragile, does not have to be silvered and is easy to mount into apparatus. Rubber has one other advantage. It is a pure van der Waals force solid whereas mica is ionic. Ion-ion interactions could in certain circumstances confuse the interpretation of results.

We thank the Avon Rubber Company Ltd. for preparing the optically smooth rubber surfaces.

[1] A. I. Bailey and J. S. Courtney-Pratt, *Proc. Roy. Soc. A*, 1955, **277**, 500.
[2] A. D. Roberts, *Eng. Materials Design*, 1968, **11** (4), 579.
[3] K. J. Mysels and M. N. Jones, *Disc. Faraday Soc.*, 1966, **42**, 43.
[4] K. Herrebrugh, *A.S.M.E. J. Lubric. Tech.*, 1968, **90**, 262.
[5] A. D. Roberts and P. D. Swales, *Brit. J. Appl. Phys.*, 1969, **2**, 1317.
[6] D. Dowson and G. R. Higginson, *Elastohydrodynamic Lubrication* (Oxford, Pergamon Press, 1966).

Smectic Model for Liquid Films on Solid Surfaces

Part 1.—Application to Monolayer Boundary Lubrication.

By E. Drauglis, A. A. Lucas* and C. M. Allen

Battelle Memorial Institute, Columbus, Ohio, U.S.A.

Received 30th April, 1970

A qualitative theory of films of hydrocarbon solutions of fatty acids described in a previous publication correlates many experimental data. According to this theory the fatty acid molecules and solvent molecules are arranged in layered structures similar to those in smectic mesophases. A simplified method for calculating the attractive energy between two such layers is described. The fatty acid molecules are assumed to behave as hard rods of constant electric polarizability and diamagnetic susceptibility. A potential describing the interaction between two rods is derived and used in a lattice sum calculation to compute the total energy of interaction of two layers as a function of average separation of molecules within a layer, relative position of the two layers and the distance between layers. Calculation of frictional force and coefficient of friction by means of this model show good agreement with a similar calculation of Cameron if the proper assumption about the relative magnitude of energy as a function of relative position is made.

---

1. INTRODUCTION

The question of whether liquid films possess properties significantly different from bulk properties when they are very close to solid surfaces has been of great interest. Griffiths [1] and Bulkley [2] are among the earliest investigators who studied the problem experimentally. Griffiths seems to have observed some surface film rigidity while Bulkley did not. Bastow and Bowden [3] studied the problem by measuring the separation of glass plates immersed in the fluid of interest and obtained the the same separation for fluids as for air. Later, these workers measured the flow of various liquids between two clean parallel glass plates.[4] No evidence was obtained that the fluid flow properties of ordinary liquids were different in thin films than in the bulk. Not unexpectedly, solutions of ammonium oleate, which can exist in a liquid crystalline phase, did show anomalies.

Positive evidence for the existence of anomalous films was obtained by Needs.[5] In his experiments he squeezed various fluids between two steel flats and observed stable residual films some 10,000 Å thick. Later, Fuks [6, 7] performed similar but more carefully controlled experiments on well-characterized hydrocarbon solutions of fatty acids as well as some oils of practical interest. For the few systems where his results could be compared with those of Needs he obtained lower values for the film thickness than Needs although his values were almost always greater than 1,000Å. Besides investigating the thickness of the films as a function of loading, Fuks studied various other properties of the films as a function of fatty acid concentration, surface

* now Chargé de Recherches au Fonds National Belge de la Recherché Scientifique, at Dept. of Physics, University of Liege, Liege, Belgium.

energy of substrates on which the films were formed and temperature. Shear stress measurements were made on various systems for various normal stresses. Many of the measurements were made on various combinations of solvent and fatty acid and correlations of the data with hydrocarbon chain length were observed. These measurements constitute much experimental data in support of the existence of multimolecular films many 1,000 Å thick.

However, there has been much criticism of this interpretation. One group of critics claims that the measurements themselves do not substantiate the presence of multimolecular films and the data can be explained as experimental artifacts caused by the presence of dirt and impurities in the fluids and asperities on the substrates. The work of Hayward and Isdale [8] provides an example of this viewpoint. A refutation of this argument is the general internal consistency of the data. A second body of criticism admits the validity of Fuks' data but claims that other phenomena such as the deposition of particles of solid fatty acid or the formation of micelles due to the presence of water could account for Fuks' results. The idea is that no continuous film having unique properties exists. Refutation of this type requires much experimental investigation of the structure and properties of the films formed in Fuks' apparatus and a reasonable theoretical explanation of not only the phenomenological data of Fuks but also the results of any new experiments on the structure and ordering properties of the films.

A reasonable approach to the development of a theory of the films is first to identify which of the various factors are the most dominant or to identify one or two physical features which must be present no matter what the complicating factors are. Then, one can construct a comparatively simple model in which only these features are considered. Such a model has been described in a recent review on boundary lubrication.[9] In this model the structure of the films is assumed to be similar to the structures found in smectic mesophases, i.e., the molecules are arranged in layers with the long axis of each molecule perpendicular to the plane of the layers. Two assumptions are necessary to account for the formation of such structures. The first is that fatty acid molecules (and straight-chain hydrocarbon solvent molecules) within a few 1,000 Å of a solid surface behave as rod-like molecules and tend to align themselves parallel to each other by means of their mutual van der Waals' forces to form ordered domains or clusters. The second assumption is that the electric dipole field caused by the formation of Fe—O bonds arising from the chemisorption of a monolayer of fatty acid and the van der Waals wall forces are sufficiently strong to align these domains so that they are all perpendicular to the substrate. If these two assumptions are valid then one could expect a coherent, continuous film having a structure similar to a smectic liquid crystal to form. As shown in ref. (9) this model explains qualitatively most of the data of Fuks.

2. ATTRACTIVE INTERACTION ENERGY BETWEEN TWO ROD-LIKE MOLECULES

Because of the complicated nature of the problem, it is not feasible to create from first principles a general theory capable of correlating all of the data of Fuks. Instead, we shall attempt to develop only a theory of the shear properties of the films based on our qualitative model. The first step is the development of methods for the calculation of the energy of interaction between two ideal smectic layers and the interaction energy of a molecule of a layer with respect to all other molecules in the layer. In principle, such calculations are straightforward because only van der Waals' forces need be considered. To obtain the interaction energy between two layers one merely sums pairwise the interaction energy between the given molecule and each molecule in the next layer and then multiplies by the number of molecules per unit area to obtain the energy per unit area.   To obtain the energy of a molecule within a layer one sums the interaction of the molecule with every other molecule within the layer.   In our calculation we shall simplify this process by deriving an analytical expression which approximates the attractive energy between two molecules.

The basis for our calculation is the London-Kirkwood expression for the attractive part of the potential due to an atom or group of atoms [10]:

$$V_{ij} = -3mc^2\alpha\chi/r_{ij}^6 \tag{2.1}$$

where V_{ij} is the attractive energy between CH_2 group i of one molecule and CH_2 group j of another molecule, r_{ij} is the distance between two such groups, m is the electronic mass $(9.109 \times 10^{-28}$ g), α is the electric polarizability of a CH_2 group $(1.777 \times 10^{-24}$ cm^3) and χ is the diamagnetic susceptibility of a CH_2 group $(19.94 \times 10^{-30}$ cm^3).

The total attractive interaction energy of a molecule with respect to all the other molecules in an adjacent layer can be obtained by summing V_{ij} for all the CH_2 groups within the molecule with all the CH_2 groups in the molecules in the adjacent layer.   (In the first approximation the end groups may be treated as CH_2 groups.) To simplify the procedure we assume that each molecule can be replaced by a rod, the electric polarizability and diamagnetic susceptibility of which is constant along the length of the rod.   An expression for the attractive interaction energy of two parallel rods can be obtained by integration over the rods.   The V_{ij} of eqn (2.1) are replaced by

$$dV = -1.5\, mc^2\alpha\chi(NN'/LL')\,dzdz'/r^6 = -A\,(NN'/LL')\,dzdz'/r^6, \tag{2.2}$$

where L and L' are the respective lengths of the molecules, N and N' are the number of CH_2 groups per molecule, dz and dz' are elements of length and r is the distance between dz and dz'.   Two distinct expressions are to be derived—the first of which gives the attractive energy for two rods in different layers as a function of interlayer separation and distance between rods within a layer and the second of which gives the attractive energy of two rods within the same layer as a function of distance between the rods.

The interaction energy for two rods in different layers may be obtained as follows. Let a rod of length L be situated on the z axis with its lower end a distance ε above the x-y plane and let a rod of length L' be situated parallel to the z axis below the x-y plane at a distance $\rho = (x^2+y^2)^{\frac{1}{2}}$ from the origin and have its upper end a distance below the x-y plane.   The interaction energy is then found by evaluating the integral obtained from eqn. (2.2):

$$V(\rho,\varepsilon) = -\left(\frac{NN'}{LL'}\right)A\iint_{\text{(lengths of both rods)}} r^{-6}dzdz'$$

$$= -\left(\frac{NN'}{LL'}\right)A\int_{\varepsilon}^{\varepsilon+L}dz\int_{-L-\varepsilon}^{-\varepsilon}[x^2+y^2+(z-z')^2]^{-3}dz'.$$

Assuming ρ not equal to zero and carrying out the integrations gives

$$V(\rho,\varepsilon) = -(A/8)(NN'/LL')\rho^{-4}[F(U_1)-F(U_2)-F(U_3)+F(U_4)], \tag{2.3}$$

where the function $F(U_i)$ is given by

$$F(U_i) = -1/(1+U_i^2)+3U_i \arctan U \tag{2.4}$$

and

$$U_1 = (2\varepsilon+L')/\rho; \quad U_2 = 2\varepsilon/\rho; \quad U_3 = (2\varepsilon+L+L')/\rho; \quad U_4 = (2\varepsilon+L)/\rho.$$

For the case $\rho = 0$, one obtains

$$V(0,\varepsilon) = -(NN'A/20LL')[(2\varepsilon+L')^{-4}-(2\varepsilon)^{-4}-(2\varepsilon+L+L')^{-4}+(2\varepsilon+L)^{-4}]. \tag{2.5}$$

The interaction energy for two rods within the same layer may be found in a similar manner.   Evaluation of the integral,

$$V(\rho) = -L^{-2}N^2A\int_0^L dz\int_0^L dz'[\rho^2+(z-z')^2]^{-3}$$

yields

$$V(\rho) = (N^2A/4\rho^4L^2)[3(L/\rho)\arctan(L/\rho)+L^2/(L^2+\rho^2)].$$

3.   MONOLAYER BOUNDARY LUBRICATION

Before an attempt is made to use these potential functions to the multilayer films of Fuks it seems prudent to apply it to a simpler system for which either experimental data or theoretical calculations are already available.   An example of such a system is that considered by Cameron in his theory of the role of van der Waals' forces in monolayer boundary lubrication.[11]   In this work the frictional force between two rubbing surfaces is assumed to arise from the van der Waals forces due to a single layer of stearic acid on each surface.   The computation of the frictional force is performed by calculating first the attractive energy of interaction of a molecule in one layer with all the other molecules in the other layer by summing the contributions to the potential of all the CH_2 groups within a radius of 8 Å of the molecule.   The r_{ij} are taken to be equal to the distances between X-ray scattering centres.   The repulsive energy is taken as equal to the repulsive energy of the hydrogen atoms of the molecule that approach closest as the two layers are pressed together.   Cameron's calculation shows that this is negligible and we therefore omit it from our calculation. When one surface is moved parallel to the other from the stable equilibrium position to the next equilibrium position, a point between these is encountered where the total interaction energy E_2 is a minimum.   If E_1 is the energy at the equilibrium point and x is the distance between equilibrium points then the mean force F needed to move a chain from one equilibrium position to the next is give by

$$E_1-E_2 = Fx. \tag{3.1}$$

To simplify the calculation Cameron argues that E_2 is negligibly small in comparison with E_1 and therefore the frictional force is given by

$$F = E_1/x. \tag{3.2}$$

This assumption has been questioned by Akhmatov, and others.[12]   Our calculations seem to verify this criticism.

To compute the attractive energy between two layers as a function of relative position and separation of the layers we carry out a simple summation procedure. We first consider a molecule of length L, the major axis of which is parallel to the

z axis, located at the point (x,y) above an infinite layer of molecules of length L. These molecules are assumed to be arranged in a square mesh of lattice constant a. The origin of the x-y coordinate system will be taken on one of the mesh points. The horizontal distance ρ_{lm} between the isolated molecule and a molecule in the layer located at the point whose coordinates are ma and la is given by

$$\rho_{lm} = [(x-ma)^2+(y-la)^2]^{\frac{1}{2}}. \tag{3.3}$$

The total attractive energy is obtained by substituting this value of ρ in eqn (2.3) and (2.5) and summing over m and l. That is, the total attractive energy is

$$W(a,\varepsilon) = \sum_{l=-\infty}^{\infty} \sum_{m=-\infty}^{\infty} V_{lm}(\rho_{lm},\varepsilon). \tag{3.4}$$

These summations must be done numerically for given values of lattice constant a and layer separation, $d = 2\varepsilon$.

Eqn (3.4) was summed numerically over a 31×31 square lattice by means of a CDC-6400 computer for several values of d, x and y. Results of this computation are given in table 1 for values of the parameters used by Cameron, i.e., $a = 4.46$ Å,

TABLE 1.—ATTRACTIVE ENERGY AS A FUNCTION OF INTERLAYER DISPLACEMENT ($y = 0.0$)

position x (in units of a)	energy of attraction (ergs/cm²)
0.0	15.824
0.1	15.708
0.2	15.409
0.3	15.053
0.4	14.774
0.5	14.669

$d = 3.09$ Å ; $a = 4.46$ Å.

$d = 3.09$ Å, $N = 18$, $L = 24.69$ Å, and a surface concentration of 5×10^{14} cm^{-2}. In this table the value of the energy for the point $(0, 0)$ is E_1 and the value of the energy for the point $(0.5a, 0)$ is E_2. Calculations for many different values of d show that for no reasonable value of d can E_2 be neglected. This is shown in table 2. In this table, $E_{min.}$ is the energy at the point $(0.5a, 0.5a)$.

TABLE 2.—DEPENDENCE OF E_1, E_2 AND E_{min} ON INTERLAYER SEPARATION

interlayer separation Å	E_{min} (ergs/cm²)	E_1 (ergs/cm²)	E_2 (ergs/cm²)
2.5	20.298	26.802	22.148
2.6	19.020	24.195	20.454
2.7	17.842	21.975	19.115
2.8	16.756	20.067	17.812
2.9	15.572	18.414	15.754
3.0	14.828	16.972	15.521
3.1	13.973	15.704	14.575
3.2	13.181	14.583	13.681
3.3	12.448	13.586	12.864
3.4	11.769	12.694	12.114
3.5	11.139	11.892	11.425
4.0	8.590	8.867	8.704

$a = 4.46$ Å.

Table 3 gives a comparison of our results with those of Cameron. The quantity F is calculated both from eqn (3.1) and (3.2) for various values of d, which are to be

TABLE 3.—COMPARISON OF HARD-ROD CALCULATION WITH CAMERON'S RESULTS

P (kgf/cm$^2 \times 10^3$)	d (Å)	F_{12}	F_1 (kgf/cm$^2 \times 10^3$)	$F_{Cameron}$
0.001	3.09	0.052	0.709	1.2
6.33	2.89	0.119	0.826	1.52
13.0	2.69	0.128	0.985	1.93
19.5	2.49	0.209	1.202	2.29

The first column gives the normal pressure, the third column gives the frictional force calculated by eqn (3.1), the fourth column gives the frictional force calculated by eqn (3.2) and the last column is the frictional force as calculated by Cameron.[11]

expected for the values of P shown, if the linear compressibility is taken as 10×10^{-12} dyn/cm^2 as in ref. (11).   Our results for F agree within a factor of two with Cameron's results if eqn (3.2) is used.   However, if eqn (3.1) is used, our results differ by about an order of magnitude both from Cameron's results and experiment.

4. DISCUSSION

The fact that our results for F show good agreement with the exact group-by-group calculation of Cameron if eqn (3.2) is used indicates that our rigid rod approximation is sufficiently accurate to encourage its further application to more complicated systems such as the multilayer model for the films of Fuks.   A simple treatment of such a model will be used by the authors in a future calculation in which the assumption is made that the molecules are arranged in regularly spaced arrays within the layers, as they are in the first chemisorbed layer.   Shear forces for such a model will be calculated as a function of interlayer separation and distance between molecules within a layer.   Comparisons will be made with Fuks' data for reasonable values of the distance parameters.   Good agreement with the data will probably not be obtained because one does not expect periodic spacing of the molecules to prevail except in the first layer.   However, such calculations will probably be useful when applied to known smectic liquid crystals in which periodicity does exist within layers.

This work was in part supported by Naval Air Systems Command, Contract No. N–00019–70–C–0139, Battelle Institute and the Columbus Laboratories of Battelle Memorial Institute.

[1] A. A. Griffiths, *Phil. Trans. A*, 1920, **221**, 163.
[2] R. Bulkley, *Bur. Stand. J. Res.*, 1931, **6**, 89.
[3] S. H. Bastow and F. P. Bowden, *Proc. Roy. Soc. A*, 1931, **134**, 404.
[4] S. H. Bastow and F. P. Bowden, *Proc. Roy. Soc. A*, 1935, **151**, 222.
[5] S. J. Needs, *Amer. Soc. Mech. Eng. Trans.*, 1940, **62**, 331.
[6] G. I. Fuks, *Research in Surface Forces*, ed. B. V. Deryagin, vol. 1, (Consultants Bureau, New York, 1964), p. 79.
[7] G. I. Fuks, *Research in Surface Forces*, 1966, **2**, 159.
[8] A. T. J. Hayward and J. D. Isdale, *Brit. J. Appl. Phys.*, 1969, 2, **2**, 251.
[9] C. M. Allen and E. Drauglis, *Wear*, 1969, **14**, 363.
[10] J. G. Kirkwood, *Phys. Z.*, 1932, **33**, 57.
[11] A. Cameron, *Amer. Soc. Lubr. Eng.*, 1960, **2**, 195.
[12] A. S. Akhmatov, *Molecular Physics of Boundary Lubrication*, Fiziko-Matematicheskoi Literatury, Moscow, 1963 (translated and published in English by Israel Program for Scientific Translations, Ltd., Jerusalem, 1966), p. 331 ff.

GENERAL DISCUSSION

Prof. J. Lyklema (*Wageningen, Netherlands*) said: With reference to the paper by Churayev *et al.*, when liquids are forced through porous media or narrow capillaries the resistance against flow is not only of a hydrodynamic but also of an electrokinetic character. The principle is that between the ends of the capillary a streaming potential is generated, which in turn gives rise to a counter electro-osmotic flow. Especially in narrow capillaries, this counter-osmotic flow can attain values that are comparable with the induced Poiseuille flow and due account of this complicating effect should be given before something definite can be said with respect to either the viscosity of the liquid or the effective radius of the capillary.

As a first approximation, the order of magnitude of the electrokinetic contribution can be assessed as follows. The streaming potential V, generated by the pressure P amounts to

$$V = \varepsilon \zeta P / 4\pi\eta\kappa \tag{1}$$

where the symbols have their usual meaning. The velocity of the counter-flow obeys

$$v_E = (\varepsilon\zeta/4\pi\eta)(V/l) = (\varepsilon\zeta/4\pi\eta)^2(P/\kappa). \tag{2}$$

If v_P is the average liquid velocity due to the Poiseuille flow, this quantity follows from

$$Q = \pi r^2 v_P = \pi r^2 P / 8\eta l, \tag{3}$$

Hence

$$v_E/v_P = \varepsilon^2\zeta^2/2\eta\pi^2\kappa r^2, \tag{4}$$

indicating that the ratio between the primary Poiseuille flow and the generated counter-flow can become appreciable in capillaries of small radius r.

For aqueous solutions at room temperature,

$$v_E/v_P = 3.8 \times 10^{-19}\, \zeta^2/\kappa r^2, \tag{5}$$

if ζ is in mV, κ in Ω^{-1} cm^{-1} and r in cm. As Churayev *et al.* do not give the quality of their water nor its pH, it is not certain what values for ζ and κ should be substituted. If $\kappa \sim 10^{-5}\, \Omega^{-1}$ cm^{-1} and $\zeta \sim -50$ mV,

$$v_E/v_P \sim 10^{-10}/r^2 = 10^{-8}/r^2\,\%, \tag{6}$$

indicating a reduction of 1 % in capillaries of 1 μ and of 100 % in capillaries of 0.1 μ.

Comparing these conclusions with the experimental results, plotted as an apparent increase in viscosity in fig. 4 of Churayev *et al.*, it appears that the reduction in flow could well be due completely to counter-osmotic flow, so that there is no indication of thick viscous layers or enhanced viscosity. The obvious experiment to ensure absence of electrokinetic counter-effects is repetition of the author's experiments in electrolyte solution instead of pure water, because addition of salts reduces ζ and increases κ (see eqn (5)). The effect of electrolyte on η is of secondary order.

Dr. N. V. Churayev, Dr. V. D. Sobolev and **Dr. Z. M. Zorin** (*Acad. Sci. Moscow*) (*communicated*): In reply to Lyklema, to evaluate the electroviscous effect, Lyklema uses the Helmholtz-Smoluchovsky eqn (1) which cannot be used for thin capillaries where overlap of the diffuse ionic atmospheres takes place. Our calculations based on formulae taking into account the overlap [1,2] show that the electroviscosity effect is considerably smaller: 2% for $\zeta = -15$ mV [3] and $\kappa = 2 \times 10^{-6} \, \Omega^{-1} \, cm^{-1}$. We do not understand why Lyklema has taken in his calculations too high a value of $\zeta \equiv -50$ mV which is well known not to be characteristic of the quartz-water interface. Moreover, the absence of considerable electroviscosity effect was proved by direct experiments with 10^{-3} N KCl solution in the same quartz capillaries. Though diffuse ionic layers were depressed we obtained the same viscosity change η/η_0 with decreasing radii of capillaries as for clean water. Thus, our experimental results cannot be explained by electroviscosity effect (or counter-osmotic flow) which was naturally taken to account and specially verified.

Dr. G. Peschel (*Universität Würzburg*) said: In the viscosity measurements by Churayev *et al.*, of water with very thin capillaries they obtained the surface zone viscosity of water structurally changed by adjacent solid walls. We also carried out experiments for determination of the surface zone viscosity of pure water. In the method developed by us we brought together two polished fused silica plates—one plate is planar, the other spherically formed with a curvature radius of 100 cm— within a very small distance under the action of a constant force. Both plates were immersed in the liquid in question. The movement of the spherical plate was registered by a special device. The theory which we developed also takes into account the surface roughness of the plates. For water, we found an enhanced surface zone viscosity which is detectable up to plate distances of about 1,500 Å and is sensibly dependent on the outer force and on the degree of coverage of the silica surfaces with hydroxyl groups. The main point is that the surface zone viscosity showed pronounced maxima at about 15, 32, 45 and 61°C; these temperatures are known to be the temperatures of higher-order transition of the water structure. The most pronounced maximum was at about 32°C. Churayev *et al.*, too, find a maximum at about this temperature. May be they could also detect the other maxima if their data were more closely spaced. The anomalous effect we found vanishes at about 70°C. The effect found by them trails off at the same temperature. In this respect our work agrees well with their evidence. Are their measurements disturbed by dust particles?

Dr. B. A. Pethica (*Unilever Res., Port Sunlight*) said: In an earlier paper, Deryaguin and his coworkers measured the shear moduli of thin liquid layers on quartz, and estimated the altered region to be ten times larger than that reported by Churayev, Sobolev and Zorin in their present paper, the latter estimate being based on viscosity measurements. Do these two experimental methods measure different aspects of the boundary layer properties, or do they refer to different surface characteristics in the quartz used?

Dr. B. Stuke (*Phys. Chem. Inst., München*) said: With regard to the paper by Churayev *et al.*, some time ago [4] we have done exactly what Lyklema has suggested and came to the same conclusion. The time was measured for a given amount of liquid to pass through filters of sintered quartz (bacteria filter with guaranteed maximum pore size, Schott u. Gen.) by pumping under manostatic conditions. Three

[1] N. V. Churayev and B. V. Deryaguin, *Dokl. Acad. Nauk.*, 1966, **169**, 396; *Kolloid Zhur.*, 1966, **28**, 751.
[2] N. V. Churayev and B. V. Deryaguin, *Research in Surface Forces*, (Nauka, Moscow, 1967), p. 295.
[3] Lars Lidström, *Acta Polytechn. Scand.*, 1968, **75**, 100.
[4] *Diplomarbeit*, L. Kolodziej, (München, 1962).

different filters were used with maximum pore sizes, 1.2, 1.7 and 3 μ. For two different electrolyte solutions with concentrations > 0.01 N the ratio of times τ_α was equal to the ratio of the corresponding viscosities η_α, i.e., $\tau_i/\tau_k = \eta_i/\eta_k$. For solutions with concentrations < 0.01 N (index 1) compared with solutions with concentrations > 0.01 N (index 2) the result was $\tau_1/\tau_2 > \eta_1/\eta_2$, depending (monotonically) on the concentration of solution 1 and pore size. The value of the ratio passing times for pure water compared with an electrolyte solution of sufficient concentration (maximum value) under conditions where $\eta_{H_2O}/\eta_2 \sim 1$ for the filter with 1.2 μ was $\tau_{H_2O}/\tau_2 \rightarrow 1.05$.

Prof. N. V. Churayev (*Moscow*) said: In reply to J. Lyklema and B. Stuke, they were of the opinion that the increase of viscosity of water in thin quartz capillaries observed by us could be connected with the electroviscosity effect. To evaluate this effect, Lyklema uses the Helmholtz-Smoluchovski equation, the applicability of which is limited by the condition of $\kappa r \gg 1$ (where $1/\kappa$ is the Debye radius of the diffusion-ion atmospheres, r is the radius of the capillary). For water ($\kappa = 10^5 \, \Omega$ cm^{-1}) and capillaries with $r = 10^{-4} - 4 \times 10^{-6}$ cm in our experiments the values of κr are in range 10-0.4. Thus, for the thinnest capillaries where the greatest change of viscosity is found, the application of the Helmholtz-Smoluchovski equation becomes impossible. For $\kappa r \approx 1$ one should take into account the overlap of the diffusion ion atmospheres that are considered theoretically in ref. (1) and (2).

Basing on the equations derived there, we may evaluate the influence of the electroviscosity effect in the case of interest. For the quartz/water interface at pH 6.5-7 $\zeta = -15 \, \text{mV}$ [3]; the specific conductivity λ of distilled water was $2 \times 10^{-6} \, \Omega^{-1} \, \text{cm}^{-1}$. The calculations show that the maximum variations of viscosity η not to exceed 2-3 %. Moreover, the highest viscosity changes would have been observed at $\kappa r \approx 2$,[1, 2] i.e., for the capillary radii $r \approx 2 \times 10^{-5}$ cm, and this was not found. All this leads to the conclusion that the results obtained could not be explained by a simple influence of electroviscosity.

However, we did not confine ourselves to theoretical estimations, but undertook additional experiments with the electrolyte solutions. In fig. 1 are the results of comparative measurements of viscosity of water and a 10^{-3} N KCl solution in the same quartz capillaries ($r = 6.9 \times 10^{-6}$ cm and $r = 8.9 \times 10^{-6}$ cm). As the κ values with electrolyte are higher by an order of magnitude than those with distilled water, one would have expected an abrupt drop of viscosity of the solution in the microcapillaries if the cause of changes in η were due to the electroviscosity effect. As the plot, however, shows, the experimental points for water and the solution lie on the same line corresponding to the increased viscosity values $\eta/\eta_0 = 1.27$ and 1.34.

An increase of 30 % viscosity of 10^{-3} N KCl solution could not have been caused by the electroviscosity effect. Calculations using eqn (5) employed by Lyklema (the equation can be applied in this case for $\kappa r \approx 5$), show that the differences in viscosities would be as small as *ca.* 2 %. Thus, the experimental data are also in contradiction to his hypothesis of the effect of electroviscosity.

The explanation for the increased viscosity of water in fine-porous bodies [4] is usually given from the standpoint of the electroviscosity effect as has been done in

[1] N. V. Churayev and B. V. Deryaguin, *Dokl. An S.S.S.R.*, 1966, **169**, 396 ; *Kolloidn. zhurn.*, 1966, **28**, 751. *Sb. Issledovaniya v oblasti poverkhnostnykh sil* (*Research in surface forces*), " Nauka ", (Moscow, 1967), p. 295.

[2] C. L. Rice and R. Whitehead, *J. Phys. Chem.*, 1965, **69**, 4017.

[3] L. Lidström, *Acta Polytechn. Scand.*, 1968, **75**, 100.

[4] J. T. Davies and E. R. Rideal, *Interfacial Phenomena* (Academic Press, 1961), p. 126.

different filters were used with maximum pore size 1.2, 1.7 and 3 μ. For two different electrolyte solutions with concentrations > 0.01 N the ratio of times τ was equal to the ratio of the corresponding viscosities η_a, i.e., $\tau_i/\tau_a = \eta_i/\eta_a$. For solutions with concentrations < 0.01 N (index 1) compared with solutions with concentrations > 0.01 N (index 2) the result was $\tau_i/\tau_a > \eta_i/\eta_a$, depending (monotonically) on the concentration of solution 1 and pore size. The value of the ratio of passing times for pure water compared with an electrolyte solution of sufficient concentration (maximum value) under conditions where $\eta_a/\eta_{a0} > 1$ for the filter with 1.2 μ was equal to 1.05.

Prof. N. V. Churayev (Moscow) said: In reply to J. Lyklema and B. Staic, they were of the opinion that the increase of viscosity of water in thin quartz capillaries observed by us could be connected with the electroviscosity effect. To evaluate the effect, Lyklema uses the Helmholtz-Smoluchowski equation the applicability of which is limited by the condition $\zeta e/\kappa > 1$ where $1/\kappa$ is the Debye radius of the diffusion-ion atmospheres, r is the radius of the capillary). For water ($\kappa = 10^6$ cm⁻¹) and capillaries with $r = 10^{-4} - 4\times10^{-5}$ cm in our experiments the values of κr are in range 10-0.4. Thus, for the thinnest capillaries where the greatest change of viscosity is found, the application of the Helmholtz-Smoluchowski equation is impossible. For $\kappa r \approx 1$ one should take into account the overlap of the diffusion-ion atmospheres that are considered theoretically in ref. (1) and (2).

Basing on the equations derived there, we may evaluate the influence of the electroviscosity effect in the case of interest. For the quartz/water mixture at pH 6.5 $\zeta = -15$ mV, the specific conductivity λ of distilled water was 2×10^{-6} Ω⁻¹ cm⁻¹. The calculations show that the maximum variations of viscosity η not to exceed 2-3%. Moreover, the highest viscosity changes would have been observed at $\kappa r \approx 1$, i.e., for the capillary radii $r \approx 2\times10^{-5}$ cm, and this was not found. All this leads to the conclusion that the results obtained could not be explained by a simple influence of electroviscosity.

However, we did not confine ourselves to theoretical estimations, but made additional experiments with the electrolyte solutions. In fig. 1 are the results of comparative measurements of viscosity of water and a 10^{-3} N KCl solution in the same quartz capillaries ($\theta = 6.9\times10^{-5}$ cm and $r = 8.9\times10^{-5}$ cm). As the values with electrolyte are higher by an order of magnitude than those with distilled water, one would have expected an abrupt drop of viscosity of the solution if the increase of changes in η were due to the electroviscosity effect. As fig. 1 shows, the experimental points for water and the corresponding to the increased viscosity values, lie

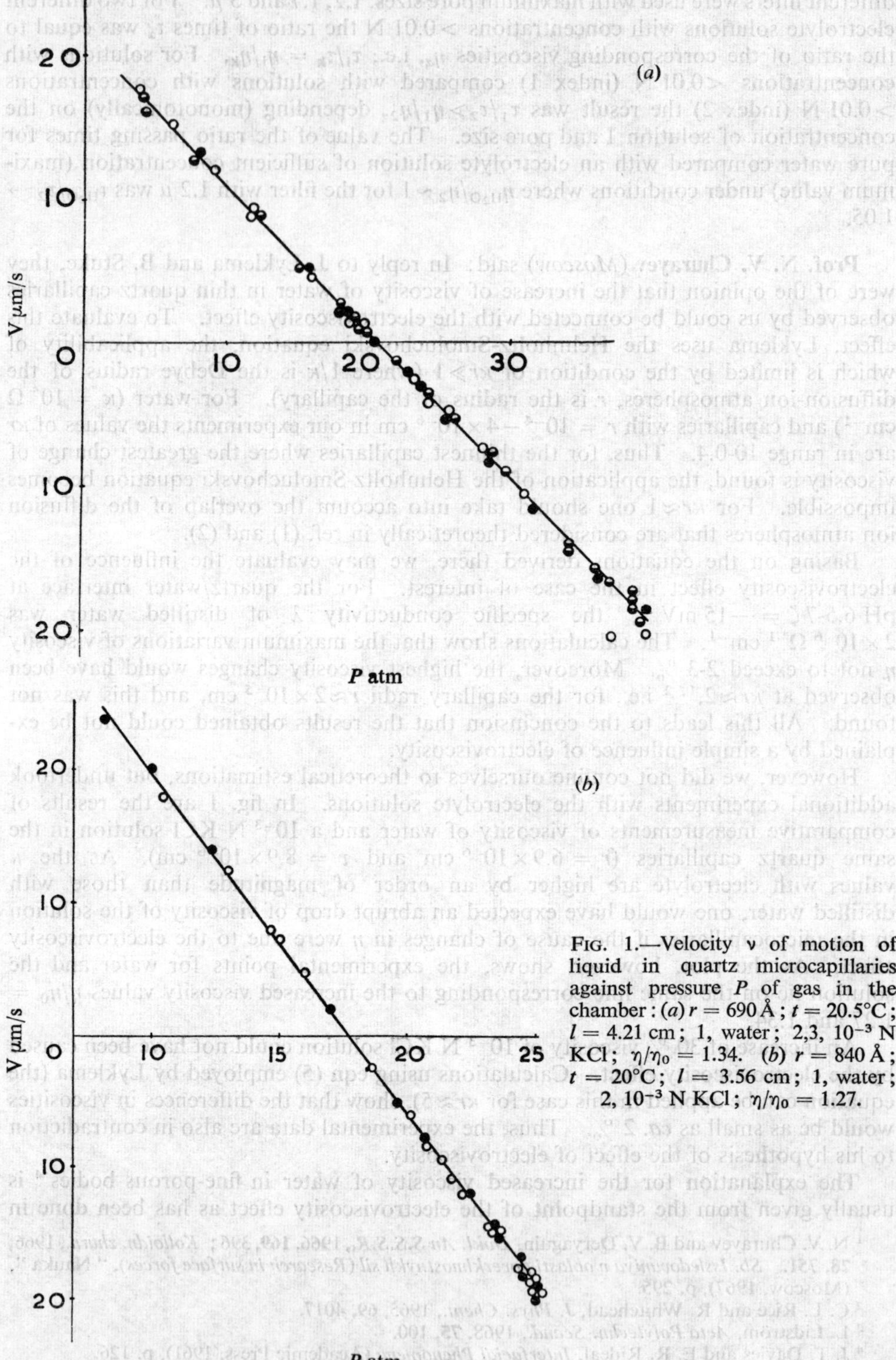

FIG. 1.—Velocity ν of motion of liquid in quartz microcapillaries against pressure P of gas in the chamber: (a) $r = 690$ Å; $t = 20.5°$C; $l = 4.21$ cm; 1, water; 2,3, 10^{-3} N KCl; $\eta/\eta_0 = 1.34$. (b) $r = 840$ Å; $t = 20°$C; $l = 3.56$ cm; 1, water; 2, 10^{-3} N KCl; $\eta/\eta_0 = 1.27$.

for a 10^{-3} N KCl solution could not have been used. Calculations using eq. (5) employed by Lyklema (the equation for $\kappa r \gg 5$) show that the differences in viscosities would be as small as ca. 2 %. Thus the experimental data are also in contradiction to his hypothesis of the effect of electroviscosity.

The explanation for the increased viscosity of water in fine-porous bodies is usually given from the standpoint of the electroviscosity effect as has been done in

1 N. V. Churayev and B. V. Deryagin, Dokl. Ak. Nauk S.S.S.R., 1966, 169, 396; Kolloid. zhur., 1966, 28, 751. Sb. Issledovaniya v oblasti poverkhnostnykh sil (Research in Surface Forces), (Moscow, 1967), p. 205.
2 C. L. Rice and R. Whitehead, J. Phys. Chem., 1965, 69, 4017.
3 L. Lidström, Acta Polytechn. Scand., 1965, 75, 100.
4 J. Davies and E. K. Rideal, Interfacial Phenomena (Academic Press, 1961), p. 126.

the paper of B. Stuke. It probably does occur in a number of systems. Nevertheless, for fine-porous bodies (generally at $\kappa r \sim 1$) if one does not consider the overlap of diffusion ion atmospheres [1,2] the influence of electroviscosity is over-estimated. Henniker,[3] while measuring the flow potential by short-circuiting the electrodes, did not detect changes in flow rate, though the viscosity of the liquid was higher by more than 20 %. Deryaguin, Zakhavayeva and Lopatina [4] have observed increases of several fold of the viscosity in thin pores of clays at a concentration of electrolyte of *ca.* 0.1 N when the diffuse layers were totally suppressed. It should be noted that the highest viscosity differences are obtained in very thin layers and pores [5,6] when (at $\kappa r \rightarrow 0$) the influence of the electroviscosity effect does not grow rapidly (a consequence of the erroneous application of the Helmholtz-Smoluchowski equation) but in contrast falls abruptly.[1,2]

All this leads to the conclusion that the real cause of the viscosity changes is the modification of the structure of liquids under the influence of surface forces. The nature of the changes depends not only on the radii of the capillaries but also on the temperature of the liquid and/or the concentration of electrolytes, that influence the structure of the boundary layers.

In reply to Pethica, Bazaron, Deryaguin and Bulgadayev [7,8] found an increase in shear modulus for the boundary layers of polar liquids with thicknesses of the order of 10^{-5} cm, from measurements by the resonance method on piezo-electric quartz at a frequency 75 kHz. This does not mean, however, that anomalous layers of the same thickness should be also observed during the viscous flow of a liquid under a constant shearing force.

The boundary layers of polar liquids seem to have complex structure.[9] The layers of water in close proximity to the surface are modified to a greater extent and have, in particular, a considerable yield value θ amounting to *ca.* 100 dyn cm^{-2} according to Nerpin and Bondarenko.[10] This result was obtained by observing changes in the filtration permeability with increase of the pressure gradient. It follows from the estimation of θ that these layers could not take part in flow in our experiments and are caused only by the " narrowing " of capillaries. The thickness of the boundary layer was determined [12] to be 80 Å, i.e., close to our value. At the same time a broader boundary region with less modified properties appears to exist. Its limits are observed, e.g., in measurements of the shear modulus [1] and the disjoining pressure.[13]

Mr. D. W. J. Osmond (*I.C.I., Ltd., Slough*) said: It has long seemed to us in industry that the thick layers of modified solvent postulated to form at interfaces, are

[1] N. V. Churayev and B. V. Deryaguin, *Dokl. An S.S.S.R.*, 1966, **169**, 396; *Kolloidn. zhurn.*, 1966, **28**, 751. *Sb. Issledovaniya v oblasti poverkhnostnykh sil* (*Research in surface forces*), "Nauka", (Moscow, 1967), p. 295.

[2] C. L. Rice and R. Whitehead, *J. Phys. Chem.*, 1965, **69**, 4017.

[3] J. C. Henniker, *J. Colloid. Sci.*, 1952, **7**, 443.

[4] B. V. Deryaguin, N. N. Zakhavayeva and A. M. Lopatina, *Sb. Issledovaniya v oblasti poverkhnostnykh sil* (*Research in surface forces*), (" Nauka ", Moscow, 1961), p. 175. *Bull. Rilem.*, 1965, **27**, 27.

[5] G. Peschel, *Mater. constr.*, 1968, **1**, 529; G. Peschel and K. H. Adlfinger, *Naturwiss.*, 1969, **56**, 558.

[6] Z. M. Tovbina, *Sb. Issledovaniya v oblasti poverkhnostnykh sil* (*Research in surface forces*), (" Nauka ", Moscow, 1967), p. 24.

[7] U. B. Bazaron, B. V. Derjaguin and A. B. Bulgadayev, *Dokl. AN S.S.S.R.*, 1965, **160**, 799; 1966, **166**, 639; *Zhurn. exper. teor. fiz.*, 1966, **51**, 969.

[8] B. V. Deryaguin, *Disc. Faraday Soc.*, 1966, **42**, 109.

[9] W. Drost-Hansen, *Ind. Eng. Chem.*, 1969, **61**, 10.

[10] S. V. Nerpin and N. F. Bondarenko, *Sb. Trudov po agrofizike*, 1969, **19**, 27.

[11] G. Peschel and K. H. Adlfinger, *Naturwissen.*, 1967, **54**, 614.

implausible except, as in Cameron's paper, as chemical or other artifacts. Our reasons for doubt are as follows. It is now a commonplace to prepare concentrated dispersions of both polar and non-polar particles in organic liquids often in the presence of fatty acids. In such dispersions the particles radius is often about 1,000 Å and the phase volume about 50 %. If layers of 1,000 Å thickness surround such particles, *seven times* as much solvent as is present would be required to form the layers. We may safely assume therefore that all of the solvent present would be heavily modified if such layers existed. Dispersions in a continuum of " modified " solvent of this type would be expected to show marked rheological anomalies, especially at low rates of shear. In practice, such dispersions are commonly mobile liquids, whose chief anomaly is intense shear *thickening* or dilatancy at very high rates of shear. It therefore seems that, either such layers do not exist, or they are so weak as to have little practical effect. On the other hand, it is well known in the Paint Industry, that fluid dispersions in inert non-aqueous media of particles of reactive materials, such as calcium carbonate, zinc oxide or basic lead carbonate, are converted on the addition of fatty acids to gelatinous dispersions. It has long been accepted, in agreement with Cameron's results, that this gelling is due to thickening of the continuous phase by metal soaps formed by chemical attack on the particle surfaces.

Dr. A. Cameron (*Mech. Eng. Dept., Imperial College*) said: Osmond's results from paint technology confirm that when the liquid or the components in the liquid do not react chemically with the particles there is no change in the viscosity. When they do react as for whiting and a fatty oil the result is a plastic putty. These are completely consonant with our findings.

Dr. H. E. Ries (*Chicago*) said: In spite of the ingenious studies on thick surface films described by Smith and Cameron and others, I believe that there is something unique about films one molecule thick—partly perhaps because I worked with Harkins and partly because of our research on monolayers extending over the last 35 years. I believe everyone agrees that the monolayer is the last line of defence in many areas—friction, wear and rust (also for emulsion and foam stability, evaporation retardation, etc.). Certainly the monolayer is held by forces stronger than those that hold any succeeding layer, and on solid surfaces it is the only layer that can be chemisorbed. More specifically, our radiotracer adsorption studies performed under carefully controlled conditions clearly demonstrate that adsorption levels off at one monolayer. Perhaps the most pertinent experiments were those in which radiostearic acid was adsorbed from n-hexadecane solutions on iron films vapour-deposited in high vacuum on the window of a Geiger tube. Although initial adsorption is relatively rapid, the equilibrium adsorption plateau, at the monolayer level extends over at least 70 h.[1] Our mixed-film studies at the water/air interface, as well as at the metal/water interface in rust-prevention experiments, also support the monolayer concept. I should thus like to ask Smith and Cameron whether some of their work suggests that the monolayer provides protection of some special significance?

Dr. A. Cameron (*Mech. Eng. Dept., Imperial College*) said: In reply to Ries, we have done nothing to displace the monolayer. Our studies are solely directed to finding under what circumstances a thick viscous layer could be formed. With

[1] D. C. Walker and H. E. Ries, Jr., *J. Colloid Sci.*, 1962, **17**, 789.

stearic acid and cetane a monolayer occurs both on platinum and on steel but it is only with *steel* that the 1,000 Å film occurs.

Dr. D. Tabor (*University of Cambridge*) said: Cameron's paper shows that when a surface, immersed in a liquid, appears to produce long-range effects this is due to chemical reaction and the formation of a relatively thick film of reaction products. These results are in close agreement with measurements made by Bowden and Moore [1] in Cambridge nearly 20 years ago, using radioactive tracer techniques. Although this seems fully confirmed for fatty acids I find it hard to believe that dimethyl silicones will produce reaction products at room temperature. Although, as Willis [2] observed, a solid surface can greatly accelerate the chemical degration of silicones the effects were negligible below about 90°C. If reactions do occur, presumably they must be due to impurities in the silicone or to contaminants in the system as a whole. I would welcome Cameron's comments on this.

Finally, may I ask a question as a tribologist rather than a surface physicist. Tribologists are interested in two questions: first, do surfaces produce long-range effects in liquids; secondly, is this long-range effect stable or is it easily disrupted by shear, i.e., is the modified film rheologically strong or not? Cameron has given an answer to the first question—could he give one to the second?

Dr. A. Cameron (*Mech. Eng. Dept., Imperial College*) said: Tabor may rest assured that we have taken full acount of the work of Bowden and Tingle, and Bowden and Moore, on the reactivity of stearic acid with metals. In the work he quotes it is interesting to note that, after 20 h, 4 molecular layers, i.e., a 100 Å, thick layer of stearic acid reacted with zinc. We find an immobilized layer 1,000 Å thick which we believe is caused by solvent meshed by soap fibrils. From normal grease technology it is reasonable to think that a soap content of about 10 % would be enough to achieve such thickening. This corresponds to 100 Å of " solid " iron stearate, the same 4 layers as Bowden and Moore found. The only difference is that they indicated this took 20 h to form at 25°C while we find our films are grown in $\frac{1}{4}$ h. We also wonder if the correction they applied for radioactive material adsorbed on the walls of their vessels was accurate. The aim of their work was quite different from ours which is simply to study the liquid layer at the surface of metals. We certainly are not competent to discuss the interesting low viscosity layer which silicones seem to form at metal surfaces, though our result confirms Deryaguin's. We have not attempted to elucidate the mechanism.

In reply to Tabor's second question, work now being carried out by flowing mercury past the films show that they are very fragile. A shear stress of about 5 dyn/cm² seems to break them down. On the other hand, stearic acid seems to promote fluid lubrication (see our ref. (9), fig. 6).

Mr. A. J. Groszek (*B.P. Co. Ltd., Sunbury*) said: In view of the results published previously by Cameron, I expected that adsorption of stearic acid from n-hexadecane on iron oxides and metals such as iron would give a higher heat of adsorption than the adsorption from a non-matching solvent such as n-heptane. We determined therefore heats of preferential adsorption of stearic acid for the two solvents in the flow-microcalorimeter. The heats of preferential adsorption obtained for three different iron oxides and for ground iron powder are shown in table 1. Stearic acid saturates the oxides at a concentration of 1 g/l., i.e., the concentration at which

[1] F. P. Bowden and A. C. Moore, *Trans. Faraday Soc.*, 1950, **47**, 900.
[2] D. Tabor and R. F. Willis, *Wear*, 1969, **13**, 413.

the integral heats of adsorption shown in table 1 were obtained.   Some of the heat evolved is due to chemisorption but a large part is caused by reversible, physical adsorption.[1]

TABLE 1.—INTEGRAL HEATS OF ADSORPTION OF OCTADECANOIC ACID ON TO IRON OXIDES AND GROUND IRON FROM n-HEPTANE AND n-HEXADECANE SOLUTIONS

| | heat of adsorption, J m^{-2} * | |
adsorbent	from n-heptane	from n-hexadecane
γ-Fe$_2$O$_3$ [1]	0.042	0.222
γ-Fe$_2$O$_3$—Fe$_3$O$_4$ [4]	0.158	0.126
α-Fe$_2$O$_3$	0.113	0.158
ground iron	0.276	0.251

* adsorption from solution containing 1 g/l. octadecanoic acid.

Perhaps the most striking result was obtained for γ-Fe$_2$O$_3$, for which the heat of adsorption from n-C$_{16}$ was considerably greater than that taking place on adsorption from n-heptane.   The result can be explained by a substantial orientation of n-hexadecane taking place near the surface saturated with stearic acid, or perhaps several layers of mixed film forming near the surface.   It is striking that the high heat of adsorption is obtained only for one of the three iron oxides examined and that there is no difference in the heats for ground iron, which emphasizes the importance of the nature of the surface in these adsorption experiments.

The effect of the chain length of fatty acids on the heat of adsorption was also obtained for the adsorption of small amounts of fatty alcohols and acids from hexadecane onto α-Fe$_2$O$_3$ at low surface coverage as shown in table 2.   The C$_{16}$ acid produces a much greater heat effect than C$_6$ acid, although an almost identical number of moles were adsorbed in both cases.   These results indicate that the matching of the chain-lengths of the solvent and solute can certainly increase the heat of adsorption and therefore stability of the adsorbed films, which agrees with the results reported by Cameron *et al.* on the relatively high effectiveness of the matched solvent/solutes in certain lubrication experiments.

TABLE 2.—PULSE ADSORPTION OF NORMAL ALCOHOLS AND CARBOXYLIC ACIDS ON TO α Fe$_2$O$_3$ FROM n-HEPTANE AND n-HEXADECANE

| | heat of adsorption, kJ mol^{-1} | |
adsorbate	from n-heptane	from n-hexadecane
acetic acid	<0.8	<0.8
hexoic acid	18.2	9.6
hexadecanoic acid	50.2	77.0
octadecanoic acid	59.2	86.0
butyl alcohol	12.1	4.2
hexyl alcohol	21.0	11.3
hexadecyl alcohol	42.7	60.7
octadecyl alcohol	67.0	75.0

What remains undecided, however, is whether the high heat of adsorption attesting to vertical orientation of hexadecane and stearic acid is confined to a close-packed monolayer or several more loosely-packed layers extending into the bulk of the liquid.   The fact that the amount of stearic acid adsorbed from n-C$_7$ and n-C$_{16}$ on α-Fe$_2$O$_3$ is the same in spite of the difference in the heats of adsorption suggests that the orientation is not confined to the monolayer.

[1] A. J. Groszek, *Ann.* 1970 *ASLE Conf.* (Chicago).

Dr. A. Cameron and **Dr. A. J. Smith** (*Mech. Eng. Dept., Imperial College*) said: With regard to the paper by Dyson, the gel-like layer formed on the surfaces of steel does not contribute to the EHL film when tested at the very high shear rates of the disc machine. There has been further work extending our results which shows that the gel is extremely fragile under shear and apparently breaks down very easily. The puzzle, however, still remains that 0.1 % of stearic acid in cetane has a profound influence on the lubrication characteristics of a 4-ball machine with 1 in. balls running at 200 rev/min. The scuffing load is considerably extended and also the electrical resistance is likewise much increased. The two seem to go hand in hand. One is therefore left with the dilemma that the film thickness measurement described here shows that there is no effect, while the scuffing tests and the electrical resistance tests show a marked effect. It would be interesting if Dyson could give any electrical resistance values for the types of system discussed by Askwith, Cameron and Crouch.

Dr. A. Dyson (*Thornton Res. Centre, Chester*) said: Cameron and Smith raise a point which is of great interest and importance but which has not yet been satisfactorily explained to my knowledge. There seems to be no necessary connection between film thickness and scuffing load. Film thicknesses calculated for the conditions obtaining just before scuffing vary over nearly an order of magnitude even if the metal surfaces and the lubricant are similar. Furthermore, additives of the " extreme pressure " type do not seem to have any effect on film thickness under conditions in which it can be measured; yet they increase the scuffing load. Recent work on the effect of very small concentrations of oxygen and of water vapour indicates that these must be regarded as " extreme pressure " agents, since scuffing loads are often very much less if oxygen and water vapour are carefully excluded from the rubbing system. We do not understand in detail the mechanism by which these materials act and I suggest that the effect of stearic acid may be essentially similar in nature.

The apparent fragility of the films formed under the conditions reported by them is interesting but, as I understand it, the observation indicates only that there is at least one point of intimate electrical contact over the rather large area of interface between the mercury and the solid metal. A coherent film could still persist over the remainder of the area. Another possibility is that the films formed by reaction with a metal surface which has been freshly exposed by abrasion are stronger than those formed on a surface protected by physically adsorbed or chemisorbed layers, such as are presumably present under the conditions of their work. I am not able to offer useful comments on electrical resistance under the conditions used by Askwith, Cameron and Crouch.

Dr. D. Tabor (*University of Cambridge*) said: Dyson's paper illustrates the gap between tribologists even of the most sophisticated kind and surface chemists. The problem is not only one of concepts but of communication and I sympathize with the difficulties he must have experienced in coping with some of the preceding papers. There is one point in his paper that may have escaped the attention of the surface chemists. In his experiments the mean contact pressure between the discs is *ca*. 3,600 atm and this is exerted on the oil film trapped between the surfaces. As a result, the viscosity of the trapped film is about 1,000 times the atmospheric viscosity. This suggests that other rheological properties of the film may be vastly different from those obtained in a normal laboratory. It is well known, e.g., that small amounts of a fatty acid can greatly improve the lubrication properties of a mineral oil. I would welcome any comments from surface chemists about the possible effects of very high pressures on the rheological properties of surface active materials adsorbed on a solid surface from a non-polar solvent.

Dr. K. J. Mysels (*R. J. Reynolds, Winston Salem*) said: Dyson's experiments were not directed to the very thin monomolecular surface layers which are bound to play an important role in lubrication. Nevertheless, the very high pressures prevailing under these lubricating conditions may have a considerable stabilizing effect upon these very thin layers. This is indicated by the experiments reported by Ash and Findenegg who found that adsorption was accompanied by contraction which could be interpreted as corresponding to the solidification of a monomolecular layer of liquid near the surface. It follows that such a layer will be stabilized by higher pressures just as the freezing point of hydrocarbons is raised by it.

Dr. H. E. Ries (*Chicago*) said: I have two questions and a comment related to the elegant studies of Roberts and Tabor: (*a*) Have the authors considered controlled deposition (Langmuir–Blodgett) of the film-forming compounds on the rubber and glass surfaces? (*b*) In the earlier work with films of calcium stearate on mica, at what surface pressures were the films transferred? Perhaps in both these cases, the monolayers should be transferred at several different surface pressures because marked differences in film structure have been demonstrated through a relatively small pressure range. For example, in our electron-microscope studies on transferred films, islands irregular in size and shape appear at low pressures, a continuous film with discontinuous open spaces at greater pressures (with the ratio of covered to uncovered area increasing with pressure), a homogeneous continuous film at still greater pressures, then ridges and folds, and finally long narrow flat platelets two molecules thick following collapse [1,2]

Dr. A. D. Roberts (*University of Cambridge*) said: In reply to Ries, an attempt was made to measure the shear strength of monolayers of sodium dodecyl sulphate and stearic acid sandwiched between rubber and glass surfaces. Monolayers were deposited on surfaces from bulk solution and excess solution " squeezed " out between the surfaces either under the influenced of a high load or by allowing the solvent around the contact region between surfaces to evaporate so that the excess solution in the sandwich would be " sucked out " by concentration difference. The separation between the surfaces was measured interferometrically. In some cases, *no* film appeared to be trapped between the surfaces: in others, the film was about 40 Å thick. This was assumed to be bimolecular film and its shear strength determined by measuring both the force required to shear the film and the area of the contact zone. By using both soft and hard rubber surfaces it was possible to change the contact pressure over the range $\frac{1}{2}$-6 atm. Results of preliminary measurements suggest that the shear strength of monolayers of sodium dodecyl sulphate and of stearic acid increase with pressure over this range. Unfortunately, the interferometric method does not indicate whether the trapped film is continuous or not. For this reason we have not quoted any specific values. However, at the higher pressure range, our shear strengths were comparable with those observed by Bailey and Courtney-Pratt in their mica experiments. The next step of this work would be to deposit monolayers by controlled deposition using the Langmuir-Blodgett technique.

[1] H. E. Ries, Jr. and W. A. Kimball, *Proc. 2nd Int. Congr. Surface Activity*, 1957, **1**, 75 ; *Nature*, 1958, **181**, 901.

[2] H. E. Ries, Jr. and D. C. Walker, *J. Colloid Sci.*, 1961, **16**, 361.

Films of calcium stearate were transferred at a surface pressure of 16 dyn/cm, castor oil being used as the piston oil. Experiments were also attempted using oleic acid as the piston oil, which exerts a pressure of about 30 dyn/cm, but in this case the higher pressure tended to crumple the film.

Dr. Th. F. Tadros (*Plant Protection Ltd., Bracknell, Berks*) said: With regard to the paper by Roberts and Tabor, in the measurement of film thickness of sodium dodecyl sulphate between rubber and glass, was the pH controlled? The electrical double-layer charge at the glass-solution interface is not only dependent on ionic strength, but also on the pH of the solution. By addition of salt the pH would vary so that the two curves of their fig. 5 can only be compared if the pH was maintained constant. Another point is with regard to the presence of specifically adsorbed Ca^{2+} ions (in the Stern plane) on glass which would cause adsorption of dodecyl sulphate ions with the head groups towards the surface and the hydrocarbon chain in solution. Mysels has commented that this seems unlikely since Ca^{2+} ions would be screened by negative silicate sites. However, from the work on quartz flotation by sodium dodecyl sulphate,[1] in the presence of bivalent ions, e.g., Ba^{2+}, dodecyl sulphate ions are probably held next to barium activated quartz by association with specifically adsorbed ions in the Stern layer. This seems to occur at high Ba^{2+} concentration in solution.

Dr. A. D. Roberts (*University of Cambridge*) said: In reply to Tadros, the salt added to films of sodium dodecyl sulphate (SDS) sandwiched between rubber and glass was sodium chloride. This is a strong electrolyte and therefore not expected to alter the solution pH significantly. Measurements made on solutions of 0.01 M SDS containing different quantities of NaCl ranging in concentration from 0.01 to 0.30 M revealed that the pH was constant at 6.0 ± 0.1. The pH of 0.01 M SDS without added NaCl was 6.14 ± 0.02.

Dr. A. Cameron (*Mech. Eng. Dept., Imperial College*) said: It is indeed gratifying to note that some work which I did in Prof. Sir Eric Rideal's laboratory in 1943 has been resurrected by Drauglis *et al.* The fact that in those days I did all the tedious summations using a slide rule naturally limited the accuracy of the results. There does seem to be some confusion in their paper in that the authors have not recognized that the reason for putting the energy E_2 much smaller than E_1 is that when the chains are in between the two equilibrium positions the repulsion energy must equal the attractive energy or be of the same magnitude. I therefore do not understand their approval of Akhmatov's criticism as Akhmatov also did not appreciate this. Could Drauglis make some comment on this point?

It is not clear why they chose a square unit cell. Also have they any intention of extending their model to include a terminal CH_3 group? They may be interested that in the Lubrication Laboratory at Imperial College this problem is being studied as completely as possible by myself and M. J. Sutcliffe. What is gratifying is that the first results show that the setting angle is 45° which agrees to considerable accuracy with the latest X-ray determinations.

Dr. D. Tabor (*University of Cambridge*) said: Drauglis has described a detailed analysis of the forces between long chain molecules. He has then deduced the frictional force between surfaces lubricated by fatty acid monolayers using the model proposed by Cameron some years ago. Such a calculation implies that Drauglis

[1] A. M. Gaudin and D. W. Fuerstenau, *A.I.M.E. Trans.*, 1955, **202**, 66.

has, in effect, calculated the bulk shear strength of a bimolecular layer of fatty acid. I find this a little disconcerting. These computations give us a value of the *theoretical* strength of the bimolecular layer; the real strength may be very different indeed if shear involves the movement of dislocations. Would Drauglis discuss this?

Dr. E. Drauglis (*Battelle Mem. Inst., Ohio*) said: Although Akhmatov was apparently unaware of Cameron's reason for neglecting E_2, (ref. (12) of our paper), we agree with his criticism because the assertion that when the chains are in between the two equilibrium positions the repulsive energy must equal the attractive energy or at least be of the same order of magnitude has not been proven. Undoubtedly, adopting this assertion makes the calculations much easier but this is not sufficient reason for doing so. We find it difficult to believe that the terminal CH_3 groups of each chain adjust their positions as the layers are squeezed together so as to make the assumption true for all values of the interlayer separation. The more elaborate calculations being carried out at his laboratory by Mr. M. J. Sutcliffe should do much to clarify this point. In our analysis, we chose a square unit cell in order to simplify the calculations. This results in a negligible error. In the future we may make our model more elaborate by adding a terminal CH_3 group to our polarizable rods, but the rod model is sufficient for our present purposes.

In reply to Tabor, our calculations and those of Cameron give the contribution of the van der Waals forces to the friction between two ideal layers of stearic acid. For two real surfaces covered with a monolayer of stearic acid one would expect that the arrangement of the molecules would not be nearly so orderly as in our models. Many dislocations and other irregularities could be expected to be present. These would lower the strength of the interaction between the layers. Since our values for the coefficient of friction are already far below the experimental values, the conclusion is that the van der Waals forces are insufficient to explain monolayer lubrication; other phenomena must be invoked. There is no question but that asperity interaction is very important in such a system. However, it is not our purpose to develop new theories of monolayer lubrication. Our only purpose in introducing Cameron's theory of monolayer lubrication was to serve as a check on our calculations based on the rigid rod model. Our primary purposes are to develop and verify our hypothesis that layered structures must exist in certain types of boundary surface films. A theory capable of quantitative correlation of the data of Fuks, and development of a better understanding of the rheology of smectic liquid crystal-like structures, are needed.

Dr. G. Frens (*Philips Res. Lab., Eindhoven*) said: Would not the study of the Brownian motion of a single particle in a very dilute suspension be a better criterion for the existence of thick, structured, layers of anomalous viscosity near a surface, than those hydrodynamic experiments which have been under discussion. If such layers did indeed exist they would cause considerable deviations from Einstein's law,

$$D = kT/6\pi\eta a = \bar{\Delta}^2/2t$$

in that they would make η (or a) appear to be higher than the values which can be determined independently. One would have to do a careful experiment so that no electrical forces or interactions between particles would interfere. But this can be done and to my knowledge no proof for thick structured layers has been found in such experiments.

AUTHOR INDEX *

* The references in heavy type indicate papers submitted for discussion.